D0857688

DISCRETE
MATHEMATICS
AND
ITS APPLICATIONS

Series Editor

Kenneth H. Rosen, Ph.D.

AT&T Laboratories
Middletown, New Jersey

Continued Titles

DISCRETE MATHEMATICS AND ITS APPLICATIONS
Series Editor KENNETH H. ROSEN

ELLIPTIC CURVES

Number Theory and Cryptography

Lawrence C. Washington

CHAPMAN & HALL/CRC

A CRC Press Company
Boca Raton   London   New York   Washington, D.C.

Library of Congress Cataloging-in-Publication Data

Washington, Lawrence C.
  Elliptic curves : number theory and cryptography / Lawrence C. Washington.
    p. cm. — (Discrete mathematics and its applications)
  Includes bibliographical references and index.
  ISBN 1-58488-365-0 (alk. paper)
    1. Curves, Elliptic. 2. Number theory. 3. Cryptography. I. Title. II. CRC Press Press series on
discrete mathematics and its applications.

QA567.2.E44W37 2003
516.3′52—dc21
                                                                        2003043972

Visit the CRC Press Web site at www.crcpress.com

© 2003 by Chapman & Hall/CRC

No claim to original U.S. Government works
International Standard Book Number 1-58488-365-0
Library of Congress Card Number 2003043972
Printed in the United States of America   3 4 5 6 7 8 9 0
Printed on acid-free paper

Preface

Over the last two or three decades, elliptic curves have been playing an increasingly important role both in number theory and in related fields such as cryptography. For example, in the 1980s, elliptic curves started being used in cryptography and elliptic curve techniques were developed for factorization and primality testing. In the 1980s and 1990s, elliptic curves played an important role in the proof of Fermat's Last Theorem. The goal of the present book is to develop the theory of elliptic curves assuming only modest backgrounds in elementary number theory and in groups and fields, approximately what would be covered in a strong undergraduate or beginning graduate abstract algebra course. In particular, we do not assume the reader has seen any algebraic geometry. Except for a few isolated sections, which can be omitted if desired, we do not assume the reader knows Galois theory. We implicitly use Galois theory for finite fields, but in this case everything can be done explicitly in terms of the Frobenius map so the general theory is not needed. The relevant facts are explained in an appendix.

The book provides an introduction to both the cryptographic side and the number theoretic side of elliptic curves. For this reason, we treat elliptic curves over finite fields early in the book, namely in Chapter 4. This immediately leads into the discrete logarithm problem and cryptography in Chapters 5, 6, and 7. The reader only interested in cryptography can subsequently skip to Chapters 10 and 11, where complex multiplication and the Weil and Tate-Lichtenbaum pairings are discussed. But surely anyone who becomes an expert in cryptographic applications will have a little curiosity as to how elliptic curves are used in number theory. Similarly, a non-applications oriented reader could skip Chapters 5, 6, and 7 and jump straight into the number theory in Chapters 8 and beyond. But the cryptographic applications are interesting and provide examples for how the theory can be used.

There are several fine books on elliptic curves already in the literature. This book in no way is intended to replace Silverman's excellent two volumes [90], [92], which are the standard references for the number theoretic aspects of elliptic curves. Instead, the present book covers some of the same material, plus applications to cryptography, from a more elementary viewpoint. It is hoped that readers of this book will subsequently find Silverman's books more accessible and will appreciate their slightly more advanced approach. The books by Knapp [47] and Koblitz [49] should be consulted for an approach to the arithmetic of elliptic curves that is more analytic than either this book or [90]. For the cryptographic aspects of elliptic curves, there is the recent book of Blake et al. [7], which gives more details on several algorithms than the present book, but contains few proofs. It should be consulted by serious students of elliptic curve cryptography. We hope that the present book provides a good introduction to and explanation of the mathematics used in that book. The books by Enge [28], Koblitz [51], [50], and Menezes [64] also treat elliptic curves from a cryptographic viewpoint and can be profitably consulted.

Notation. The symbols $\mathbf{Z}$, $\mathbf{F}_q$, $\mathbf{Q}$, $\mathbf{R}$, $\mathbf{C}$ denote the integers, the finite field with q elements, the rationals, the reals, and the complex numbers, respectively. We have used $\mathbf{Z}_n$ (rather than $\mathbf{Z}/n\mathbf{Z}$) to denote the integers mod n. However, when p is a prime and we are working with $\mathbf{Z}_p$ as a field, rather than as a group or ring, we use $\mathbf{F}_p$ in order to remain consistent with the notation $\mathbf{F}_q$. Note that $\mathbf{Z}_p$ does not denote the p-adic integers. This choice was made for typographic reasons since the integers mod p are used frequently, while a symbol for the p-adic integers is used only in a few examples in Chapter 13 (where we use $\mathcal{O}_p$). The p-adic rationals are denoted by $\mathbf{Q}_p$. If K is a field, then $\overline{K}$ denotes an algebraic closure of K. If R is a ring, then $R^\times$ denotes the invertible elements of R. When K is a field, $K^\times$ is therefore the multiplicative group of nonzero elements of K. Throughout the book, the letters K and E are generally used to denote a field and an elliptic curve (except in Chapter 9, where K is used a few times for an elliptic integral).

Acknowledgments. The author thanks Bob Stern of CRC Press for suggesting that this book be written and for his encouragement, and the editorial staff at CRC Press for their help during the preparation of the book. Ed Eikenberg, Jim Owings, Susan Schmoyer, Brian Conrad, and Sam Wagstaff made many suggestions that greatly improved the manuscript. Of course, there is always room for more improvement. Please send suggestions and corrections to the author (lcw@math.umd.edu). Corrections will be listed on the web site for the book (www.math.umd.edu/~lcw/ellipticcurves.html).

To Susan and Patrick

Contents

Chapter 1

Introduction

Suppose a collection of cannonballs is piled in a square pyramid with one ball on the top layer, four on the second layer, nine on the third layer, etc. If the pile collapses, is it possible to rearrange the balls into a square array?

Figure 1.1

A Pyramid of Cannonballs

If the pyramid has three layers, then this cannot be done since there are $1 + 4 + 9 = 14$ balls, which is not a perfect square. Of course, if there is only one ball, it forms a height one pyramid and also a one-by-one square. If there are no cannonballs, we have a height zero pyramid and a zero-by-zero square. Besides theses trivial cases, are there any others? We propose to find another example, using a method that goes back to Diophantus (around 250 A.D.).

If the pyramid has height x, then there are

$$1^2 + 2^2 + 3^2 + \cdots + x^2 = \frac{x(x+1)(2x+1)}{6}$$

balls (see Exercise 1.1). We want this to be a perfect square, which means that we want to find a solution to

$$y^2 = \frac{x(x+1)(2x+1)}{6}$$

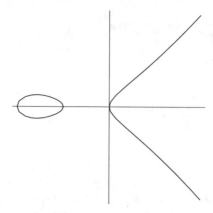

Figure 1.2

$y^2 = x(x+1)(2x+1)/6$

in positive integers x, y. An equation of this type represents an **elliptic curve**. The graph is given in Figure 1.2.

The method of Diophantus uses the points we already know to produce new points. Let's start with the points $(0,0)$ and $(1,1)$. The line through these two points is $y = x$. Intersecting with the curve gives the equation

$$x^2 = \frac{x(x+1)(2x+1)}{6} = \frac{1}{3}x^3 + \frac{1}{2}x^2 + \frac{1}{6}x.$$

Rearranging yields

$$x^3 - \frac{3}{2}x^2 + \frac{1}{2}x = 0.$$

Fortunately, we already know two roots of this equation: $x = 0$ and $x = 1$. This is because the roots are the x-coordinates of the intersections between the line and the curve. We could factor the polynomial to find the third root, but there is a better way. Note that for any numbers a, b, c, we have

$$(x - a)(x - b)(x - c) = x^3 - (a + b + c)x^2 + (ab + ac + bc)x - abc.$$

Therefore, when the coefficient of x^3 is 1, the negative of the coefficient of x^2 is the sum of the roots.

In our case, we have roots $0, 1$, and x, so

$$0 + 1 + x = \frac{3}{2}.$$

Therefore, $x = 1/2$. Since the line was $y = x$, we have $y = 1/2$, too. It's hard to say what this means in terms of piles of cannonballs, but at least we have found another point on the curve. In fact, we automatically have even one more point, namely $(1/2, -1/2)$, because of the symmetry of the curve.

Let's repeat the above procedure using the points $(1/2, -1/2)$ and $(1, 1)$. Why do we use these points? We are looking for a point of intersection somewhere in the first quadrant, and the line through these two points seems to be the best choice. The line is easily seen to be $y = 3x - 2$. Intersecting with the curve yields

$$(3x - 2)^2 = \frac{x(x + 1)(2x + 1)}{6}.$$

This can be rearranged to obtain

$$x^3 - \frac{51}{2}x^2 + \cdots = 0.$$

(By the above trick, we will not need the lower terms.) We already know the roots $1/2$ and 1, so we obtain

$$\frac{1}{2} + 1 + x = \frac{51}{2},$$

or $x = 24$. Since $y = 3x - 2$, we find that $y = 70$. This means that

$$1^2 + 2^2 + 3^2 + \cdots + 24^2 = 70^2.$$

If we have 4900 cannonballs, we can arrange them in a pyramid of height 24, or put them in a 70-by-70 square. If we keep repeating the above procedure, for example, using the point just found as one of our points, we'll obtain infinitely many rational solutions to our equation. However, it can be shown that $(24, 70)$ is the only solution to our problem in positive integers other than the trivial solution with $x = 1$. This requires more sophisticated techniques and we omit the details.

Here is another example of Diophantus's method. Is there a right triangle with rational sides with area equal to 5? The smallest Pythagorean triple $(3,4,5)$ yields a triangle with area 6, so we see that we cannot restrict our attention to integers. Now look at the triangle with sides $(8, 15, 17)$. This yields a triangle with area 60. If we divide the sides by 2, we end up with a triangle with sides $(4, 15/2, 17/2)$ and area 15. So it is possible to have non-integral sides but integral area.

Let the triangle we are looking for have sides a, b, c, as in Figure 1.3. Since the area is $ab/2 = 5$, we are looking for rational numbers a, b, c such that

$$a^2 + b^2 = c^2, \qquad ab = 10.$$

A little manipulation yields

$$\left(\frac{a + b}{2}\right)^2 = \frac{a^2 + 2ab + b^2}{4} = \frac{c^2 + 20}{4} = \left(\frac{c}{2}\right)^2 + 5,$$

$$\left(\frac{a - b}{2}\right)^2 = \frac{a^2 - 2ab + b^2}{4} = \frac{c^2 - 20}{4} = \left(\frac{c}{2}\right)^2 - 5.$$

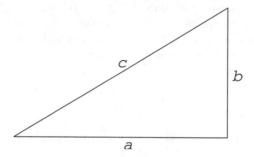

Figure 1.3

Let $x = (c/2)^2$. Then we have

$$x - 5 = ((a - b)/2)^2 \quad \text{and} \quad x + 5 = ((a + b)/2)^2.$$

We are therefore looking for a rational number x such that

$$x - 5, \quad x, \quad x + 5$$

are simultaneously squares of rational numbers. Another way to say this is that we want three squares of rational numbers to be in an arithmetical progression with difference 5.

Suppose we have such a number x. Then the product $(x - 5)(x)(x + 5) = x^3 - 25x$ must also be a square, so we need a rational solution to

$$y^2 = x^3 - 25x.$$

As above, this is the equation of an elliptic curve. Of course, if we have such a rational solution, we are not guaranteed that there will be a corresponding rational triangle (see Exercise 1.2). However, once we have a rational solution with $y \neq 0$, we can use it to obtain another solution that does correspond to a rational triangle (see Exercise 1.2). This is what we'll do below.

For future use, we record that

$$x = \left(\frac{c}{2}\right)^2, \quad y = ((x - 5)(x)(x + 5))^{1/2} = \frac{(a - b)(c)(a + b)}{8} = \frac{(a^2 - b^2)c}{8}.$$

There are three "obvious" points on the curve: $(-5, 0), (0, 0), (5, 0)$. These do not help us much. They do not yield triangles and the line through any two of them intersects the curve in the remaining point. A small search yields the point $(-4, 6)$. The line through this point and any one of the three other points yields nothing useful. The only remaining possibility is to take the line through $(-4, 6)$ and itself, namely, the tangent line to the curve at the $(-4, 6)$. Implicit differentiation yields

$$2yy' = 3x^2 - 25, \quad y' = \frac{3x^2 - 25}{2y} = \frac{23}{12}.$$

The tangent line is therefore

$$y = \frac{23}{12}x + \frac{41}{3}.$$

Intersecting with the curve yields

$$\left(\frac{23}{12}x + \frac{41}{3}\right)^2 = x^3 - 25x,$$

which implies

$$x^3 - \left(\frac{23}{12}\right)^2 x^2 + \cdots = 0.$$

Since the line is tangent to the curve at $(-4, 6)$, the root $x = -4$ is a double root. Therefore the sum of the roots is

$$-4 - 4 + x = \left(\frac{23}{12}\right)^2.$$

We obtain $x = 1681/144 = (41/12)^2$. The equation of the line yields $y = 62279/1728$.

Since $x = (c/2)^2$, we obtain $c = 41/6$. Therefore,

$$\frac{62279}{1728} = y = \frac{(a^2 - b^2)c}{8} = \frac{41(a^2 - b^2)}{48}.$$

This yields

$$a^2 - b^2 = \frac{1519}{36}.$$

Since

$$a^2 + b^2 = c^2 = (41/6)^2,$$

we solve to obtain $a^2 = 400/9$ and $b^2 = 9/4$. We obtain a triangle (see Figure 1.4) with

$$a = \frac{20}{3}, \quad b = \frac{3}{2}, \quad c = \frac{41}{6},$$

which has area 5. This is, of course, the $(40, 9, 41)$ triangle rescaled by a factor of 6.

There are infinitely many other solutions. These can be obtained by successively repeating the above procedure, for example, starting with the point just found (see Exercise 1.4).

The question of which integers n can occur as areas of right triangles with integer sides is known as the **congruent number problem**. Another formulation, as we saw above, is whether there are three rational squares in arithmetic progression with difference n. It appears in Arab manuscripts around 900 A.D. A conjectural answer to the problem was proved by Tunnell in the 1980s [102]. Recall that an integer n is called squarefree if n is not

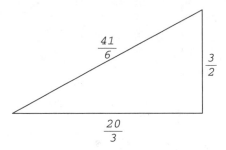

Figure 1.4

a multiple of any perfect square other than 1. For example, 5 and 15 are squarefree, while 24 and 75 are not.

CONJECTURE 1.1

Let n be an odd, squarefree, positive integer. Then n can be expressed as the area of a right triangle with rational sides if and only if the number of integer solutions to

$$2x^2 + y^2 + 8z^2 = n$$

with z even equals the number of solutions with z odd.

   Let $n = 2m$ with m odd, squarefree, and positive. Then n can be expressed as the area of a right triangle with rational sides if and only if the number of integer solutions to

$$4x^2 + y^2 + 8z^2 = m$$

with z even equals the number of integer solutions with z odd.

Tunnell [102] proved that if there is a triangle with area n, then the number of odd solutions equals the number of even solutions. However, the proof of the converse, namely that the condition on the number of solutions implies the existence of a triangle of area n, uses the Conjecture of Birch and Swinnerton-Dyer, which is not yet proved (see Chapter 12).

For example, consider $n = 5$. There are no solutions to $2x^2 + y^2 + 8z^2 = 5$. Since $0 = 0$, the condition is trivially satisfied and the existence of a triangle of area 5 is predicted. Now consider $n = 1$. The solutions to $2x^2 + y^2 + 8z^2 = 1$ are $(x, y, z) = (0, 1, 0)$ and $(0, -1, 0)$, and both have z even. Since $2 \neq 0$, there is no rational right triangle of area 1. This was first proved by Fermat by his method of descent (see Chapter 8).

For a nontrivial example, consider $n = 41$. The solutions to $2x^2 + y^2 + 8z^2 = 41$ are

$$(\pm 4, \pm 3, 0), (\pm 4, \pm 1, \pm 1), (\pm 2, \pm 5, \pm 1), (\pm 2, \pm 1, \pm 2), (0, \pm 3, \pm 2)$$

(all possible combinations of plus and minus signs are allowed). There are 32 solutions in all. There are 16 solutions with z even and 16 with z odd. Therefore, we expect a triangle with area 41. The same method as above, using the tangent line at the point $(-9, 120)$ to the curve $y^2 = x^3 - 41^2 x$, yields the triangle with sides $(40/3, 123/20, 881/60)$ and area 41.

For much more on the congruent number problem, see [49].

Finally, let's consider the quartic Fermat equation. We want to show that

$$a^4 + b^4 = c^4 \tag{1.1}$$

has no solutions in nonzero integers a, b, c. This equation represents the easiest case of Fermat's Last Theorem, which asserts that the sum of two nonzero nth powers of integers cannot be a nonzero nth power when $n \geq 3$. This general result was proved by Wiles (using work of Frey, Ribet, Serre, Mazur, Taylor, ...) in 1994 using properties of elliptic curves. We'll discuss some of these ideas in Chapter 13, but, for the moment, we restrict our attention to the much easier case of $n = 4$. The first proof in this case was due to Fermat.

Suppose $a^4 + b^4 = c^4$ with $a \neq 0$. Let

$$x = 2\frac{b^2 + c^2}{a^2}, \quad y = 4\frac{b(b^2 + c^2)}{a^3}$$

(see Example 2.2). A straightforward calculation shows that

$$y^2 = x^3 - 4x.$$

In Chapter 8 we'll show that the only rational solutions to this equation are

$$(x, y) = (0, 0), \ (2, 0), \ (-2, 0).$$

These all correspond to $b = 0$, so there are no nontrivial integer solutions of (1.1).

The cubic Fermat equation also can be changed to an elliptic curve. Suppose that $a^3 + b^3 = c^3$ and $abc \neq 0$. Since $a^3 + b^3 = (a + b)(a^2 - ab + b^2)$, we must have $a + b \neq 0$. Let

$$x = 12\frac{c}{a + b}, \quad y = 36\frac{a - b}{a + b}.$$

Then

$$y^2 = x^3 - 432.$$

(Where did this change of variables come from? See Section 2.5.2.) It can be shown (but this is not easy) that the only rational solutions to this equation are $(x, y) = (12, \pm 36)$. The case $y = 36$ yields $a - b = a + b$, so $b = 0$. Similarly, $y = -36$ yields $a = 0$. Therefore, there are no solutions to $a^3 + b^3 = c^3$ when $abc \neq 0$.

Exercises

1.1 Use induction to show that

$$1^2 + 2^2 + 3^2 + \cdots + x^2 = \frac{x(x+1)(2x+1)}{6}$$

for all integers $x \geq 0$.

1.2 (a) Show that if x, y are rational numbers satisfying $y^2 = x^3 - 25x$ and x is a square of a rational number, then this does not imply that $x + 5$ and $x - 5$ are squares. (*Hint:* Let $x = 25/4$.)
(b) Let n be an integer. Show that if x, y are rational numbers satisfying $y^2 = x^3 - n^2x$, and $x \neq 0, \pm n$, then the tangent line to this curve at (x, y) intersects the curve in a point (x_1, y_1) such that x_1, $x_1 - n$, $x_1 + n$ are squares of rational numbers. (For a more general statement, see Theorem 8.14.) This shows that the method used in the text is guaranteed to produce a triangle of area n if we can find a starting point with $x \neq 0, \pm n$.

1.3 Diophantus did not work with analytic geometry and certainly did not know how to use implicit differentiation to find the slope of the tangent line. Here is how he could find the tangent to $y^2 = x^3 - 25x$ at the point $(-4, 6)$. It appears that Diophantus regarded this simply as an algebraic trick. Newton seems to have been the first to recognize the connection with finding tangent lines.
(a) Let $x = -4 + t$, $y = 6 + mt$. Substitute into $y^2 = x^3 - 25x$. This yields a cubic equation in t that has $t = 0$ as a root.
(b) Show that choosing $m = 23/12$ makes $t = 0$ a double root.
(c) Find the nonzero root t of the cubic and use this to produce $x = 1681/144$ and $y = 62279/1728$.

1.4 Use the tangent line at $(x, y) = (1681/144, 62279/1728)$ to find another right triangle with area 5.

1.5 Show that the change of variables $x_1 = 12x + 6$, $y_1 = 72y$ changes the curve $y_1^2 = x_1^3 - 36x_1$ to $y^2 = x(x+1)(2x+1)/6$.

Chapter 2

The Basic Theory

2.1 Weierstrass Equations

For most situations in this book, an **elliptic curve** E is the graph of an equation of the form

$$y^2 = x^3 + Ax + B,$$

where A and B are constants. This will be referred to as the **Weierstrass equation** for an elliptic curve. We will need to specify what set A, B, x, and y belong to. Usually, they will be taken to be elements of a field, for example, the real numbers $\mathbf{R}$, the complex numbers $\mathbf{C}$, the rational numbers $\mathbf{Q}$, one of the finite fields $\mathbf{F}_p \,(= \mathbf{Z}_p)$ for a prime p, or one of the finite fields $\mathbf{F}_q$, where $q = p^k$ with $k \geq 1$. In fact, for almost all of this book, the reader who is not familiar with fields may assume that a field means one of the fields just listed. If K is a field with $A, B \in K$, then we say that E **is defined over** K. Throughout this book, E and K will implicitly be assumed to denote an elliptic curve and a field over which E is defined.

If we want to consider points with coordinates in some field $L \supseteq K$, we write $E(L)$. By definition, this set always contains the point ∞ defined later in this section:

$$E(L) = \{\infty\} \cup \left\{ (x, y) \in L \times L \,|\, y^2 = x^3 + Ax + B \right\}.$$

It is not possible to draw meaningful pictures of elliptic curves over most fields. However, for intuition, it is useful to think in terms of graphs over the real numbers. These have two basic forms, depicted in Figure 2.1.

The cubic $y^2 = x^3 - x$ in the first case has three distinct real roots. In the second case, the cubic $y^2 = x^3 + x$ has only one real root.

What happens if there is a multiple root? We don't allow this. Namely, we assume that

$$4A^3 + 27B^2 \neq 0.$$

If the roots of the cubic are r_1, r_2, r_3, then it can be shown that the discriminant of the cubic is

$$((r_1 - r_2)(r_1 - r_3)(r_2 - r_3))^2 = -(4A^3 + 27B^2).$$

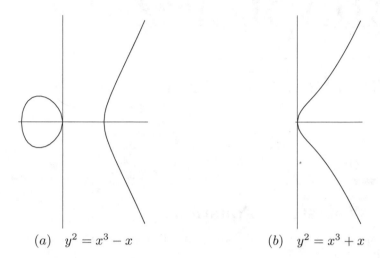

(a) $y^2 = x^3 - x$ (b) $y^2 = x^3 + x$

Figure 2.1

Therefore, the roots of the cubic must be distinct. However, the case where the roots are not distinct is still interesting and will be discussed in Section 2.9.

In order to have a little more flexibility, we also allow somewhat more general equations of the form

$$y^2 + a_1 xy + a_3 y = x^3 + a_2 x^2 + a_4 x + a_6, \tag{2.1}$$

where $a_1, \ldots, a_6$ are constants. This more general form (we'll call it the **generalized Weierstrass equation**) is useful when working with fields of characteristic 2 and characteristic 3. If the characteristic of the field is not 2, then we can divide by 2 and complete the square:

$$\left(y + \frac{a_1 x}{2} + \frac{a_3}{2}\right)^2 = x^3 + \left(a_2 + \frac{a_1^2}{4}\right)x^2 + \left(a_4 + \frac{a_1 a_3}{2}\right)x + \left(\frac{a_3^2}{4} + a_6\right),$$

which can be written as

$$y_1^2 = x^3 + a_2' x^2 + a_4' x + a_6',$$

with $y_1 = y + a_1 x/2 + a_3/2$ and with some constants a_2', a_4', a_6'. If the characteristic is also not 3, then we can let $x_1 = x + a_2'/3$ and obtain

$$y_1^2 = x_1^3 + A x_1 + B,$$

for some constants A, B.

In most of this book, we will develop the theory using the Weierstrass equation, occasionally pointing out what modifications need to be made in characteristics 2 and 3. In Section 2.7, we discuss the case of characteristic 2 in more detail, since the formulas for the (non-generalized) Weierstrass equation do not apply. In contrast, these formulas are correct in characteristic 3 for curves of the form $y^2 = x^3 + Ax + B$, but there are curves that are not of this form. The general case for characteristic 3 can be obtained by using the present methods to treat curves of the form $y^2 = x^3 + Cx^2 + Ax + B$.

Finally, suppose we start with an equation

$$cy^2 = dx^3 + ax + b$$

with $c, d \neq 0$. Multiply both sides of the equation by $c^3 d^2$ to obtain

$$(c^2 dy)^2 = (cdx)^3 + (ac^2 d)(cdx) + (bc^3 d^2).$$

The change of variables

$$y_1 = c^2 dy, \quad x_1 = cdx$$

yields an equation in Weierstrass form.

Later in this chapter, we will meet other types of equations that can be transformed into Weierstrass equations for elliptic curves. These will be useful in certain contexts.

For technical reasons, it is useful to add a **point at infinity** to an elliptic curve. In Section 2.3, this concept will be made rigorous. However, it is easiest to regard it as a point (∞, ∞), usually denoted simply by ∞, sitting at the top of the y-axis. For computational purposes, it will be a formal symbol satisfying certain computational rules. For example, a line is said to pass through ∞ exactly when this line is vertical (i.e., $x =$constant). The point ∞ might seem a little unnatural, but we will see that including it has very useful consequences.

We now make one more convention regarding ∞. It not only is at the top of the y-axis, it is also at the bottom of the y-axis. Namely, we think of the ends of the y-axis as wrapping around and meeting (perhaps somewhere in the back behind the page) in the point ∞. This might seem a little strange. However, if we are working with a field other than the real numbers, for example, a finite field, then there might not be any meaningful ordering of the elements and therefore distinguishing a top and a bottom of the y-axis might not make sense. In fact, in this situation, the ends of the y-axis do not have meaning until we introduce projective coordinates in Section 2.3. This is why it is best to regard ∞ as a formal symbol satisfying certain properties. Also, we have arranged that two vertical lines meet at ∞. By symmetry, if they meet at the top of the y-axis, they should also meet at the bottom. But two lines should intersect in only one point, so the "top ∞" and the "bottom ∞" need to be the same. In any case, this will be a useful property of ∞.

2.2 The Group Law

As we saw in Chapter 1, we could start with two points, or even one point, on an elliptic curve, and produce another point. We now examine this process in more detail.

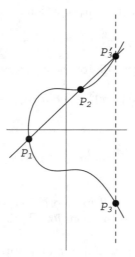

Figure 2.2

Adding Points on an Elliptic Curve

Start with two points

$$P_1 = (x_1, y_1), \quad P_2 = (x_2, y_2)$$

on an elliptic curve E given by the equation $y^2 = x^3 + Ax + B$. Define a new point P_3 as follows. Draw the line L through P_1 and P_2. We'll see below that L intersects E in a third point P_3'. Reflect P_3' across the x-axis (i.e., change the sign of the y-coordinate) to obtain P_3. We define

$$P_1 + P_2 = P_3.$$

Examples below will show that this is not the same as adding coordinates of the points. It might be better to denote this operation by $P_1 +_E P_2$, but we opt for the simpler notation since we will never be adding points by adding coordinates.

Assume first that $P_1 \neq P_2$ and that neither point is ∞. Draw the line L through P_1 and P_2. Its slope is

$$m = \frac{y_2 - y_1}{x_2 - x_1}.$$

If $x_1 = x_2$, then L is vertical. We'll treat this case later, so let's assume that $x_1 \neq x_2$. The equation of L is then

$$y = m(x - x_1) + y_1.$$

To find the intersection with E, substitute to get

$$(m(x - x_1) + y_1)^2 = x^3 + Ax + B.$$

This can be rearranged to the form

$$0 = x^3 - m^2 x^2 + \cdots .$$

The three roots of this cubic correspond to the three points of intersection of L with E. Generally, solving a cubic is not easy, but in the present case we already know two of the roots, namely x_1 and x_2, since P_1 and P_2 are points on both L and E. Therefore, we could factor the cubic to obtain the third value of x. But there is an easier way. As in Chapter 1, if we have a cubic polynomial $x^3 + ax^2 + bx + c$ with roots r, s, t, then

$$x^3 + ax^2 + bx + c = (x - r)(x - s)(x - t) = x^3 - (r + s + t)x^2 + \cdots .$$

Therefore,

$$r + s + t = -a.$$

If we know two roots r, s, then we can recover the third as $t = -a - r - s$.

In our case, we obtain

$$x = m^2 - x_1 - x_2$$

and

$$y = m(x - x_1) + y_1.$$

Now, reflect across the x-axis to obtain the point $P_3 = (x_3, y_3)$:

$$x_3 = m^2 - x_1 - x_2, \qquad y_3 = m(x_1 - x_3) - y_1.$$

In the case that $x_1 = x_2$ but $y_1 \neq y_2$, the line through P_1 and P_2 is a vertical line, which therefore intersects E in ∞. Reflecting ∞ across the x-axis yields the same point ∞ (this is why we put ∞ at both the top and the bottom of the y-axis). Therefore, in this case $P_1 + P_2 = \infty$.

Now consider the case where $P_1 = P_2 = (x_1, y_1)$. When two points on a curve are very close to each other, the line through them approximates a tangent line. Therefore, when the two points coincide, we take the line L through them to be the tangent line. Implicit differentiation allows us to find the slope m of L:

$$2y \frac{dy}{dx} = 3x^2 + A, \quad \text{so} \quad m = \frac{dy}{dx} = \frac{3x_1^2 + A}{2y_1}.$$

If $y_1 = 0$ then the line is vertical and we set $P_1 + P_2 = \infty$, as before. (*Technical point:* if $y_1 = 0$, then the numerator $3x_1^2 + A \neq 0$. See Exercise 2.2.) Therefore, assume that $y_1 \neq 0$. The equation of L is

$$y = m(x - x_1) + y_1,$$

as before. We obtain the cubic equation

$$0 = x^3 - m^2 x^2 + \cdots .$$

This time, we know only one root, namely x_1, but it is a double root since L is tangent to E at P_1. Therefore, proceeding as before, we obtain

$$x_3 = m^2 - 2x_1, \qquad y_3 = m(x_1 - x_3) - y_1.$$

Finally, suppose $P_2 = \infty$. The line through P_1 and ∞ is a vertical line that intersects E in the point P_1' that is the reflection of P_1 across the x-axis. When we reflect P_1' across the x-axis to get $P_3 = P_1 + P_2$, we are back at P_1. Therefore

$$P_1 + \infty = P_1$$

for all points P_1 on E. Of course, we extend this to include $\infty + \infty = \infty$.

Let's summarize the above discussion:

GROUP LAW
Let E be an elliptic curve defined by $y^2 = x^3 + Ax + B$. Let $P_1 = (x_1, y_1)$ and $P_2 = (x_2, y_2)$ be points on E with $P_1, P_2 \neq \infty$. Define $P_1 + P_2 = P_3 = (x_3, y_3)$ as follows:

1. If $x_1 \neq x_2$, then

$$x_3 = m^2 - x_1 - x_2, \qquad y_3 = m(x_1 - x_3) - y_1, \qquad \text{where } m = \frac{y_2 - y_1}{x_2 - x_1}.$$

2. If $x_1 = x_2$ but $y_1 \neq y_2$, then $P_1 + P_2 = \infty$.

3. If $P_1 = P_2$ and $y_1 \neq 0$, then

$$x_3 = m^2 - 2x_1, \qquad y_3 = m(x_1 - x_3) - y_1, \qquad \text{where } m = \frac{3x_1^2 + A}{2y_1}.$$

4. If $P_1 = P_2$ and $y_1 = 0$, then $P_1 + P_2 = \infty$.

Moreover, define

$$P + \infty = P$$

for all points P on E.

Note that when P_1 and P_2 have coordinates in a field L that contains A and B, then $P_1 + P_2$ also has coordinates in L. Therefore $E(L)$ is closed under the above addition of points.

This addition of points might seem a little unnatural. Later (in Chapters 9 and 11), we'll interpret it as corresponding to some very natural operations, but, for the present, let's show that it has some nice properties.

THEOREM 2.1

The addition of points on an elliptic curve E satisfies the following properties:

1. *(commutativity)* $P_1 + P_2 = P_2 + P_1$ *for all P_1, P_2 on E.*

2. *(existence of identity)* $P + \infty = P$ *for all points P on E.*

3. *(existence of inverses) Given P on E, there exists P' on E with $P+P' = \infty$. This point P' will usually be denoted $-P$.*

4. *(associativity) $(P_1 + P_2) + P_3 = P_1 + (P_2 + P_3)$ for all P_1, P_2, P_3 on E.*

In other words, the points on E form an additive abelian group with ∞ as the identity element.

PROOF   The commutativity is obvious, either from the formulas or from the fact that the line through P_1 and P_2 is the same as the line through P_2 and P_1. The identity property of ∞ holds by definition. For inverses, let P' be the reflection of P across the x-axis. Then $P + P' = \infty$.

Finally, we need to prove associativity. This is by far the most subtle and non-obvious property of the addition of points on E. It is possible to define many laws of composition satisfying (1), (2), (3) for points on E, either simpler or more complicated than the one being considered. But it is very unlikely that such a law will be associative. In fact, it is rather surprising that the law of composition that we have defined is associative. After all, we start with two points P_1 and P_2 and perform a certain procedure to obtain a third point $P_1 + P_2$. Then we repeat the procedure with $P_1 + P_2$ and P_3 to obtain $(P_1 + P_2) + P_3$. If we instead start by adding P_2 and P_3, then computing $P_1 + (P_2 + P_3)$, there seems to be no obvious reason that this should give the same point as the other computation.

The associative law can be verified by calculation with the formulas. There are several cases, depending on whether or not $P_1 = P_2$, and whether or not $P_3 = (P_1 + P_2)$, etc., and this makes the proof rather messy. However, we prefer a different approach, which we give in Section 2.4.   ∎

Warning: For the Weierstrass equation, if $P = (x, y)$, then $-P = (x, -y)$. For the generalized Weierstrass equation (2.1), this is no longer the case. If $P = (x, y)$ is on the curve described by (2.1), then (see Exercise 2.5)

$$-P = (x, \ -a_1 x - a_3 - y).$$

Example 2.1

The calculations of Chapter 1 can now be interpreted as adding points on elliptic curves. On the curve

$$y^2 = \frac{x(x+1)(2x+1)}{6},$$

we have

$$(0,0) + (1,1) = (\frac{1}{2}, -\frac{1}{2}), \quad (\frac{1}{2}, -\frac{1}{2}) + (1,1) = (24, -70).$$

On the curve

$$y^2 = x^3 - 25x,$$

we have

$$2(-4,6) = (-4,6) + (-4,6) = \left(\frac{1681}{144}, -\frac{62279}{1728} \right).$$

We also have

$$(0,0) + (-5,0) = (5,0), \quad 2(0,0) = 2(-5,0) = 2(5,0) = \infty.$$

☐

The fact that the points on an elliptic curve form an abelian group is be-hind most of the interesting properties and applications. The question arises: what can we say about the groups of points that we obtain? Here are some examples.

1. An elliptic curve over a finite field has only finitely many points with coordinates in that finite field. Therefore, we obtain a finite abelian group in this case. Properties of such groups, and applications to cryp-tography, will be discussed in later chapters.

2. If E is an elliptic curve defined over $\mathbf{Q}$, then $E(\mathbf{Q})$ is a finitely generated abelian group. This is the Mordell-Weil theorem, which we prove in Chapter 8. Such a group is isomorphic to $\mathbf{Z}^r \oplus F$ for some $r \geq 0$ and some finite group F. The integer r is called the **rank** of $E(\mathbf{Q})$. Determining r is fairly difficult in general. It is not known whether r can be arbitrarily large. At present, there are elliptic curves known with rank at least 24. The finite group F is easy to compute using the Lutz-Nagell theorem of Chapter 8. Moreover, a deep theorem of Mazur says that there are only finitely many possibilities for F, as E ranges over all elliptic curves defined over $\mathbf{Q}$.

3. An elliptic curve over the complex numbers $\mathbf{C}$ is isomorphic to a torus. This will be proved in Chapter 9. The usual way to obtain a torus is as $\mathbf{C}/\mathcal{L}$, where $\mathcal{L}$ is a lattice in $\mathbf{C}$. The usual addition of complex numbers induces a group law on $\mathbf{C}/\mathcal{L}$ that corresponds to the group law on the elliptic curve under the isomorphism between the torus and the elliptic curve.

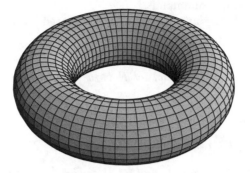

Figure 2.3

An Elliptic Curve over C

4. If E is defined over $\mathbf{R}$, then $E(\mathbf{R})$ is isomorphic to the unit circle S^1 or to $S^1 \oplus \mathbf{Z}_2$. The first case corresponds to the case where the cubic polynomial $x^3 + Ax + B$ has only one real root (think of the ends of the graph in Figure 2.1(b) as being hitched together at the point ∞ to get a loop). The second case corresponds to the case where the cubic has three real roots. The closed loop in Figure 2.1(a) is the set $S^1 \oplus \{1\}$, while the open-ended loop can be closed up using ∞ to obtain the set $S^1 \oplus \{0\}$. If we have an elliptic curve E defined over $\mathbf{R}$, then we can consider its complex points $E(\mathbf{C})$. These form a torus, as in (3) above. The real points $E(\mathbf{R})$ are obtained by intersecting the torus with a plane. If the plane passes through the hole in the middle, we obtain a curve as in Figure 2.1(a). If it does not pass through the hole, we obtain a curve as in Figure 2.1(b) (see Section 9.3).

If P is a point on an elliptic curve and k is a positive integer, then kP denotes $P + P + \cdots + P$ (with k summands). If $k < 0$, then $kP = (-P) + (-P) + \cdots (-P)$, with $|k|$ summands. To compute kP for a large integer k, it is inefficient to add P to itself repeatedly. It is much faster to use **successive doubling**. For example, to compute $19P$, we compute

$$2P, \quad 4P = 2P + 2P, \quad 8P = 4P + 4P, \quad 16P = 8P + 8P, \quad 19P = 16P + 2P + P.$$

This method allows us to compute kP for very large k, say of several hundred digits, very quickly. The only difficulty is that the size of the coordinates of the points increases very rapidly if we are working over the rational numbers (see Theorem 8.18). However, when we are working over a finite field, for example $\mathbf{F}_p$, this is not a problem because we can continually reduce mod p and thus keep the numbers involved relatively small. Note that the associative law allows us to make these computations without worrying about what order we use to combine the summands.

The method of successive doubling can be stated in general as follows:

INTEGER TIMES A POINT

Let k be a positive integer and let P be a point on an elliptic curve. The following procedure computes kP.

1. *Start with $a = k$, $B = \infty$, $C = P$.*

2. *If a is even, let $a = a/2$, and let $B = B$, $C = 2C$.*

3. *If a is odd, let $a = a - 1$, and let $B = B + C$, $C = C$.*

4. *If $a \neq 0$, go to step 2.*

5. *Output B.*

The output B is kP (see Exercise 2.4).

On the other hand, if we are working over a large finite field and are given points P and kP, it is very difficult to determine the value of k. This is called the **discrete logarithm problem** for elliptic curves and is the basis for the cryptographic applications that will be discussed in Chapter 6.

2.3   Projective Space and the Point at Infinity

We all know that parallel lines meet at infinity. Projective space allows us to make sense out of this statement and also to interpret the point at infinity on an elliptic curve.

Let K be a field. Two-dimensional **projective space** $\mathbf{P}_K^2$ over K is given by equivalence classes of triples (x, y, z) with $x, y, z \in K$ and at least one of x, y, z nonzero. Two triples (x_1, y_1, z_1) and (x_2, y_2, z_2) are said to be **equivalent** if there exists a nonzero element $\lambda \in K$ such that

$$(x_1, y_1, z_1) = (\lambda x_2, \lambda y_2, \lambda z_2).$$

We write $(x_1, y_1, z_1) \sim (x_2, y_2, z_2)$. The equivalence class of a triple only depends on the ratios of x to y to z. Therefore, the equivalence class of (x, y, z) is denoted $(x : y : z)$.

If $(x : y : z)$ is a point with $z \neq 0$, then $(x : y : z) = (x/z : y/z : 1)$. These are the "finite" points in $\mathbf{P}_K^2$. However, if $z = 0$ then dividing by z should be thought of as giving ∞ in either the x or y coordinate, and therefore the points $(x : y : 0)$ are called the **"points at infinity"** in $\mathbf{P}_K^2$. The point at infinity on an elliptic curve will soon be identified with one of these points at infinity in $\mathbf{P}_K^2$.

The two-dimensional **affine plane** over K is often denoted

$$\mathbf{A}_K^2 = \{(x, y) \in K \times K\}.$$

We have an inclusion

$$\mathbf{A}_K^2 \hookrightarrow \mathbf{P}_K^2$$

given by

$$(x, y) \mapsto (x : y : 1).$$

In this way, the affine plane is identified with the finite points in $\mathbf{P}_K^2$. Adding the points at infinity to obtain $\mathbf{P}_K^2$ can be viewed as a way of "compactifying" the plane (see Exercise 2.6).

A polynomial is **homogeneous** of degree n if it is a sum of terms of the form $ax^i y^j z^k$ with $a \in K$ and $i + j + k = n$. For example, $F(x, y, z) = 2x^3 - 5xyz + 7yz^2$ is homogeneous of degree 3. If a polynomial F is homogeneous of degree n then $F(\lambda x, \lambda y, \lambda z) = \lambda^n F(x, y, z)$ for all $\lambda \in K$. It follows that if F is homogeneous of some degree, and $(x_1, y_1, z_1) \sim (x_2, y_2, z_2)$, then $F(x_1, y_1, z_1) = 0$ if and only if $F(x_2, y_2, z_2) = 0$. Therefore, a zero of F in $\mathbf{P}_K^2$ does not depend on the choice of representative for the equivalence class, so the set of zeros of F in $\mathbf{P}_K^2$ is well defined.

If $F(x, y, z)$ is an arbitrary polynomial in x, y, z, then we cannot talk about a point in $\mathbf{P}_K^2$ where $F(x, y, z) = 0$ since this depends on the representative (x, y, z) of the equivalence class. For example, let $F(x, y, z) = x^2 + 2y - 3z$. Then $F(1, 1, 1) = 0$, so we might be tempted to say that F vanishes at $(1 : 1 : 1)$. But $F(2, 2, 2) = 2$ and $(1 : 1 : 1) = (2 : 2 : 2)$. To avoid this problem, we need to work with homogeneous polynomials.

If $f(x, y)$ is a polynomial in x and y, then we can make it homogeneous by inserting appropriate powers of z. For example, if $f(x, y) = y^2 - x^3 - Ax - B$, then we obtain the homogeneous polynomial $F(x, y, z) = y^2 z - x^3 - Axz^2 - Bz^3$. If F is homogeneous of degree n then

$$F(x, y, z) = z^n f(\frac{x}{z}, \frac{y}{z})$$

and

$$f(x, y) = F(x, y, 1).$$

We can now see what it means for two parallel lines to meet at infinity. Let

$$y = mx + b_1, \qquad y = mx + b_2$$

be two nonvertical parallel lines with $b_1 \neq b_2$. They have the homogeneous forms

$$y = mx + b_1 z, \qquad y = mx + b_2 z.$$

(The preceding discussion considered only equations of the form $f(x, y) = 0$ and $F(x, y, z) = 0$; however, there is nothing wrong with rearranging these equations to the form "homogeneous of degree n = homogeneous of degree n.") When we solve the simultaneous equations to find their intersection, we obtain

$$z = 0 \quad \text{and} \quad y = mx.$$

Since we cannot have all of x, y, z being 0, we must have $x \neq 0$. Therefore, we can rescale by dividing by x and find that the intersection of the two lines is

$$(x : mx : 0) = (1 : m : 0).$$

Similarly, if $x = c_1$ and $x = c_2$ are two vertical lines, they intersect in the point $(0 : 1 : 0)$. This is one of the points at infinity in $\mathbf{P}^2_K$.

Now let's look at the elliptic curve E given by $y^2 = x^3 + Ax + B$. Its homogeneous form is $y^2 z = x^3 + Axz^2 + Bz^3$. The points (x, y) on the original curve correspond to the points $(x : y : 1)$ in the projective version. To see what points on E lie at infinity, set $z = 0$ and obtain $0 = x^3$. Therefore $x = 0$, and y can be any nonzero number (recall that $(0 : 0 : 0)$ is not allowed). Rescale by y to find that $(0 : y : 0) = (0 : 1 : 0)$ is the only point at infinity on E. As we saw above, $(0 : 1 : 0)$ lies on every vertical line, so every vertical line intersects E at this point at infinity. Moreover, since $(0 : 1 : 0) = (0 : -1 : 0)$, the "top" and the "bottom" of the y-axis are the same.

There are situations where using projective coordinates speeds up computations on elliptic curves (see [20]). However, in this book we almost always work in affine (non-projective) coordinates and treat the point at infinity as a special case when needed. An exception is the proof of associativity of the group law given in the Section 2.4, where it will be convenient to have the point at infinity treated like any other point $(x : y : z)$.

2.4   Proof of Associativity

In this section, we prove the associativity of addition of points on an elliptic curve. The reader who is willing to believe this result may skip this section without missing anything that is needed in the rest of the book. However, as corollaries of the proof, we will obtain two results that are not about elliptic curves but which are interesting in their own right, namely the theorems of Pappus and Pascal.

First, we need to discuss lines in $\mathbf{P}^2_K$. The standard way to describe a line is by a linear equation: $ax + by + cz = 0$. Sometimes it is useful to give a parametric description:

$$x = a_1 u + b_1 v$$
$$y = a_2 u + b_2 v \qquad (2.2)$$
$$z = a_3 u + b_3 v$$

where u, v run through K, and at least one of u, v is nonzero. For example, if $a \neq 0$, the line

$$ax + by + cz = 0$$

can be described by

$$x = -(b/a)u - (c/a)v, y = u, z = v.$$

Suppose all the vectors (a_i, b_i) are multiples of each other, say $(a_i, b_i) = \lambda_i(a_1, b_1)$. Then $(x, y, z) = x(1, \lambda_2, \lambda_3)$ for all u, v such that $x \neq 0$. So we get a point, rather than a line, in projective space. Therefore, we need a condition on the coefficients $a_1, \ldots, b_3$ that ensure that we actually get a line. It is not hard to see that we must require the matrix

$$\begin{pmatrix} a_1 & b_1 \\ a_2 & b_2 \\ a_3 & b_3 \end{pmatrix}$$

to have rank 2 (cf. Exercise 2.8).

If $(u_1, v_1) = \lambda(u_2, v_2)$ for some $\lambda \in K^\times$, then (u_1, v_1) and (u_2, v_2) yield equivalent triples (x, y, z). Therefore, we can regard (u, v) as running through points $(u : v)$ in 1-dimensional projective space $\mathbf{P}_K^1$. Consequently, a line corresponds to a copy of the projective line $\mathbf{P}_K^1$ embedded in the projective plane.

We need to quantify the order to which a line intersects a curve at a point. The following gets us started.

LEMMA 2.2

Let $G(u, v)$ be a nonzero homogeneous polynomial and let $(u_0 : v_0) \in \mathbf{P}_K^1$. Then there exists an integer $k \geq 0$ and a polynomial $H(u, v)$ with $H(u_0, v_0) \neq 0$ such that

$$G(u, v) = (v_0 u - u_0 v)^k H(u, v).$$

PROOF Suppose $v_0 \neq 0$. Let m be the degree of G. Let $g(u) = G(u, v_0)$. By factoring out as large a power of $u - \frac{u_0}{v_0}$ as possible, we can write $g(u) = (v_0 u - u_0)^k h(u)$ for some k and for some polynomial h of degree $m - k$ with $h(u_0) \neq 0$. Let $H(u, v) = v^{m-k} h(u/v)$, so $H(u, v)$ is homogeneous of degree $m - k$. Then

$$G(u, v) = v^m g(u/v) = (v_0 u - u_0 v)^k H(u, v),$$

as desired.

If $v_0 = 0$, then $u_0 \neq 0$. Reversing the roles of u and v yields the proof in this case.   ∎

Let $f(x, y) = 0$ (where f is a polynomial) describe a curve C in the affine plane and let

$$x = a_1 t + b_1, y = a_2 t + b_2$$

be a line L written in terms of the parameter t. Let

$$\tilde{f}(t) = f(a_1 t + b_1, a_2 t + b_2).$$

Then L intersects C when $t = t_0$ if $\tilde{f}(t_0) = 0$. If $(t - t_0)^2$ divides $\tilde{f}(t)$, then L is tangent to C (if the point corresponding to t_0 is nonsingular. See Lemma 2.5). More generally, we say that L intersects C to order n at the point (x, y) corresponding to $t = t_0$ if $(t - t_0)^n$ is the highest power of $(t - t_0)$ that divides $\tilde{f}(t)$.

The homogeneous version of the above is the following. Let $F(x, y, z)$ be a homogeneous polynomial, so $F = 0$ describes a curve C in $\mathbf{P}_K^2$. Let L be a line given parametrically by (2.2) and let

$$\tilde{F}(u, v) = F(a_1 u + b_1 v, a_2 u + b_2 v, a_3 u + b_3 v).$$

We say that L **intersects** C **to order** n at the point $P = (x_0 : y_0 : z_0)$ corresponding to $(u : v) = (u_0 : v_0)$ if $(v_0 u - u_0 v)^n$ is the highest power of $(v_0 u - u_0 v)$ dividing $\tilde{F}(u, v)$. We denote this by

$$\mathrm{ord}_{L,P}(F) = n.$$

If $\tilde{F}$ is identically 0, then we let $\mathrm{ord}_{L,P}(F) = \infty$. It is not hard to show that $\mathrm{ord}_{L,P}(F)$ is independent of the choice of parameterization of the line L. Note that $v = v_0 = 1$ corresponds to the non-homogeneous situation above, and the definitions coincide (at least when $z \neq 0$). The advantage of the homogeneous formulation is that it allows us to treat the points at infinity along with the finite points in a uniform manner.

LEMMA 2.3

Let L_1 and L_2 be lines intersecting in a point P, and, for $i = 1, 2$, let $L_i(x, y, z)$ be a linear polynomial defining L_i. Then $\mathrm{ord}_{L_1, P}(L_2) = 1$ unless $L_1(x, y, z) = \alpha L_2(x, y, z)$ for some constant α, in which case $\mathrm{ord}_{L_1, P}(L_2) = \infty$.

PROOF   When we substitute the parameterization for L_1 into $L_2(x, y, z)$, we obtain $\tilde{L}_2$, which is a linear expression in u, v. Let P correspond to $(u_0 : v_0)$. Since $\tilde{L}_2(u_0, v_0) = 0$, it follows that $\tilde{L}_2(u, v) = \beta(v_0 u - u_0 v)$ for some constant β. If $\beta \neq 0$, then $\mathrm{ord}_{L_1, P}(L_2) = 1$. If $\beta = 0$, then all points on

L_1 lie on L_2. Since two points in $\mathbf{P}_K^2$ determine a line, and L_1 has at least three points ($\mathbf{P}_K^1$ always contains the points $(1:0), (0:1), (1:1)$), it follows that L_1 and L_2 are the same line. Therefore $L_1(x,y,z)$ is proportional to $L_2(x,y,z)$. ∎

Usually, a line that intersects a curve to order at least 2 is tangent to the curve. However, consider the curve C defined by

$$F(x,y,z) = y^2 z - x^3 = 0.$$

Let

$$x = au, \quad y = bu, \quad z = v$$

be a line through the point $P = (0:0:1)$. Note that P corresponds to $(u:v) = (0:1)$. We have $\tilde{F}(u,v) = u^2(b^2 v - a^3 u)$, so every line through P intersects C to order at least 2. The line with $b = 0$, which is the best choice for the tangent at P, intersects C to order 3. The affine part of C is the curve $y^2 = x^3$, which is pictured in Figure 2.7. The point $(0,0)$ is a singularity of the curve, which is why the intersections at P have higher orders than might be expected. This is a situation we usually want to avoid.

DEFINITION 2.4 *A curve C in $\mathbf{P}_K^2$ defined by $F(x,y,z) = 0$ is said to be* **nonsingular** *at a point P if at least one of the partial derivatives F_x, F_y, F_z is nonzero at P.*

For example, consider an elliptic curve defined by $F(x,y,z) = y^2 z - x^3 - Axz^2 - Bz^3 = 0$, and assume the characteristic of our field K is not 2 or 3. We have

$$F_x = -3x^2 - Az^2, \quad F_y = 2yz, \quad F_z = y^2 - 2Axz - 3Bz^2.$$

Suppose $P = (x:y:z)$ is a singular point. If $z = 0$, then $F_x = 0$ implies $x = 0$ and $F_z = 0$ implies $y = 0$, so $P = (0:0:0)$, which is impossible. Therefore $z \neq 0$, so we may take $z = 1$ (and therefore ignore it). If $F_y = 0$, then $y = 0$. Since $(x:y:1)$ lies on the curve, x must satisfy $x^3 + Ax + B = 0$. If $F_x = -(3x^2 + A) = 0$, then x is a root of a polynomial and a root of its derivative, hence a double root. Since we assumed that the cubic polynomial has no multiple roots, we have a contradiction. Therefore an elliptic curve has no singular points. Note that this is true even if we are considering points with coordinates in $\overline{K}$ (= algebraic closure of K). In general, by a **nonsingular curve** we mean a curve with no singular points in $\overline{K}$.

If we allow the cubic polynomial to have a multiple root x, then it is easy to see that the curve has a singularity at $(x:0:1)$. This case will be discussed in Section 2.9.

If P is a nonsingular point of a curve $F(x,y,z) = 0$, then the tangent line at P is

$$F_x(P)x + F_y(P)y + F_z(P)z = 0.$$

For example, if $F(x, y, z) = y^2 z - x^3 - Axz^2 - Bz^3 = 0$, then the **tangent line** at $(x_0 : y_0 : z_0)$ is

$$(-3x_0^2 - Az_0^2)x + 2y_0 z_0 y + (y_0^2 - 2Ax_0 z_0 - 3Bz_0^2)z = 0.$$

If we set $z_0 = z = 1$, then we obtain

$$(-3x_0^2 - A)x + 2y_0 y + (y_0^2 - 2Ax_0 - 3B) = 0.$$

Using the fact that $y_0^2 = x_0^3 + Ax_0 + B$, we can rewrite this as

$$(-3x_0^2 - A)(x - x_0) + 2y_0(y - y_0) = 0.$$

This is the tangent line in affine coordinates that we used in obtaining the formulas for adding a point to itself on an elliptic curve. Now let's look at the point at infinity on this curve. We have $(x_0 : y_0 : z_0) = (0 : 1 : 0)$. The tangent line is given by $0x + 0y + z = 0$, which is the "line at infinity" in $\mathbf{P}_K^2$. It intersects the elliptic curve only in the point $(0 : 1 : 0)$. This corresponds to the fact that $\infty + \infty = \infty$ on an elliptic curve.

LEMMA 2.5

Let $F(x, y, z) = 0$ define a curve C. If P is a nonsingular point of C, then there is exactly one line in $\mathbf{P}_K^2$ that intersects C to order at least 2, and it is the tangent to C at P.

PROOF   Let L be a line intersecting C to order $k \geq 1$. Parameterize L by (2.2) and substitute into F. This yields $\tilde{F}(u, v)$. Let $(u_0 : v_0)$ correspond to P. Then $\tilde{F} = (v_0 u - u_0 v)^k H(u, v)$ for some $H(u, v)$ with $H(u_0, v_0) \neq 0$. Therefore,

$$\tilde{F}_u(u, v) = kv_0(v_0 u - u_0 v)^{k-1} H(u, v) + (v_0 u - u_0 v)^k H_u(u, v)$$

and

$$\tilde{F}_v(u, v) = -ku_0(v_0 u - u_0 v)^{k-1} H(u, v) + (v_0 u - u_0 v)^k H_v(u, v).$$

It follows that $k \geq 2$ if and only if $\tilde{F}_u(u_0, v_0) = \tilde{F}_v(u_0, v_0) = 0$.

Suppose $k \geq 2$. The chain rule yields

$$\tilde{F}_u = a_1 F_x + a_2 F_y + a_3 F_z = 0, \quad \tilde{F}_v = b_1 F_x + b_2 F_y + b_3 F_z = 0 \qquad (2.3)$$

at P. Recall that since the parameterization (2.2) yields a line, the vectors (a_1, a_2, a_3) and (b_1, b_2, b_3) must be linearly independent.

Suppose L' is another line that intersects C to order at least 2. Then we obtain another set of equations

$$a_1' F_x + a_2' F_y + a_3' F_z = 0, \quad b_1' F_x + b_2' F_y + b_3' F_z = 0$$

at P.

If the vectors $\mathbf{a}' = (a_1', a_2', a_3')$ and $\mathbf{b}' = (b_1', b_2', b_3')$ span the same plane in K^3 as $\mathbf{a} = (a_1, a_2, a_3)$ and $\mathbf{b} = (b_1, b_2, b_3)$, then

$$\mathbf{a}' = \alpha\mathbf{a} + \beta\mathbf{b}, \quad \mathbf{b}' = \gamma\mathbf{a} + \delta\mathbf{b}$$

for some invertible matrix $\begin{pmatrix} \alpha & \beta \\ \gamma & \delta \end{pmatrix}$. Therefore,

$$u\mathbf{a}' + v\mathbf{b}' = (u\alpha + v\gamma)\mathbf{a} + (u\beta + v\delta)\mathbf{b} = u_1\mathbf{a} + v_1\mathbf{b}$$

for a new choice of parameters u_1, v_1. This means that L and L' are the same line.

If L and L' are different lines, then $\mathbf{a}, \mathbf{b}$ and $\mathbf{a}', \mathbf{b}'$ span different planes, so the vectors $\mathbf{a}, \mathbf{b}, \mathbf{a}', \mathbf{b}'$ must span all of K^3. Since (F_x, F_y, F_z) has dot product 0 with these vectors, it must be the 0 vector. This means that P is a singular point, contrary to our assumption.

Finally, we need to show that the tangent line intersects the curve to order at least 2. Suppose, for example, that $F_x \neq 0$ at P. The cases where $F_y \neq 0$ and $F_z \neq 0$ are similar. The tangent line can be given the parameterization

$$x = -(F_y/F_x)u - (F_z/F_x)v, \quad y = u, \quad z = v,$$

so

$$a_1 = -F_y/F_x, \; b_1 = -F_z/F_x, \; a_2 = 1, \; b_2 = 0, \; a_3 = 0 \; b_3 = 1$$

in the notation of (2.2). Substitute into (2.3) to obtain

$$\tilde{F}_u = (-F_y/F_x)F_x + F_y = 0, \quad \tilde{F}_v = (-F_z/F_x)F_x + F_z = 0.$$

By the discussion at the beginning of the proof, this means that the tangent line intersects the curve to order $k \geq 2$.   ∎

The associativity of elliptic curve addition will follow easily from the next result. The proof can be simplified if the points P_{ij} are assumed to be distinct. The cases where points are equal correspond to situations where tangent lines are used in the definition of the group law. Correspondingly, this is where it is more difficult to verify the associativity by direct calculation with the formulas for the group law.

THEOREM 2.6

Let $C(x, y, z)$ be a homogeneous cubic polynomial, and let C be the curve in $\mathbf{P}_K^2$ described by $C(x, y, z) = 0$. Let ℓ_1, ℓ_2, ℓ_3 and m_1, m_2, m_3 be lines in $\mathbf{P}_K^2$ such that $\ell_i \neq m_j$ for all i, j. Let P_{ij} be the point of intersection of ℓ_i and m_j. Suppose P_{ij} is a nonsingular point on the curve C for all $(i, j) \neq (3, 3)$. In addition, we require that if, for some i, there are $k \geq 2$ of the points

P_{i1}, P_{i2}, P_{i3} equal to the same point, then ℓ_i intersects C to order at least k at this point. Also, if, for some j, there are $k \geq 2$ of the points P_{1j}, P_{2j}, P_{3j} equal to the same point, then m_j intersects C to order at least k at this point. Then P_{33} also lies on the curve C.

PROOF   Express ℓ_1 in the parametric form (2.2). Then $C(x,y,z)$ becomes $\tilde{C}(u,v)$. The line ℓ_1 passes through P_{11}, P_{12}, P_{13}. Let $(u_1 : v_1), (u_2 : v_2), (u_3 : v_3)$ be the parameters on ℓ_1 for these points. Since these points lie on C, we have $\tilde{C}(u_i, v_i) = 0$ for $i = 1, 2, 3$.

Let m_j have equation $m_j(x, y, z) = a_j x + b_j y + c_j z = 0$. Substituting the parameterization for ℓ_1 yields $\tilde{m}_j(u, v)$. Since P_{ij} lies on m_j, we have $\tilde{m}_j(u_j, v_j) = 0$ for $j = 1, 2, 3$. Since $\ell_1 \neq m_j$ and since the zeros of $\tilde{m}_j$ yield the intersections of ℓ_1 and m_j, the function $\tilde{m}_j(u, v)$ vanishes only at P_{1j}, so the linear form $\tilde{m}_j$ is nonzero. Therefore, the product $\tilde{m}_1(u, v)\tilde{m}_2(u, v)\tilde{m}_3(u, v)$ is a nonzero cubic homogeneous polynomial. We need to relate this product to $\tilde{C}$.

LEMMA 2.7

Let $R(u, v)$ and $S(u, v)$ be homogeneous polynomials of degree 3, with $S(u, v)$ not identically 0, and suppose there are three points $(u_i : v_i)$, $i = 1, 2, 3$, at which R and S vanish. Moreover, if k of these points are equal to the same point, we require that R and S vanish to order at least k at this point (that is, $(v_i u - u_i v)^k$ divides R and S). Then there is a constant $\alpha \in K$ such that $R = \alpha S$.

PROOF   First, observe that a nonzero cubic homogeneous polynomial $S(u, v)$ can have at most 3 zeros $(u : v)$ in $\mathbf{P}_K^1$ (counting multiplicities). This can be proved as follows. Factor off the highest possible power of v, say v^k. Then $S(u, v)$ vanishes to order k at $(1 : 0)$, and $S(u, v) = v^k S_0(u, v)$ with $S_0(1, 0) \neq 0$. Since $S_0(u, 1)$ is a polynomial of degree $3 - k$, the polynomial $S_0(u, 1)$ can have at most $3 - k$ zeros, counting multiplicities (it has exactly $3 - k$ if K is algebraically closed). All points $(u : v) \neq (1 : 0)$ can be written in the form $(u : 1)$, so $S_0(u, v)$ has at most $3 - k$ zeros. Therefore, $S(u, v)$ has at most $k + (3 - k) = 3$ zeros in $\mathbf{P}_K^1$.

It follows easily that the condition that $S(u, v)$ vanish to order at least k could be replaced by the condition that $S(u, v)$ vanish to order exactly k. However, it is easier to check "at least" than "exactly." Since we are allowing the possibility that $R(u, v)$ is identically 0, this remark does not apply to R.

Let $(u_0, : v_0)$ be any point in $\mathbf{P}_K^1$ not equal to any of the $(u_i : v_i)$. (*Technical point*: If K has only two elements, then $\mathbf{P}_K^1$ has only three elements. In this case, enlarge K to $GF(4)$. The α we obtain is forced to be in K since it is the ratio of a coefficient of R and a coefficient of S, both of which are in K.) Since S can have at most three zeros, $S(u_0, v_0) \neq 0$. Let $\alpha = R(u_0, v_0)/S(u_0, v_0)$.

Then $R(u, v) - \alpha S(u, v)$ is a cubic homogeneous polynomial that vanishes at the four points $(u_i : v_i)$, $i = 0, 1, 2, 3$. Therefore $R - \alpha S$ must be identically zero. ∎

Returning to the proof of the theorem, we note that $\tilde{C}$ and $\tilde{m}_1 \tilde{m}_2 \tilde{m}_3$ vanish at the points $(u_i : v_i)$, $i = 1, 2, 3$. Moreover, if k of the points P_{1j} are the same point, then k of the linear functions vanish at this point, so the product $\tilde{m}_1(u, v)\tilde{m}_2(u, v)\tilde{m}_3(u, v)$ vanishes to order at least k. By assumption, $\tilde{C}$ vanishes to order at least k in this situation. By the lemma, there exists a constant α such that

$$\tilde{C} = \alpha \tilde{m}_1 \tilde{m}_2 \tilde{m}_3.$$

Let

$$C_1(x, y, z) = C(x, y, z) - \alpha m_1(x, y, z)m_2(x, y, z)m_3(x, y, z).$$

The line ℓ_1 can be described by a linear equation $\ell_1(x, y, z) = ax + by + cz = 0$. At least one coefficient is nonzero, so let's assume $a \neq 0$. The other cases are similar. The parameterization of the line ℓ_1 can be taken to be

$$x = -(b/a)u - (c/a)v, \quad y = u, \quad z = v. \tag{2.4}$$

Then $\tilde{C}_1(u, v) = C_1(-(b/a)u - (c/a)v, u, v)$. Write $C_1(x, y, z)$ as a polynomial in x with polynomials in y, z as coefficients. Writing

$$x^n = (1/a^n)\left((ax + by + cz) - (by + cz)\right)^n = (1/a^n)\left((ax + by + cz)^n + \cdots\right),$$

we can rearrange $C_1(x, y, z)$ to be a polynomial in $ax + by + cz$ whose coefficients are polynomials in y, z:

$$C_1(x, y, z) = a_3(y, z)(ax + by + cz)^3 + \cdots + a_0(y, z). \tag{2.5}$$

Substituting (2.4) into (2.5) yields

$$0 = \tilde{C}_1(u, v) = a_0(u, v),$$

since $ax + by + cz$ vanishes identically when x, y, z are written in terms of u, v. Therefore $a_0(y, z) = a_0(u, v)$ is the zero polynomial. It follows from (2.5) that $C_1(x, y, z)$ is a multiple of $\ell_1(x, y, z) = ax + by + cz$.

Similarly, there exists a constant β such that $C(x, y, z) - \beta \ell_1 \ell_2 \ell_3$ is a multiple of m_1.

Let

$$D(x, y, z) = C - \alpha m_1 m_2 m_3 - \beta \ell_1 \ell_2 \ell_3.$$

Then $D(x, y, z)$ is a multiple of ℓ_1 and a multiple of m_1.

LEMMA 2.8
$D(x, y, z)$ is a multiple of $\ell_1(x, y, z)m_1(x, y, z)$.

PROOF   Write $D = m_1 D_1$. We need to show that ℓ_1 divides D_1. We could quote some result about unique factorization, but instead we proceed as follows. Parameterize the line ℓ_1 via (2.4) (again, we are considering the case $a \neq 0$). Substituting this into the relation $D = m_1 D_1$ yields $\tilde{D} = \tilde{m}_1 \tilde{D}_1$. Since ℓ_1 divides D, we have $\tilde{D} = 0$. Since $m_1 \neq \ell_1$, we have $\tilde{m}_1 \neq 0$. Therefore $\tilde{D}_1(u, v)$ is the zero polynomial. As above, this implies that $D_1(x, y, z)$ is a multiple of ℓ_1, as desired.   ∎

By the lemma,

$$D(x, y, z) = \ell_1 m_1 \ell,$$

where $\ell(x, y, z)$ is linear. By assumption, $C = 0$ at P_{22}, P_{23}, P_{32}. Also, $\ell_1 \ell_2 \ell_3$ and $m_1 m_2 m_3$ vanish at these points. Therefore, $D(x, y, z)$ vanishes at these points. Our goal is to show that D is identically 0.

LEMMA 2.9
$\ell(P_{22}) = \ell(P_{23}) = \ell(P_{32}) = 0$.

PROOF   First suppose that $P_{13} \neq P_{23}$. If $\ell_1(P_{23}) = 0$, then P_{23} is on the line ℓ_1 and also on ℓ_2 and m_3 by definition. Therefore, P_{23} equals the intersection P_{13} of ℓ_1 and m_3. Since P_{23} and P_{13} are for the moment assumed to be distinct, this is a contradiction. Therefore $\ell_1(P_{23}) \neq 0$. Since $D(P_{23}) = 0$, it follows that $m_1(P_{23})\ell(P_{23}) = 0$.

Suppose now that $P_{13} = P_{23}$. Then, by the assumption in the theorem, m_3 is tangent to C at P_{23}, so $\mathrm{ord}_{m_3, P_{23}}(C) \geq 2$. Since $P_{13} = P_{23}$ and P_{23} lies on m_3, we have $\mathrm{ord}_{m_3, P_{23}}(\ell_1) = \mathrm{ord}_{m_3, P_{23}}(\ell_2) = 1$. Therefore, $\mathrm{ord}_{m_3, P_{23}}(\alpha \ell_1 \ell_2 \ell_3) \geq 2$. Also, $\mathrm{ord}_{m_3, P_{23}}(\beta m_1 m_2 m_3) = \infty$. Therefore, $\mathrm{ord}_{m_3, P_{23}}(D) \geq 2$, since D is a sum of terms, each of which vanishes to order at least 2. But $\mathrm{ord}_{m_3, P_{23}}(\ell_1) = 1$, so we have

$$\mathrm{ord}_{m_3, P_{23}}(m_1 \ell) \geq \mathrm{ord}_{m_3, P_{23}}(D) - \mathrm{ord}_{m_3, P_{23}}(\ell_1) \geq 1.$$

Therefore $m_1(P_{23})\ell(P_{23}) = 0$.

In both cases, we have $m_1(P_{23})\ell(P_{23}) = 0$.

If $m_1(P_{23}) \neq 0$, then $\ell(P_{23}) = 0$, as desired.

If $m_1(P_{23}) = 0$, then P_{23} lies on m_1, and also on ℓ_2 and m_3, by definition. Therefore, $P_{23} = P_{21}$, since ℓ_2 and m_1 intersect in a unique point. By assumption, ℓ_2 is therefore tangent to C at P_{23}. Therefore, $\mathrm{ord}_{\ell_2, P_{23}}(C) \geq 2$. As above, $\mathrm{ord}_{\ell_2, P_{23}}(D) \geq 2$, so

$$\mathrm{ord}_{\ell_2, P_{23}}(\ell_1 \ell) \geq 1.$$

If in this case we have $\ell_1(P_{23}) = 0$, then P_{23} lies on ℓ_1, ℓ_2, m_3. Therefore $P_{13} = P_{23}$. By assumption, the line m_3 is tangent to C at P_{23}. Since P_{23} is a nonsingular point of C, Lemma 2.5 says that $\ell_2 = m_3$, contrary to hypothesis. Therefore, $\ell_1(P_{23}) \neq 0$, so $\ell(P_{23}) = 0$.

Similarly, $\ell(P_{22}) = \ell(P_{32}) = 0.$ ∎

If $\ell(x, y, z)$ is identically 0, then D is identically 0. Therefore, assume that $\ell(x, y, z)$ is not zero and hence it defines a line ℓ.

First suppose that P_{23}, P_{22}, P_{32} are distinct. Then ℓ and ℓ_2 are lines through P_{23} and P_{22}. Therefore $\ell = \ell_2$. Similarly, $\ell = m_2$. Therefore $\ell_2 = m_2$, contradiction.

Now suppose that $P_{32} = P_{22}$. Then m_2 is tangent to C at P_{22}. As before,

$$\mathrm{ord}_{m_2, P_{22}}(\ell_1 m_1 \ell) \geq 2.$$

We want to show that this forces ℓ to be the same line as m_2.

If $m_1(P_{22}) = 0$, then P_{22} lies on m_1, m_2, ℓ_2. Therefore, $P_{21} = P_{22}$. This means that ℓ_2 is tangent to C at P_{22}. By Lemma 2.5, $\ell_2 = m_2$, contradiction. Therefore, $m_1(P_{22}) \neq 0$.

If $\ell_1(P_{22}) \neq 0$, then $\mathrm{ord}_{m_2, P_{22}}(\ell) \geq 2$. This means that ℓ is the same line as m_2.

If $\ell_1(P_{22}) = 0$, then $P_{22} = P_{32}$ lies on $\ell_1, \ell_2, \ell_3, m_2$, so $P_{12} = P_{22} = P_{32}$. Therefore $\mathrm{ord}_{m_2, P_{22}}(C) \geq 3$. By the reasoning above, we now have $\mathrm{ord}_{m_2, P_{22}}(\ell_1 m_1 \ell) \geq 3$. Since we have proved that $m_1(P_{22}) \neq 0$, we have $\mathrm{ord}_{m_2, P_{22}}(\ell) \geq 2$. This means that ℓ is the same line as m_2.

So now we have proved, under the assumption that $P_{32} = P_{22}$, that ℓ is the same line as m_2. By Lemma 2.9, P_{23} lies on ℓ, and therefore on m_2. It also lies on ℓ_2 and m_3. Therefore, $P_{22} = P_{23}$. This means that ℓ_2 is tangent to C at P_{22}. Since $P_{32} = P_{22}$ means that m_2 is also tangent to C at P_{22}, we have $\ell_2 = m_2$, contradiction. Therefore, $P_{32} \neq P_{22}$ (under the assumption that $\ell \neq 0$).

Similarly, $P_{23} \neq P_{22}$.

Finally, suppose $P_{23} = P_{32}$. Then P_{23} lies on ℓ_2, ℓ_3, m_2, m_3. This forces $P_{22} = P_{32}$, which we have just shown is impossible.

Therefore, all possibilities lead to contradictions. It follows that $\ell(x, y, z)$ must be identically 0. Therefore $D = 0$, so

$$C = \alpha \ell_1 \ell_2 \ell_3 + \beta m_1 m_2 m_3.$$

Since ℓ_3 and m_3 vanish at P_{33}, we have $C(P_{33}) = 0$, as desired. This completes the proof of Theorem 2.6. ∎

REMARK 2.10 Note that we proved the stronger result that

$$C = \alpha \ell_1 \ell_2 \ell_3 + \beta m_1 m_2 m_3$$

for some constants α, β. Since there are 10 coefficients in an arbitrary homogeneous cubic polynomial in three variables and we have required that C vanish at eight points (when the P_{ij} are distinct), it is not surprising that the set of possible polynomials is a two-parameter family. When the P_{ij} are not distinct, the tangency conditions add enough restrictions that we still obtain a two-parameter family.   ∎

We can now prove the associativity of addition for an elliptic curve. Let P, Q, R be points on E. Define the lines

$$\ell_1 = \overline{PQ}, \quad \ell_2 = \overline{\infty, Q+R}, \quad \ell_3 = \overline{R, P+Q}$$

$$m_1 = \overline{QR}, \quad m_2 = \overline{\infty, P+Q}, \quad m_3 = \overline{P, Q+R}.$$

We have the following intersections:

	ℓ_1	ℓ_2	ℓ_3
m_1	Q	$-(Q+R)$	R
m_2	$-(P+Q)$	∞	$P+Q$
m_3	P	$Q+R$	X

Assume for the moment that the hypotheses of the theorem are satisfied. Then all the points in the table, including X, lie on E. The line ℓ_3 has three points of intersection with E, namely $R, P+Q$, and X. By the definition of addition, $X = -((P+Q)+R)$. Similarly, m_3 intersects C in 3 points, which means that $X = -(P+(Q+R))$. Therefore, after reflecting across the x-axis, we obtain $(P+Q)+R = P+(Q+R)$, as desired.

It remains to verify the hypotheses of the theorem, namely that the orders of intersection are correct and that the lines ℓ_i are distinct from the lines m_j.

First we want to dispense with cases where ∞ occurs. The problem is that we treated ∞ as a special case in the definition of the group law. However, as pointed out earlier, the tangent line at ∞ intersects the curve only at ∞ (and intersects to order 3 at ∞). It follows that if two of the entries in a row or column of the above table of intersections are equal to ∞, then so is the third, and the line intersects the curve to order 3. Therefore, this hypothesis is satisfied.

It is also possible to treat directly the cases where some of the intersection points $P, Q, R, \pm(P+Q), \pm(Q+R)$ are ∞. In the cases where at least one of P, Q, R is ∞, associativity is trivial.

If $P+Q = \infty$, then $(P+Q)+R = \infty+R = R$. On the other hand, the sum $Q+R$ is computed by first drawing the line L through Q and R, which intersects E in $-(Q+R)$. Since $P+Q = \infty$, the reflection of Q across the x-axis is P. Therefore, the reflection L' of L passes through $P, -R$, and $Q+R$. The sum $P+(Q+R)$ is found by drawing the line through P and $Q+R$, which is L'. We have just observed that the third point of intersection of L' with E is $-R$. Reflecting yields $P+(Q+R) = R$, so associativity holds in this case.

Similarly, associativity holds when $Q+R = \infty$.

Finally, we need to consider what happens if some line ℓ_i equals some line m_j, since then Theorem 2.6 does not apply.

First, observe that if P, Q, R are collinear, then associativity is easily verified directly.

Second, suppose that $P, Q, Q + R$ are collinear. Then $P + (Q + R) = -Q$. Also, $P + Q = -(Q + R)$, so $(P + Q) + R = -(Q + R) + R$. The second equation of the following shows that associativity holds in this case.

LEMMA 2.11

Let P_1, P_2 be points on an elliptic curve. Then $(P_1 + P_2) - P_2 = P_1$ and $-(P_1 + P_2) + P_2 = -P_1$.

PROOF   The two relations are reflections of each other, so it suffices to prove the second one. The line L through P_1 and P_2 intersects the elliptic curve in $-(P_1 + P_2)$. Regarding L as the line through $-(P_1 + P_2)$ and P_2 yields $-(P_1 + P_2) + P_2 = -P_1$, as claimed.   ■

Suppose that $\ell_i = m_j$ for some i, j. We consider the various cases. By the above discussion, we may assume that all points in the table of intersections are finite, except for ∞ and possibly X. Note that each ℓ_i and each m_j meets E in three points (counting multiplicity), one of which is P_{ij}. If the two lines coincide, then the other two points must coincide in some order.

1. $\ell_1 = m_1$: Then P, Q, R are collinear, and associativity follows.

2. $\ell_1 = m_2$: In this case, P, Q, ∞ are collinear, so $P + Q = \infty$; associativity follows by the direct calculation made above.

3. $\ell_2 = m_1$: Similar to the previous case.

4. $\ell_1 = m_3$: Then $P, Q, Q + R$ are collinear; associativity was proved above.

5. $\ell_3 = m_1$: Similar to the previous case.

6. $\ell_2 = m_2$: Then $P + Q$ must be $\pm(Q + R)$. If $P + Q = Q + R$, then commutativity plus the above lemma yields

$$P = (P + Q) - Q = (Q + R) - Q = R.$$

Therefore,

$$(P+Q)+R = R+(P+Q) = P+(P+Q) = P+(R+Q) = P+(Q+R).$$

If $P + Q = -(Q + R)$, then

$$(P + Q) + R = -(Q + R) + R = -Q$$

and

$$P + (Q + R) = P - (P + Q) = -Q,$$

so associativity holds.

7. $\ell_2 = m_3$: In this case, the line m_3 through P and $(Q + R)$ intersects E in ∞, so $P = -(Q + R)$. Since $-(Q + R)$, Q, R are collinear, we have that P, Q, R are collinear and associativity holds.

8. $\ell_3 = m_2$: Similar to the previous case.

9. $\ell_3 = m_3$: Since ℓ_3 cannot intersect E in 4 points (counting multiplicities), it is easy to see that $P = R$ or $P = P + Q$ or $Q + R = P + Q$. The case $P = R$ was treated in the case $\ell_2 = m_2$. Assume $P = P + Q$. Adding $-P$ and applying Lemma 2.11 yields $\infty = Q$, in which case associativity immediately follows. If $Q + R = P + Q$, then adding $-Q$ and applying Lemma 2.11 yields $P = R$, which we have already treated.

If $\ell_i \neq m_j$ for all i, j, then the hypotheses of the theorem are satisfied, so the addition is associative, as proved above. This completes the proof of the associativity of elliptic curve addition. $\blacksquare$

REMARK 2.12   Note that for most of the proof, we did not use the Weierstrass equation for the elliptic curve. In fact, any nonsingular cubic curve would suffice. The identity O for the group law needs to be a point whose tangent line intersects to order 3. Three points sum to 0 if they lie on a straight line. Negation of a point P is accomplished by taking the line through O and P. The third point of intersection is then $-P$. Associativity of this group law follows just as in the Weierstrass case.

2.4.1   The Theorems of Pappus and Pascal

Theorem 2.6 has two other nice applications outside the realm of elliptic curves.

THEOREM 2.13 (Pascal's Theorem)

Let $ABCDEF$ be a hexagon inscribed in a conic section (ellipse, parabola, or hyperbola), where A, B, C, D, E, F are distinct points in the affine plane. Let X be the intersection of $\overline{AB}$ and $\overline{DE}$, let Y be the intersection of $\overline{BC}$ and $\overline{EF}$, and let Z be the intersection of $\overline{CD}$ and $\overline{FA}$. Then X, Y, Z are collinear (see Figure 2.4).

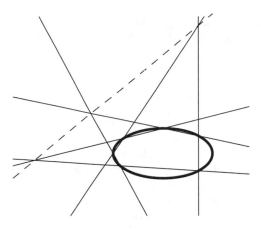

Figure 2.4
Pascal's Theorem

REMARK 2.14   (1) A conic is given by an equation $q(x,y) = ax^2 + bxy + cy^2 + dx + ey + f = 0$ with at least one of a, b, c nonzero. Usually, it is assumed that $b^2 - 4ac \neq 0$; otherwise, the conic degenerates into a product of two linear factors, and the graph is the union of two lines. The present theorem holds even in this case, as long as the points A, C, E lie on one of the lines, B, D, F lie on the other, and none is the intersection of the two lines.

(2) Possibly $\overline{AB}$ and $\overline{DE}$ are parallel, for example. Then X is an infinite point in $\mathbf{P}_K^2$.

(3) Note that X, Y, Z will always be distinct. This is easily seen as follows: First observe that X, Y, Z cannot lie on the conic since a line can intersect the conic in at most two points; the points A, B, C, D, E, F are assumed to be distinct and therefore exhaust all possible intersections. If $X = Y$, then $\overline{AB}$ and $\overline{BC}$ meet in both B and Y, and therefore the lines are equal. But this means that $A = C$, contradiction. Similarly, $X \neq Z$ and $Y \neq Z$.   ∎

PROOF   Define the following lines:

$$\ell_1 = \overline{EF}, \; \ell_2 = \overline{AB}, \; \ell_3 = \overline{CD}, \; m_1 = \overline{BC}, \; m_2 = \overline{DE}, \; m_3 = \overline{FA}.$$

We have the following table of intersections:

	ℓ_1	ℓ_2	ℓ_3
m_1	Y	B	C
m_2	E	X	D
m_3	F	A	Z

Let $q(x,y) = 0$ be the affine equation of the conic. In order to apply Theorem 2.6, we change $q(x, y)$ to its homogeneous form $Q(x, y, z)$. Let $\ell(x, y, z)$ be a linear form giving the line through X and Y. Then

$$C(x, y, z) = Q(x, y, z)\ell(x, y, z)$$

is a homogeneous cubic polynomial. The curve $C = 0$ contains all of the points in the table, with the possible exception of Z. It is easily checked that the only singular points of C are the points of intersection of $Q = 0$ and $\ell = 0$, and the intersection of the two lines comprising $Q = 0$ in the case of a degenerate conic. Since none of these points occur among the points we are considering, the hypotheses of Theorem 2.6 are satisfied. Therefore, $C(Z) = 0$. Since $Q(Z) \neq 0$, we must have $\ell(Z) = 0$, so Z lies on the line through X and Y. Therefore, X, Y, Z are collinear. This completes the proof of Pascal's theorem.   ∎

COROLLARY 2.15 (Pappus's Theorem)

Let ℓ and m be two distinct lines in the plane. Let A, B, C be distinct points of ℓ and let A', B', C' be distinct points of m. Assume that none of these points is the intersection of ℓ and m. Let X be the intersection of $\overline{AB'}$ and $\overline{A'B}$, let Y be the intersection of $\overline{B'C}$ and $\overline{BC'}$, and let Z be the intersection of $\overline{CA'}$ and $\overline{C'A}$. Then X, Y, Z are collinear (see Figure 2.5).

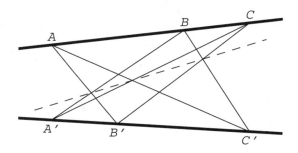

Figure 2.5

Pappus's Theorem

PROOF   This is the case of a degenerate conic in Theorem 2.13. The "hexagon" is $AB'CA'BC'$.   ∎

2.5   Other Equations for Elliptic Curves

In this book, we are mainly using the Weierstrass equation for an elliptic curve. However, elliptic curves arise in various other guises, and it is worthwhile to discuss these briefly.

2.5.1   Legendre Equation

This is a variant on the Weierstrass equation. Its advantage is that it allows us to express all elliptic curves over an algebraically closed field (of characteristic not 2) in terms of one parameter.

PROPOSITION 2.16
Let K be a field of characteristic not 2 and let

$$y^2 = x^3 + ax^2 + bx + c = (x - e_1)(x - e_2)(x - e_3)$$

be an elliptic curve E over K with $e_1, e_2, e_3 \in K$. Let

$$x_1 = (e_2 - e_1)^{-1}(x - e_1), \quad y_1 = (e_2 - e_1)^{-3/2}y, \quad \lambda = \frac{e_3 - e_1}{e_2 - e_1}.$$

Then $\lambda \neq 0, 1$ and

$$y_1^2 = x_1(x_1 - 1)(x_1 - \lambda).$$

PROOF   This is a straightforward calculation.   ∎

The parameter λ for E is not unique. In fact, each of

$$\{\lambda, \frac{1}{\lambda}, 1 - \lambda, \frac{1}{1 - \lambda}, \frac{\lambda}{\lambda - 1}, \frac{\lambda - 1}{\lambda}\}$$

yields a Legendre equation for E. They correspond to the six permutations of the roots e_1, e_2, e_3. It can be shown that these are the only values of λ corresponding to E, so the map $\lambda \mapsto E$ is six-to-one, except where $\lambda = -1, 1/2, 2$, or $\lambda^2 - \lambda + 1 = 0$ (in these situations, the above set collapses; see Exercise 2.9).

2.5.2   Cubic Equations

It is possible to start with a cubic equation $C(x, y) = 0$, over a field K of characteristic not 2 or 3, that has a point with $x, y \in K$ and find a change of variables that transforms the equation to Weierstrass form (although possibly $4A^3 + 27B^2 = 0$). The procedure is fairly complicated (see [21] or [68]), so we restrict our attention to a specific example.

Consider the cubic Fermat equation

$$x^3 + y^3 + z^3 = 0.$$

The fact that this equation has no rational solutions with $xyz \neq 0$ was conjectured by the Arabs in the 900s and represents a special case of Fermat's Last Theorem, which asserts that the sum of two nonzero nth powers of integers cannot be a nonzero nth power when $n \geq 3$. The first proof in the case $n = 3$ was probably due to Fermat. We'll discuss some of the ideas for the proof in the general case in Chapter 13.

Suppose that $x^3 + y^3 + z^3 = 0$ and $xyz \neq 0$. Since $x^3 + y^3 = (x + y)(x^2 - xy + y^2)$, we must have $x + y \neq 0$. Write

$$\frac{x}{z} = u + v, \quad \frac{y}{z} = u - v.$$

Then $(u + v)^3 + (u - v)^3 + 1 = 0$, so $2u^3 + 6uv^2 + 1 = 0$. Divide by u^3 (since $x + y \neq 0$, we have $u \neq 0$) and rearrange to obtain

$$6(v/u)^2 = -(1/u)^3 - 2.$$

Let

$$x_1 = \frac{-6}{u} = -12\frac{z}{x + y}, \quad y_1 = \frac{36v}{u} = 36\frac{x - y}{x + y}.$$

Then

$$y_1^2 = x_1^3 - 432.$$

It can be shown (this is somewhat nontrivial) that the only rational solutions to this equation are $(x_1, y_1) = (12, \pm 36)$, and ∞. The case $y_1 = 36$ yields $x - y = x + y$, so $y = 0$. Similarly, $y_1 = -36$ yields $x = 0$. The point with $(x_1, y_1) = \infty$ corresponds to $x = -y$, which means that $z = 0$. Therefore, there are no solutions to $x^3 + y^3 + z^3 = 0$ when $xyz \neq 0$.

2.5.3   Quartic Equations

Occasionally, we will meet curves defined by equations of the form

$$v^2 = au^4 + bu^3 + cu^2 + du + e, \tag{2.6}$$

with $a \neq 0$. If we have a point (p, q) lying on the curve with $p, q \in K$, then the equation (when it is nonsingular) can be transformed into a Weierstrass equation by a change of variables that uses rational functions with coefficients in the field K. Note that an elliptic curve E defined over a field K always has a point in $E(K)$, namely ∞ (whose projective coordinates $(0 : 1 : 0)$ certainly lie in K). Therefore, if we are going to transform a curve C into Weierstrass form in such a way that all coefficients of the rational functions describing the transformation lie in K, then we need to start with a point on C that has coordinates in K.

There are curves of the form (2.6) that do not have points with coordinates in K. This phenomenon will be discussed in more detail in Chapter 8.

Suppose we have a curve defined by an equation (2.6) and suppose we have a point (p, q) lying on the curve. By changing x to $x + p$, we may assume $p = 0$, so the point has the form $(0, q)$.

First, suppose $q = 0$. If $d = 0$, then the curve has a singularity at $(u, v) = (0, 0)$. Therefore, assume $d \neq 0$. Then

$$(\frac{v}{u^2})^2 = d(\frac{1}{u})^3 + c(\frac{1}{u})^2 + b(\frac{1}{u}) + a.$$

This can be easily transformed into a Weierstrass equation in d/u and dv/u^2.

The harder case is when $q \neq 0$. We have the following result.

THEOREM 2.17
Let K be a field of characteristic not 2. Consider the equation

$$v^2 = au^4 + bu^3 + cu^2 + du + q^2$$

with $a, b, c, d, q \in K$. Let

$$x = \frac{2q(v + q) + du}{u^2}, \quad y = \frac{4q^2(v + q) + 2q(du + cu^2) - (d^2u^2/2q)}{u^3}.$$

Define

$$a_1 = d/q, \quad a_2 = c - (d^2/4q^2), \quad a_3 = 2qb, \quad a_4 = -4q^2a, \quad a_6 = a_2a_4.$$

Then

$$y^2 + a_1xy + a_3y = x^3 + a_2x^2 + a_4x + a_6.$$

The inverse transformation is

$$u = \frac{2q(x + c) - (d^2/2q)}{y}, \quad v = -q + \frac{u(ux - d)}{2q}.$$

The point $(u, v) = (0, q)$ corresponds to the point $(x, y) = \infty$ and $(u, v) = (0, -q)$ corresponds to $(x, y) = (-a_2, a_1a_2 - a_3)$.

PROOF   Most of the proof is a "straightforward" calculation that we omit. For the image of the point $(0, -q)$, see [21, pp. 105-107].   ∎

Example 2.2

Consider the equation

$$v^2 = u^4 + 1. \tag{2.7}$$

Then $a = 1$, $b = c = d = 0$, and $q = 1$. If

$$x = \frac{2(v+1)}{u^2}, \quad y = \frac{4(v+1)}{u^3},$$

then we obtain the elliptic curve E given by

$$y^2 = x^3 - 4x.$$

The inverse transformation is

$$u = 2x/y, \quad v = -1 + (2x^3/y^2).$$

The point $(u, v) = (0, 1)$ corresponds to ∞ on E, and $(u, v) = (0, -1)$ corresponds to $(0, 0)$. We will show in Chapter 8 that

$$E(\mathbf{Q}) = \{\infty, (0,0), (2,0), (-2,0)\}.$$

These correspond to $(u, v) = (0, 1), (0, -1)$, and points at infinity. Therefore, the only finite rational points on the quartic curve are $(u, v) = (0, \pm 1)$. It is easy to deduce from this that the only integer solutions to

$$a^4 + b^4 = c^2$$

satisfy $ab = 0$. This yields Fermat's Last Theorem for exponent 4. We will discuss this in more detail in Chapter 8.

It is worth considering briefly the situation at infinity in u, v. If we make the equation (2.7) homogeneous, we obtain

$$F(u, v, w) = v^2 w^2 - u^4 - w^4 = 0.$$

The points at infinity have $w = 0$. To find them, we set $w = 0$ and get $0 = u^4$, which means $u = 0$. We thus find only the point $(u : v : w) = (0 : 1 : 0)$. But we have two points, namely $(2, 0)$ and $(-2, 0)$ in the corresponding Weierstrass model. The problem is that $(u : v : w) = (0 : 1 : 0)$ is a singular point in the quartic model. At this point we have

$$F_u = F_v = F_w = 0.$$

What is happening is that the curve intersects itself at the point $(u : v : w) = (0 : 1 : 0)$. One branch of the curve is $v = +u^2\sqrt{1 + (1/u)^4}$ and the other is $v = -u^2\sqrt{1 + (1/u)^4}$. For simplicity, let's work with real or complex numbers. If we substitute the first of these expressions into $x = 2(v + 1)/u^2$ and take the limit as $u \to \infty$, we obtain

$$x = \frac{2(v+1)}{u^2} = \frac{2(1 - u^2\sqrt{1 + (1/u)^4})}{u^2} \to -2.$$

If we use the other branch, we find $x \to +2$. So the transformation that changes the quartic equation into the Weierstrass equation has pulled apart the two branches (the technical term is "resolved the singularities") at the singular point. ▯

2.5.4   Intersection of Two Quadratic Surfaces

The intersection of two quadratic surfaces in three-dimensional space, along with a point on this intersection, is usually an elliptic curve. Rather than work in full generality, we'll consider pairs of equations of the form

$$au^2 + bv^2 = e, \quad cu^2 + dw^2 = f,$$

where a, b, c, d, e, f are nonzero elements of a field K of characteristic not 2. Each separate equation may be regarded as a surface in uvw-space, and they intersect in a curve. We'll show that if we have a point P in the intersection, then we can transform this curve into an elliptic curve in Weierstrass form.

Before analyzing the intersection of these two surfaces, let's consider the first equation by itself. It can be regarded as giving a curve C in the uv-plane. Let $P = (u_0, v_0)$ be a point on C. Let L be the line through P with slope m:

$$u = u_0 + t, \quad v = v_0 + mt.$$

We want to find the other point where L intersects C. See Figure 2.6. Substitute into the equation for C and use the fact that $au_0^2 + bv_0^2 = e$ to

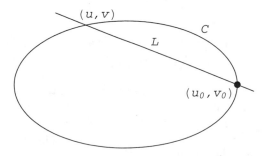

Figure 2.6

obtain

$$a(2u_0 t + t^2) + b(2v_0 mt + m^2 t^2) = 0.$$

Since $t = 0$ corresponds to (u_0, v_0), we factor out t and obtain

$$t = -\frac{2au_0 + 2bv_0 m}{a + bm^2}.$$

Therefore,

$$u = u_0 - \frac{2au_0 + 2bv_0 m}{a + bm^2}, \quad v = v_0 - \frac{2amu_0 + 2bv_0 m^2}{a + bm^2}.$$

We make the convention that $m = \infty$ yields $(u_0, -v_0)$, which is what we get if we are working with real numbers and let $m \to \infty$. Also, possibly the denominator $a + bm^2$ vanishes, in which case we get points "at infinity" in the uv-projective plane (see Exercise 2.10).

Note that if (u, v) is any point on C with coordinates in K, then the slope m of the line through (u, v) and P is in K (or is infinite). We have therefore obtained a bijection, modulo a few technicalities, between values of m (including ∞) and points on C (including points at infinity). The main point is that we have obtained a parameterization of the points on C. A similar procedure works for any conic section containing a point with coordinates in K.

Which value of m corresponds to the original point (u_0, v_0)? Let m be the slope of the tangent line at (u_0, v_0). The second point of intersection of the tangent line with the curve is again the point (u_0, v_0), so this slope is the desired value of m. The value $m = 0$ yields the point $(-u_0, v_0)$. This can be seen from the formulas, or from the fact that the line through $(-u_0, v_0)$ and (u_0, v_0) has slope 0.

We now want to intersect C, regarded as a "cylinder" in uvw-space, with the surface $cu^2 + dw^2 = f$. Substitute the expression just obtained for u to obtain

$$dw^2 = f - c\left(u_0 - \frac{2au_0 + 2bv_0 m}{a + bm^2}\right)^2.$$

This may be rewritten as

$$d(w(a + bm^2))^2 = (a + bm^2)^2 f - c(bu_0 m^2 - 2bv_0 m - au_0)^2$$
$$= (b^2 f - cb^2 u_0^2)m^4 + \cdots.$$

This may now be changed to Weierstrass form by the procedure given earlier. Note that the leading coefficient $b^2 f - cb^2 u_0^2$ equals $b^2 dw_0^2$. If $w_0 = 0$, then fourth degree polynomial becomes a cubic polynomial, so the equation just obtained is easily put into Weierstrass form. The leading term of this cubic polynomial vanishes if and only if $v_0 = 0$. But in this case, the point $(u_0, v_0, w_0) = (u_0, 0, 0)$ is a singular point of the uvw curve – a situation that we should avoid (see Exercise 2.11).

The procedure for changing "square = degree four polynomial" into Weierstrass form requires a point satisfying this equation. We could let m be the slope of the tangent line at (u_0, v_0), which corresponds to the point (u_0, v_0). The formula of Theorem 2.17 then requires that we shift the value of m to obtain $m = 0$. Instead, it's easier to use $m = 0$ directly, since this value corresponds to $(-u_0, v_0)$, as pointed out above.

Example 2.3

Consider the intersection

$$u^2 + v^2 = 2, \quad u^2 + 4w^2 = 5.$$

Let $(u_0, v_0, w_0) = (1, 1, 1)$. First, we parameterize the solutions to $u^2 + v^2 = 2$. Let $u = 1 + t, v = 1 + mt$. This yields

$$(1 + t)^2 + (1 + mt)^2 = 2,$$

which yields $t(2 + 2m) + t^2(1 + m^2) = 0$. Discarding the solution $t = 0$, we obtain $t = -(2 + 2m)/(1 + m^2)$, hence

$$u = 1 - \frac{2 + 2m}{1 + m^2} = \frac{m^2 - 2m - 1}{1 + m^2}, \quad v = 1 - m\frac{2 + 2m}{1 + m^2} = \frac{1 - 2m - m^2}{1 + m^2}.$$

Note that $m = -1$ corresponds to $(u, v) = (1, 1)$ (this is because the tangent at this point has slope $m = -1$). Substituting into $u^2 + 4w^2 = 5$ yields

$$4(w(1 + m^2))^2 = 5(1 + m^2)^2 - (m^2 - 2m - 1)^2 = 4m^4 + 4m^3 + 8m^2 - 4m + 4.$$

Letting $r = w(1 + m^2)$ yields

$$r^2 = m^4 + m^3 + 2m^2 - m + 1.$$

In Theorem 2.17, we use $q = 1$. The formulas then change this curve to the Weierstrass equation

$$y^2 - xy + 2y = x^3 + \frac{7}{4}x^2 - 4x - 7.$$

Completing the square yields

$$y_1^2 = x^3 + 2x^2 - 5x - 6,$$

where $y_1 = y + 1 - \frac{1}{2}x$. $\square$

2.6   The j-invariant

Let E be the elliptic curve given by $y^2 = x^3 + Ax + B$, where A, B are elements of a field K of characteristic not 2 or 3. If we let

$$x_1 = \mu^2 x, \quad y_1 = \mu^3 y, \tag{2.8}$$

with $\mu \in \overline{K}^{\times}$, then we obtain

$$y_1^2 = x_1^3 + A_1 x_1 + B_1,$$

with

$$A_1 = \mu^4 A, B_1 = \mu^6 B.$$

(In the generalized Weierstrass equation $y^2 + a_1 xy + a_3 y = x^3 + a_2 x^2 + a_4 x + a_6$, this change of variables yields new coefficients $\mu^i a_i$. This explains the numbering of the coefficients.)

Define the **j-invariant** of E to be

$$j = j(E) = 1728 \frac{4A^3}{4A^3 + 27B^2}.$$

Note that the denominator is the negative of the discriminant of the cubic, hence is nonzero by assumption. The change of variables (2.8) leaves j unchanged. The converse is true, too.

THEOREM 2.18

Let $y_1^2 = x_1^3 + A_1 x_1 + B_1$ and $y_2^2 = x_2^3 + A_2 x_2 + B_2$ be two elliptic curves with j-invariants j_1 and j_2, respectively. If $j_1 = j_2$, then there exists $\mu \neq 0$ in $\overline{K}$ (= algebraic closure of K) such that

$$A_2 = \mu^4 A_1, \quad B_2 = \mu^6 B_1.$$

The transformation

$$x_2 = \mu^2 x_1, \quad y_2 = \mu^3 y_1$$

takes one equation to the other.

PROOF   First, assume that $A_1 \neq 0$. Since this is equivalent to $j_1 \neq 0$, we also have $A_2 \neq 0$. Choose μ such that $A_2 = \mu^4 A_1$. Then

$$\frac{4A_2^3}{4A_2^3 + 27B_2^2} = \frac{4A_1^3}{4A_1^3 + 27B_1^2} = \frac{4\mu^{-12} A_2^3}{4\mu^{-12} A_2^3 + 27B_1^2} = \frac{4A_2^3}{4A_2^3 + 27\mu^{12} B_1^2},$$

which implies that

$$B_2^2 = (\mu^6 B_1)^2.$$

Therefore $B_2 = \pm \mu^6 B_1$. If $B_2 = \mu^6 B_1$, we're done. If $B_2 = -\mu^6 B_1$, then change μ to $i\mu$ (where $i^2 = -1$). This preserves the relation $A_2 = \mu^4 A_1$ and also yields $B_2 = \mu^6 B_1$.

If $A_1 = 0$, then $A_2 = 0$. Since $\Delta \neq 0$, we have $B_1, B_2 \neq 0$. Choose μ such that $B_2 = \mu^6 B_1$.   ∎

There are two special values of j that arise quite often:

1. $j = 0$: In this case, the elliptic curve E has the form $y^2 = x^3 + B$.

2. $j = 1728$: In this case, the elliptic curve has the form $y^2 = x^3 + Ax$.

The first one, with $B = -432$, was obtained in Section 2.5.2 from the Fermat equation $x^3 + y^3 + z^3 = 0$. The second curve, once with $A = -25$ and once with $A = -4$, appeared in Chapter 1.

The curves with $j = 0$ and with $j = 1728$ have automorphisms (bijective group homomorphisms from the curve to itself) other than the one defined by $(x, y) \mapsto (x, -y)$, which is an automorphism for any elliptic curve in Weierstrass form.

1. $y^2 = x^3 + B$ has the automorphism $(x, y) \mapsto (\zeta x, -y)$, where ζ is a nontrivial cube root of 1.

2. $y^2 = x^3 + Ax$ has the automorphism $(x, y) \mapsto (-x, iy)$, where $i^2 = -1$.

(See Exercise 2.12.)

Note that the j-invariant tells us when two curves are isomorphic over an algebraically closed field. However, if we are working with a non-algebraically closed field K, then it is possible to have two curves with the same j-invariant that cannot be transformed into each other using rational functions with coefficients in K. For cxample, both $y^2 = x^3 - 25x$ and $y^2 = x^3 - 4x$ have $j = 1728$. The first curve has infinitely points with coordinates in $\mathbf{Q}$, for example, all integer multiples of $(-4, 6)$ (see Section 8.4). The only rational points on the second curve are ∞, $(2, 0)$, $(-2, 0)$, and $(0, 0)$ (see Section 8.4). Therefore, we cannot change one curve into the other using only rational functions defined over $\mathbf{Q}$. Of course, we can use the field $\mathbf{Q}(\sqrt{10})$ to change one curve to the other via $(x, y) \mapsto (\mu^2 x, \mu^3 y)$, where $\mu = \sqrt{10}/2$.

If two different elliptic curves defined over a field K have the same j-invariant, then we say that the two curves are **twists** of each other.

Finally, we note that j is the j-invariant of

$$y^2 = x^3 + \frac{3j}{1728 - j}x + \frac{2j}{1728 - j} \tag{2.9}$$

when $j \neq 0, 1728$. Since $y^2 = x^3 + 1$ and $y^2 = x^3 + x$ have j-invariants 0 and 1728, we find the j-invariant gives a bijection between elements of K and $\overline{K}$-isomorphism classes of elliptic curves defined over K (that is, each $j \in K$ corresponds to an elliptic curve defined over K, and any two elliptic curves defined over K and with the same j-invariant can be transformed into each other by a change of variables (2.8) defined over $\overline{K}$.

If the characteristic of K is 2 or 3, the j-invariant can also be defined, and results similar to the above one hold. See Section 2.7 and Exercise 2.13.

2.7   Elliptic Curves in Characteristic 2

Since we have been using the Weierstrass equation rather than the generalized Weierstrass equation in most of the preceding sections, the formulas given do not apply when the field K has characteristic 2. In this section, we sketch what happens in this case.

Note that the Weierstrass equation is singular. Let $f(x, y) = y^2 - x^3 - Ax - B$. Then $f_y = 2y = 0$, since $2 = 0$ in characteristic 2. Let x_0 be a root (possibly in some extension of K) of $f_x = -3x^2 - A = 0$ and let y_0 be the square root of $x_0^3 + Ax_0 + B$. Then (x_0, y_0) lies on the curve and $f_x(x_0, y_0) = f_y(x_0, y_0) = 0$.

Therefore, we work with the generalized Weierstrass equation for an elliptic curve E:

$$y^2 + a_1 xy + a_3 y = x^3 + a_2 x^2 + a_4 x + a_6.$$

If $a_1 \neq 0$, then the change of variables

$$x = a_1^2 x_1 + (a_3/a_1), \quad y = a_1^3 y_1 + a_1^{-3}(a_1^2 a_4 + a_3^2)$$

changes the equation to the form

$$y_1^2 + x_1 y_1 = x_1^3 + a_2' x_1^2 + a_6'.$$

This curve is nonsingular if and only if $a_6' \neq 0$. The j-invariant in this case is $1/a_6'$.

If $a_1 = 0$, we let $x = x_1 + a_2$, $y = y_1$ to obtain an equation of the form

$$y_1^2 + a_3' y_1 = x_1^3 + a_4' x_1 + a_6'.$$

This curve is nonsingular if and only if $a_3' \neq 0$. The j-invariant is 0.

Let's return to the generalized Weierstrass equation and look for points at infinity. Make the equation homogeneous:

$$y^2 z + a_1 xyz + a_3 yz^2 = x^3 + a_2 x^2 z + a_4 xz^2 + a_6 z^3.$$

Now set $z = 0$ to obtain $0 = x^3$. Therefore, $\infty = (0 : 1 : 0)$ is the only point at infinity on E, just as with the standard Weierstrass equation. A line L through (x_0, y_0) and ∞ is a vertical line $x = x_0$. If (x_0, y_0) lies on E then the other point of intersection of L and E is $(x_0, -a_1 x_0 - a_3 - y_0)$. See Exercise 2.5.

We can now describe addition of points. Of course, $P + \infty = P$, for all points P. Three points P, Q, R add to ∞ if and only if they are collinear. The negation of a point is given by

$$-(x, y) = (x, -a_1 x - a_3 - y).$$

To add two points P_1 and P_2, we therefore proceed as follows. Draw the line L through P_1 and P_2 (take the tangent if $P_1 = P_2$). It will intersect E in a third point P_3'. Now compute $P_3 = -P_3'$ by the formula just given (do not simply reflect across the x-axis). Then $P_1 + P_2 = P_3$.

The proof that this addition law is associative is the same as that given in Section 2.4. The points on E, including ∞, therefore form an abelian group.

Since we will need it later, let's look at the formula for doubling a point in characteristic 2. To keep the formulas from becoming too lengthy, we'll treat separately the two cases obtained above.

1. $y^2 + xy = x^3 + a_2 x^2 + a_6$. Rewrite this as $y^2 + xy + x^3 + a_2 x^2 + a_6 = 0$ (remember, we are in characteristic 2). Implicit differentiation yields

$$xy' + (y + x^2) = 0$$

(since $2 = 0$ and $3 = 1$). Therefore the slope of the line L through $P = (x_0, y_0)$ is $m = (y_0 + x_0^2)/x_0$. The line is

$$y = m(x - x_0) + y_0 = mx + b$$

for some b. Substitute to find the intersection (x_1, y_1) of L and E:

$$0 = (mx + b)^2 + x(mx + b) + x^3 + a_2 x^2 + a_6 = x^3 + (m^2 + m + a_2)x^2 + \cdots .$$

The sum $x_0 + x_0 + x_1$ of the roots is $(m^2 + m + a_2)$, so we obtain

$$x_1 = m^2 + m + a_2 = \frac{y_0^2 + x_0^4 + x_0 y_0 + x_0^3 + a_2 x_0^2}{x_0^2} - \frac{x_0^4 + a_6}{x_0^2}$$

(since $y_0^2 = x_0 y_0 + x_0^3 + a_2 x_0^2 + a_6$). The y-coordinate of the intersection is $y_1 = m(x_1 - x_0) + y_0$. The point (x_1, y_1) equals $-2P$. Therefore $2P = (x_2, y_2)$, with

$$x_2 = (x_0^4 + a_6)/x_0^2, \quad y_2 = -x_1 - y_1 = x_1 + y_1.$$

2. $y^2 + a_3 y = x^3 + a_4 x + a_6$. Rewrite this as $y^2 + a_3 y + x^3 + a_4 x + a_6 = 0$. Implicit differentiation yields

$$a_3 y' + (x^2 + a_4) = 0.$$

Therefore the tangent line L is

$$y = m(x - x_0) + y_0, \quad \text{with} \quad m = \frac{x_0^2 + a_4}{a_3}.$$

Substituting and solving, as before, finds the point of intersection (x_1, y_1) of L and E, where

$$x_1 = m^2 = \frac{x_0^4 + a_4^2}{a_3^2}$$

and $y_1 = m(x_1 - x_0) + y_0$. Therefore, $2P = (x_2, y_2)$ with

$$x_2 = (x_0^4 + a_4^2)/a_3^2, \quad y_2 = a_3 + y_1.$$

2.8   Endomorphisms

The main purpose of this section is to prove Proposition 2.20, which will be used in the proof of Hasse's theorem in Chapter 4. We'll also prove a few technical results on separable endomorphisms. The reader willing to believe that every endomorphism used in this book is separable, except for powers of the Frobenius map and multiplication by multiples of p in characteristic p, can safely omit the technical parts of this section.

By an **endomorphism** of E, we mean a homomorphism $\alpha : E(\overline{K}) \to E(\overline{K})$ that is given by rational functions. In other words, $\alpha(P_1+P_2) = \alpha(P_1)+\alpha(P_2)$, and there are rational functions (quotients of polynomials) $R_1(x,y), R_2(x,y)$ with coefficients in $\overline{K}$ such that

$$\alpha(x,y) = (R_1(x,y),\ R_2(x,y))$$

for all $(x,y) \in E(\overline{K})$. There are a few technicalities when the rational functions are not defined at a point. These will be dealt with below. Of course, since α is a homomorphism, we have $\alpha(\infty) = \infty$. We will also assume that α is nontrivial; that is, there exists some (x,y) such that $\alpha(x,y) \neq \infty$. The trivial endomorphism that maps every point to ∞ will be denoted by 0.

Example 2.4

Let E be given by $y^2 = x^3 + Ax + B$ and let $\alpha(P) = 2P$. Then α is a homomorphism and

$$\alpha(x,y) = (R_1(x,y),\ R_2(x,y)),$$

where

$$R_1(x,y) = \left(\frac{3x^2 + A}{2y}\right)^2 - 2x$$

$$R_2(x,y) = \left(\frac{3x^2 + A}{2y}\right)\left(3x - \left(\frac{3x^2 + A}{2y}\right)^2\right) - y.$$

Since α is a homomorphism given by rational functions it is an endomorphism of E.   ☐

It will be useful to have a standard form for the rational functions describing an endomorphism. For simplicity, we assume that our elliptic curve is given in Weierstrass form. Let $R(x,y)$ be any rational function. Since $y^2 = x^3+Ax+B$ for all $(x,y) \in E(\overline{K})$, we can replace any even power of y by a polynomial in x and replace any odd power of y by y times a polynomial in x and obtain a rational function that gives the same function as $R(x, y)$ on points in $E(\overline{K})$. Therefore, we may assume that

$$R(x, y) = \frac{p_1(x) + p_2(x)y}{p_3(x) + p_4(x)y}.$$

Moreover, we can rationalize the denominator by multiplying the numerator and denominator by $p_3 - p_4 y$ and then replacing y^2 by $x^3 + Ax + B$. This yields

$$R(x, y) = \frac{q_1(x) + q_2(x)y}{q_3(x)}. \tag{2.10}$$

Consider an endomorphism given by

$$\alpha(x, y) = (R_1(x, y), R_2(x, y)),$$

as above. Since α is a homomorphism,

$$\alpha(x, -y) = \alpha(-(x, y)) = -\alpha(x, y).$$

This means that

$$R_1(x, -y) = R_1(x, y) \quad \text{and} \quad R_2(x, -y) = -R_2(x, y).$$

Therefore, if R_1 is written in the form (2.10), then $q_2(x) = 0$, and if R_2 is written in the form (2.10), then the corresponding $q_1(x) = 0$. Therefore, we may assume that
$$\alpha(x, y) = (r_1(x), r_2(x)y)$$
with rational functions $r_1(x), r_2(x)$.

We can now say what happens when one of the rational functions is not defined at a point. Write

$$r_1(x) = p(x)/q(x)$$

with polynomials $p(x)$ and $q(x)$ that do not have a common factor. If $q(x) = 0$ for some point (x, y), then we assume that $\alpha(x, y) = \infty$. If $q(x) \neq 0$, then Exercise 2.14 shows that $r_2(x)$ is defined; hence the rational functions defining α are defined.

We define the **degree** of α to be

$$\deg(\alpha) = \text{Max}\{\deg p(x), \deg q(x)\}$$

if α is nontrivial. When $\alpha = 0$, let $\deg(0) = 0$. Define $\alpha \neq 0$ to be a **separable** endomorphism if the derivative $r_1'(x)$ is not identically zero. This is equivalent to saying that at least one of $p'(x)$ and $q'(x)$ is not identically zero. See Exercise 2.17. (In characteristic 0, a nonconstant polynomial will have nonzero derivative. In characteristic $p > 0$, the polynomials with zero derivative are exactly those of the form $g(x^p)$.)

Example 2.5

We continue with the previous example, where $\alpha(P) = 2P$. We have

$$R_1(x, y) = \left(\frac{3x^2 + A}{2y}\right)^2 - 2x.$$

The fact that $y^2 = x^3 + Ax + B$, plus a little algebraic manipulation, yields

$$r_1(x) = \frac{x^4 - 2Ax^2 - 8Bx + A^2}{4(x^3 + Ax + B)}.$$

(This is the same as the expression in terms of division polynomials that will be given in Section 3.2.) Therefore, $\deg(\alpha) = 4$. The polynomial $q'(x) = 4(3x^2 + A)$ is not zero (including in characteristic 3, since if $A = 0$ then $x^3 + B$ has multiple roots, contrary to assumption). Therefore α is separable.
□

Example 2.6

Let's repeat the previous example, but in characteristic 2. We'll use the formulas from Section 2.7 for doubling a point. First, let's look at $y^2 + xy = x^3 + a_2x^2 + a_6$. We have

$$\alpha(x, y) = (r_1(x), R_2(x, y))$$

with $r_1(x) = (x^4 + a_6)/x^2$. Therefore $\deg(\alpha) = 4$. Since $p'(x) = 4x^3 = 0$ and $q'(x) = 2x = 0$, the endomorphism α is not separable.

Similarly, in the case $y^2 + a_3y = x^3 + a_4x + a_6$, we have $r_1(x) = (x^4 + a_4^2)/a_3^2$. Therefore, $\deg(\alpha) = 4$, but α is not separable.   □

In general, in characteristic p, the map $\alpha(Q) = pQ$ has degree p^2 and is not separable. The statement about the degree is Corollary 3.7. The fact that α is not separable is proved in Section 2.8.

An important example of an endomorphism is the **Frobenius map**. Suppose E is defined over the finite field $\mathbf{F}_q$. Let

$$\phi_q(x, y) = (x^q, y^q).$$

The Frobenius map ϕ_q plays a crucial role in the theory of elliptic curves over $\mathbf{F}_q$.

LEMMA 2.19

Let E be defined over $\mathbf{F}_q$. Then ϕ_q is an endomorphism of E of degree q, and ϕ_q is not separable.

PROOF   Since $\phi_q(x, y) = (x^q, y^q)$, the map is given by rational functions (in fact, by polynomials) and the degree is q. The main point is that $\phi_q :$ $E(\overline{\mathbf{F}}_q) \to E(\overline{\mathbf{F}}_q)$ is a homomorphism. Let $(x_1, y_1), (x_2, y_2) \in E(\overline{\mathbf{F}}_q)$ with $x_1 \neq x_2$. The sum is (x_3, y_3), with

$$x_3 = m^2 - x_1 - x_2, \qquad y_3 = m(x_1 - x_3) - y_1, \qquad \text{where } m = \frac{y_2 - y_1}{x_2 - x_1}$$

(we are working with the Weierstrass form here; the proof for the generalized Weierstrass form is essentially the same). Raise everything to the qth power to obtain

$$x_3^q = m'^2 - x_1^q - x_2^q, \qquad y_3^q = m'(x_1^q - x_3^q) - y_1^q, \qquad \text{where } m' = \frac{y_2^q - y_1^q}{x_2^q - x_1^q}.$$

This says that

$$\phi_q(x_3, y_3) = \phi_q(x_1, y_1) + \phi_q(x_2, y_2).$$

The cases where $x_1 = x_2$ or where one of the points is ∞ are checked similarly. However, there is one subtlety that arises when adding a point to itself. The formula says that $2(x_1, y_1) = (x_3, y_3)$, with

$$x_3 = m^2 - 2x_1, \qquad y_3 = m(x_1 - x_3) - y_1, \qquad \text{where } m = \frac{3x_1^2 + A}{2y_1}.$$

When this is raised to the qth power, we obtain

$$x_3^q = m'^2 - 2x_1^q, \qquad y_3^q = m'(x_1^q - x_3^q) - y_1^q, \qquad \text{where } m' = \frac{3^q(x_1^q)^2 + A^q}{2^q y_1^q}.$$

Since $2, 3, A \in \mathbf{F}_q$, we have $2^q = 2, 3^q = 3, A^q = A$. This means that we obtain the formula for doubling the point (x_1^q, y_1^q) on E (if A^q didn't equal A, we would be working on a new elliptic curve with A^q in place of A).

Since ϕ_q is a homomorphism given by rational functions, it is an endomorphism of E. Since $q = 0$ in $\mathbf{F}_q$, the derivative of x^q is identically zero. Therefore, ϕ_q is not separable.   ∎

The following result will be crucial in the proof of Hasse's theorem in Chapter 4 and in the proof of Theorem 3.2.

PROPOSITION 2.20

Let $\alpha \neq 0$ be a separable endomorphism of an elliptic curve E. Then

$$\deg \alpha = \#Ker(\alpha),$$

where $Ker(\alpha)$ is the kernel of the homomorphism $\alpha : E(\overline{K}) \to E(\overline{K})$.
If $\alpha \neq 0$ is not separable, then

$$\deg \alpha > \#Ker(\alpha).$$

PROOF   Write $\alpha(x, y) = (r_1(x), yr_2(x))$ with $r_1(x) = p(x)/q(x)$, as above. Then $r_1' \neq 0$, so $p'q - pq'$ is not the zero polynomial.

Let S be the set of $x \in \overline{K}$ such that $(pq' - p'q)(x)\, q(x) = 0$. Let $(a, b) \in E(\overline{K})$ be such that

1. $a \neq 0$, $b \neq 0$, $(a, b) \neq \infty$,

2. $\deg(p(x) - aq(x)) = \mathrm{Max}\{\deg(p), \deg(q)\} = \deg(\alpha)$,

3. $a \notin r_1(S)$, and

4. $(a, b) \in \alpha(E(\overline{K}))$.

Since $pq' - p'q$ is not the zero polynomial, S is a finite set, hence its image under α is finite. The function $r_1(x)$ is easily seen to take on infinitely many distinct values as x runs through $\overline{K}$. Since, for each x, there is a point $(x, y) \in E(\overline{K})$, we see that $\alpha(E(\overline{K}))$ is an infinite set. Therefore, such an (a, b) exists.

We claim that there are exactly $\deg(\alpha)$ points $(x_1, y_1) \in E(\overline{K})$ such that $\alpha(x_1, y_1) = (a, b)$. For such a point, we have

$$\frac{p(x_1)}{q(x_1)} = a, \quad y_1 r_2(x_1) = b.$$

Since $(a, b) \neq \infty$, we must have $q(x_1) \neq 0$. By Exercise 2.14, $r_2(x_1)$ is defined. Since $b \neq 0$ and $y_1 r_2(x_1) = b$, we must have $y_1 = b/r_2(x_1)$. Therefore, x_1 determines y_1 in this case, so we only need to count values of x_1.

By assumption (2), $p(x) - aq(x) = 0$ has $\deg(\alpha)$ roots, counting multiplicities. We therefore must show that $p - aq$ has no multiple roots. Suppose that x_0 is a multiple root. Then

$$p(x_0) - aq(x_0) = 0 \quad \text{and} \quad p'(x_0) - aq'(x_0) = 0.$$

Multiplying the equations $p = aq$ and $aq' = p'$ yields

$$ap(x_0)q'(x_0) = ap'(x_0)q(x_0).$$

Since $a \neq 0$, this implies that x_0 is a root of $pq' - p'q$, so $x_0 \in S$. Therefore, $a = r_1(x_0) \in r_1(S)$, contrary to assumption. It follows that $p - aq$ has no multiple roots, and therefore has $\deg(\alpha)$ distinct roots.

Since there are exactly $\deg(\alpha)$ points (x_1, y_1) with $\alpha(x_1, y_1) = (a, b)$, the kernel of α has $\deg(\alpha)$ elements.

Of course, since α is a homomorphism, for each $(a, b) \in \alpha(E(\overline{K}))$, there are exactly $\deg(\alpha)$ points (x_1, y_1) with $\alpha_1(x_1, y_1) = (a, b)$. The assumptions on (a, b) were made during the proof to obtain this result for at least one point, which suffices.

If α is not separable, then the steps of the above proof hold, except that $p' - aq'$ is always the zero polynomial, so $p(x) - aq(x) = 0$ always has multiple roots and therefore has fewer than $\deg(\alpha)$ solutions.   ∎

THEOREM 2.21

Let E be an elliptic curve defined over a field K. Let $\alpha \neq 0$ be an endomorphism of E. Then $\alpha : E(\overline{K}) \to E(\overline{K})$ is surjective.

REMARK 2.22   We definitely need to be working with $\overline{K}$ instead of K in the theorem. For example, the Mordell-Weil theorem (Theorem 8.17) implies that multiplication by 2 cannot be surjective on $E(\mathbf{Q})$ if there is a point in $E(\mathbf{Q})$ of infinite order. Intuitively, working with an algebraically closed field allows us to solve the equations defining α in order to find the inverse image of a point.   ∎

PROOF   Let $(a, b) \in E(\overline{K})$. Since $\alpha(\infty) = \infty$, we may assume that $(a, b) \neq \infty$. Let $r_1(x) = p(x)/q(x)$ be as above. If $p(x) - aq(x)$ is not a constant polynomial, then it has a root x_0. Since p and q have no common roots, $q(x_0) \neq 0$. Choose $y_0 \in \overline{K}$ to be either square root of $x_0^3 + Ax_0 + B$. Then $\alpha(x_0, y_0)$ is defined (Exercise 2.14) and equals (a, b') for some b'. Since $b'^2 = a^3 + Aa + B = b^2$, we have $b = \pm b'$. If $b' = b$, we're done. If $b' = -b$, then $\alpha(x_0, -y_0) = (a, -b') = (a, b)$.

We now need to consider the case when $p - aq$ is constant. Since $E(\overline{K})$ is infinite and the kernel of α is finite, only finitely many points of $E(\overline{K})$ can map to a point with a given x-coordinate. Therefore, either $p(x)$ or $q(x)$ is not constant. If p and q are two nonconstant polynomials, then there is at most one constant a such that $p - aq$ is constant (if a' is another such number, then $(a' - a)q = (p - aq) - (p - a'q)$ is constant and $(a' - a)p = a'(p - aq) - a(p - a'q)$ is constant, which implies that p and q are constant). Therefore, there are at most two points, (a, b) and $(a, -b)$ for some b, that are not in the image of α. Let (a_1, b_1) be any other point. Then $\alpha(P_1) = (a_1, b_1)$ for some P_1. We can choose (a_1, b_1) such that $(a_1, b_1) + (a, b) \neq (a, \pm b)$, so there exists P_2 with $\alpha(P_2) = (a_1, b_1) + (a, b)$. Then $\alpha(P_2 - P_1) = (a, b)$, and $\alpha(P_1 - P_2) = (a, -b)$. Therefore, α is surjective.   ∎

For later applications, we need a convenient criterion for separability. If (x, y) is a variable point on $y^2 = x^3 + Ax + B$, then we can differentiate y with respect to x:

$$2yy' = 3x^2 + A.$$

Similarly, we can differentiate a rational function $f(x, y)$ with respect to x:

$$\frac{d}{dx} f(x, y) = f_x(x, y) + f_y(x, y)y',$$

where f_x and f_y denote the partial derivatives.

LEMMA 2.23

Let E be the elliptic curve $y^2 = x^3 + Ax + B$. Fix a point (u, v) on E. Write

$$(x, y) + (u, v) = (f(x, y), g(x, y)),$$

where $f(x, y)$ and $g(x, y)$ are rational functions of x, y (the coefficients depend on (u, v)). Then

$$\frac{\frac{d}{dx} f(x, y)}{g(x, y)} = \frac{1}{y}.$$

PROOF   The addition formulas give

$$f(x, y) = \left(\frac{y - v}{x - u} \right)^2 - x - u$$

$$g(x, y) = \frac{-(y - v)^3 + x(y - v)(x - u)^2 + 2u(y - v)(x - u)^2 - v(x - u)^3}{(x - u)^3}$$

$$\frac{d}{dx} f(x, y) = \frac{2y'(y - v)(x - u) - 2(y - v)^2 - (x - u)^3}{(x - u)^3}.$$

A straightforward but lengthy calculation, using the fact that $2yy' = 3x^2 + A$, yields

$$(x - u)^3 (y \frac{d}{dx} f(x, y) - g(x, y))$$
$$= v(Au + u^3 - v^2 - Ax - x^3 + y^2) + y(-Au - u^3 + v^2 + Ax + x^3 - y^2).$$

Since (u, v) and (x, y) are on E, we have $v^2 = u^3 + Au + B$ and $y^2 = x^3 + Ax + B$. Therefore, the above expression becomes

$$v(-B + B) + y(B - B) = 0.$$

Therefore, $y \frac{d}{dx} f(x, y) = g(x, y)$.   ∎

REMARK 2.24   Lemma 2.23 is perhaps better stated in terms of differentials. It says that the differential dx/y is translation invariant. In fact, it is the unique translation invariant differential, up to scalar multiples, for E. See [90].   ∎

LEMMA 2.25

Let $\alpha_1, \alpha_2, \alpha_3$ be nonzero endomorphisms of an elliptic curve E with $\alpha_1 + \alpha_2 = \alpha_3$. Write

$$\alpha_j(x, y) = (R_{\alpha_j}(x), y S_{\alpha_j}(x)).$$

Suppose there are constants $c_{\alpha_1}, c_{\alpha_2}$ such that

$$\frac{R'_{\alpha_1}(x)}{S_{\alpha_1}(x)} = c_{\alpha_1}, \quad \frac{R'_{\alpha_2}(x)}{S_{\alpha_2}(x)} = c_{\alpha_2}.$$

Then

$$\frac{R'_{\alpha_3}(x)}{S_{\alpha_3}(x)} = c_{\alpha_1} + c_{\alpha_2}.$$

PROOF   Let (x_1, y_1) and (x_2, y_2) be variable points on E. Write·

$$(x_3, y_3) = (x_1, y_1) + (x_2, y_2),$$

where

$$(x_1, y_1) = \alpha_1(x, y), \quad (x_2, y_2) = \alpha_2(x, y).$$

Then x_3 and y_3 are rational functions of x_1, y_1, x_2, y_2, which in turn are rational functions of x, y. By Lemma 2.23, with $(u, v) = (x_2, y_2)$,

$$\frac{\partial x_3}{\partial x_1} = \frac{y_3}{y_1}.$$

Similarly,

$$\frac{\partial x_3}{\partial x_2} = \frac{y_3}{y_2}.$$

By assumption,

$$\frac{\partial x_j}{\partial x} = c_{\alpha_j}\frac{y_j}{y}$$

for $j = 1, 2$. By the chain rule,

$$\begin{aligned}
\frac{dx_3}{dx} &= \frac{\partial x_3}{\partial x_1}\frac{\partial x_1}{\partial x} + \frac{\partial x_3}{\partial x_2}\frac{\partial x_2}{\partial x} \\
&= \frac{y_3}{y_1}\frac{y_1}{y}c_{\alpha_1} + \frac{y_3}{y_2}\frac{y_2}{y}c_{\alpha_2} \\
&= (c_{\alpha_1} + c_{\alpha_2})\frac{y_3}{y}.
\end{aligned}$$

Dividing by y_3/y yields the result.   ∎

REMARK 2.26   In terms of differentials (see the previous Remark), we have $(dx/y)\circ\alpha$ is a translation-invariant differential on E. Therefore it must be a scalar multiple $c_\alpha dx/y$ of dx/y. It follows that every nonzero endomorphism α satisfies the hypotheses of Lemma 2.23.   ∎

PROPOSITION 2.27
Let E be an elliptic curve defined over a field K, and let n be a nonzero integer. Suppose that multiplication by n on E is given by

$$n(x, y) = (R_n(x), yS_n(x))$$

for all $(x, y) \in E(\overline{K})$, where R_n and S_n are rational functions. Then

$$\frac{R'_n(x)}{S_n(x)} = n.$$

Therefore, multiplication by n is separable if and only if n is not a multiple of the characteristic p of the field.

PROOF Since $R_{-n} = R_n$ and $S_{-n} = -S_n$, we have $R'_{-n}/S_{-n} = -R'_n/S_n$. Therefore, the result for positive n implies the result for negative n.

Note that the first part of the proposition is trivially true for $n = 1$. If it is true for n, then Lemma 2.25 implies that it is true for $n + 1$, which is the sum of n and 1. Therefore, $\frac{R'_n(x)}{S(x)} = n$ for all n.

We have $R'_n(x) \neq 0$ if and only if $n = R'_n(x)/S_n(x) \neq 0$, which is equivalent to p not dividing n. Since the definition of separability is that $R'_n \neq 0$, this proves the second part of the proposition. ∎

Finally, we use Lemma 2.25 to prove a result that will be needed in Section 11.2. Let E be an elliptic curve defined over a finite field $\mathbf{F}_q$. The Frobenius endomorphism ϕ_q is defined by $\phi_q(x, y) = (x^q, y^q)$. It is an endomorphism of E by Lemma 2.19.

PROPOSITION 2.28

Let E be an elliptic curve defined over $\mathbf{F}_q$, where q is a power of the prime p. Let r and s be integers, not both 0. The endomorphism $r\phi_q + s$ is separable if and only if $p \nmid s$.

PROOF Write the multiplication by r endomorphism as

$$r(x, y) = (R_r(x), yS_r(x)).$$

Then

$$(R_{r\phi_q}(x), yS_{r\phi_q}(x)) = (r\phi_q)(x, y) = (R_r^q(x), \, y^q S_r^q(x))$$
$$= \left(R_r^q(x), \, y(x^3 + Ax + B)^{(q-1)/2} S_r^q(x) \right).$$

Therefore,

$$c_{r\phi_q} = R'_{r\phi_q}/S_{r\phi_q} = qR_r^{q-1}R'_r/S_{r\phi_q} = 0.$$

Also, $c_s = R'_s/S_s = s$ by Proposition 2.27. By Lemma 2.25,

$$R'_{r\phi_q+s}/S_{r\phi_q+s} = c_{r\phi_q+s} = c_{r\phi_q} + c_s = 0 + s = s.$$

Therefore, $R'_{r\phi_q+s} \neq 0$ if and only if $p \nmid s$. ∎

2.9   Singular Curves

We have been working with $y^2 = x^3 + Ax + B$ under the assumption that $x^3 + Ax + B$ has distinct roots. However, it is interesting to see what happens when there are multiple roots. It will turn out that elliptic curve addition becomes either addition of elements in K or multiplication of elements in $K^\times$ or in a quadratic extension of K. This means that an algorithm for a group $E(K)$ arising from elliptic curves, such as one to solve a discrete logarithm problem (see Chapter 5), will probably also apply to these more familiar situations. See also Chapter 7. Moreover, as we'll discuss briefly at the end of this section, singular curves arise naturally when elliptic curves defined over the integers are reduced modulo various primes.

We first consider the case where $x^3 + Ax + B$ has a triple root at $x = 0$, so the curve has the equation

$$y^2 = x^3.$$

The point $(0,0)$ is the only singular point on the curve (see Figure 2.7). Since

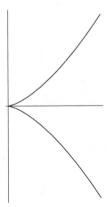

Figure 2.7

$y^2 = x^3$

any line through this point intersects the curve in at most one other point, $(0,0)$ causes problems if we try to include it in our group. So we leave it out. The remaining points, which we denote $E_{ns}(K)$, form a group, with the group law defined in the same manner as when the cubic has distinct roots. The only thing that needs to be checked is that the sum of two points cannot be $(0,0)$. But since a line through $(0,0)$ has at most one other intersection point with the curve, a line through two nonsingular points cannot pass through $(0,0)$ (this will also follow from the proof of the theorem below).

THEOREM 2.29

Let E be the curve $y^2 = x^3$ and let $E_{ns}(K)$ be the nonsingular points on this curve with coordinates in K, including the point $\infty = (0 : 1 : 0)$. The map

$$E_{ns}(K) \to K, \quad (x, y) \mapsto \frac{x}{y}, \quad \infty \mapsto 0$$

is a group isomorphism between $E_{ns}(K)$ and K, regarded as an additive group.

PROOF   Let $t = x/y$. Then $x = (y/x)^2 = 1/t^2$ and $y = x/t = 1/t^3$. Therefore we can express all of the points in $E_{ns}(K)$ in terms of the parameter t. Let $t = 0$ correspond to $(x, y) = \infty$. It follows that the map of the theorem is a bijection. (Note that $1/t$ is the slope of the line through $(0, 0)$ and (x, y), so this parameterization is obtained similarly to the one obtained for quadratic curves in Section 2.5.4.)

Suppose $(x_1, y_1) + (x_2, y_2) = (x_3, y_3)$. We must show that $t_1 + t_2 = t_3$, where $t_i = x_i/y_i$. If $(x_1, y_1) \neq (x_2, y_2)$, the addition formulas say that

$$x_3 = \left(\frac{y_2 - y_1}{x_2 - x_1} \right)^2 - x_1 - x_2.$$

Substituting $x_i = 1/t_i^2$ yields

$$t_3^{-2} = \left(\frac{t_2^{-3} - t_1^{-3}}{t_2^{-2} - t_1^{-2}} \right)^2 - t_1^{-2} - t_2^{-2}.$$

A straightforward calculation simplifies this to

$$t_3^{-2} = (t_1 + t_2)^{-2}.$$

Similarly,

$$-y_3 = \left(\frac{y_2 - y_1}{x_2 - x_1} \right)(x_3 - x_1) + y_1$$

may be rewritten in terms of the t_i to yield

$$t_3^{-3} = (t_1 + t_2)^{-3}.$$

Taking the ratio of the expressions for t_3^{-2} and t_3^{-3} gives

$$t_3 = t_1 + t_2,$$

as desired.

If $(x_1, y_1) = (x_2, y_2)$, the proof is similar. Finally, the cases where one or more of the points $(x_i, y_i) = \infty$ are easily checked.   ∎

We now consider the case where $x^3 + Ax + B$ has a double root. By translating x, we may assume that this root is 0 and the curve E has the equation

$$y^2 = x^2(x + a)$$

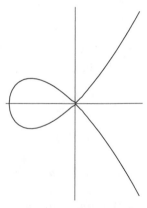

Figure 2.8

$y^2 = x^3 + x^2$

for some $a \neq 0$. The point $(0,0)$ is the only singularity (see Figure 2.8). Let $E_{ns}(K)$ be the nonsingular points on E with coordinates in K, including the point ∞. Let $\alpha^2 = a$ (so α might lie in an extension of K). The equation for E may be rewritten as

$$\left(\frac{y}{x}\right)^2 = a + x.$$

When x is near 0, the right side of this equation is approximately a. Therefore, E is approximated by $(y/x)^2 = a$, or $y/x = \pm\alpha$ near $x = 0$. This means that the two "tangents" to E at $(0,0)$ are

$$y = \alpha x \quad \text{and} \quad y = -\alpha x$$

(for a different way to obtain these tangents, see Exercise 2.15).

THEOREM 2.30

Let E be the curve $y^2 = x^2(x + a)$ with $0 \neq a \in K$. Let $E_{ns}(K)$ be the nonsingular points on E with coordinates in K. Let $\alpha^2 = a$. Consider the map

$$\psi : (x, y) \mapsto \frac{y + \alpha x}{y - \alpha x}, \quad \infty \mapsto 1.$$

1. *If $\alpha \in K$, then ψ gives an isomorphism from $E_{ns}(K)$ to $K^\times$, considered as a multiplicative group.*

2. *If $\alpha \notin K$, then ψ gives an isomorphism*

$$E_{ns}(K) \simeq \{u + \alpha v \mid u, v \in K,\ u^2 - av^2 = 1\},$$

where the right hand side is a group under multiplication.

PROOF   Let

$$t = \frac{y + \alpha x}{y - \alpha x}.$$

This may be solved for y/x to obtain

$$\frac{y}{x} = \alpha \frac{t + 1}{t - 1}.$$

Since $x + a = (y/x)^2$, we obtain

$$x = \frac{4\alpha^2 t}{(t - 1)^2} \quad \text{and} \quad y = \frac{4\alpha^3 t(t + 1)}{(t - 1)^3}$$

(the second is obtained from the first using $y = x(y/x)$). Therefore, (x, y) determines t and t determines (x, y), so the map ψ is injective, and is a bijection in case (1).

In case (2), rationalize the denominator by multiplying the numerator and denominator of $(y + \alpha x)/(y - \alpha x)$ by $y + \alpha x$ to obtain an expression of the form $u + \alpha v$:

$$\frac{(y + \alpha x)}{(y - \alpha x)} = u + \alpha v.$$

We can change the sign of α throughout this equation and preserve the equality. Now multiply the resulting expression by the original to obtain

$$u^2 - av^2 = (u + \alpha v)(u - \alpha v) = \frac{(y + \alpha x)}{(y - \alpha x)} \frac{(y - \alpha x)}{(y + \alpha x)} = 1.$$

Conversely, suppose $u^2 - av^2 = 1$. Let

$$x = \left(\frac{u + 1}{v}\right)^2 - a, \quad y = \left(\frac{u + 1}{v}\right) x.$$

Then (x, y) is on the curve E and

$$\psi(x, y) = \frac{(y/x) + \alpha}{(y/x) - \alpha} = \frac{u + 1 + \alpha v}{u + 1 - \alpha v} = u + \alpha v$$

(the last equality uses the fact that $u^2 - av^2 = 1$). Therefore, ψ is surjective, hence is a bijection in case (2), too.

It remains to show that ψ is a homomorphism. Suppose $(x_1, y_1) + (x_2, y_2) = (x_3, y_3)$. Let

$$t_i = \frac{y_i + \alpha x_i}{y_i - \alpha x_i}.$$

We must show that $t_1 t_2 = t_3$.

When $(x_1, y_1) \neq (x_2, y_2)$, we have

$$x_3 = \left(\frac{y_2 - y_1}{x_2 - x_1}\right)^2 - a - x_1 - x_2.$$

Substituting $x_i = \dfrac{4\alpha t_i}{(t_i - 1)^2}$ and $y_i = \dfrac{4\alpha^3 t_i(t_i + 1)}{(t_i - 1)^3}$ and simplifying yields

$$\frac{4t_3}{(t_3 - 1)^2} = \frac{4t_1 t_2}{(t_1 t_2 - 1)^2}. \tag{2.11}$$

Similarly,

$$-y_3 = x = \left(\frac{y_2 - y_1}{x_2 - x_1}\right)(x_3 - x_1) + y_1$$

yields

$$\frac{4\alpha^3 t_3(t_3 + 1)}{(t_3 - 1)^3} = \frac{4\alpha^3 t_1 t_2(t_1 t_2 + 1)}{(t_1 t_2 - 1)^3}.$$

The ratio of this equation and (2.11) yields

$$\frac{t_3 - 1}{t_3 + 1} = \frac{t_1 t_2 - 1}{t_1 t_2 + 1}.$$

This simplifies to yield

$$t_1 t_2 = t_3,$$

as desired.

The case where $(x_1, y_1) = (x_2, y_2)$ is similar, and the cases where one or more of the points is ∞ are trivial. This completes the proof. ∎

One situation where the above singular curves arise naturally is when we are working with curves with integral coefficients and reduce modulo various primes. For example, let E be $y^2 = x(x + 35)(x - 55)$. Then we have

$$E \bmod 5: \quad y^2 \equiv x^3,$$
$$E \bmod 7: \quad y^2 \equiv x^2(x + 1),$$
$$E \bmod 11: \quad y^2 \equiv x^2(x + 2).$$

The first case is treated in Theorem 2.29 and is called **additive reduction**. The second case is **split multiplicative reduction** and is covered by Theorem 2.30(1). In the third case, $\alpha \notin \mathbf{F}_{11}$, so we are in the situation of Theorem 2.30(2). This is called **non-split multiplicative reduction**. For all primes $p \geq 13$, the cubic polynomial has distinct roots mod p, so $E \bmod p$ is nonsingular. This situation is called **good reduction**.

2.10 Elliptic Curves mod n

In a few situations, we'll need to work with elliptic curves mod n, where n is composite. We'll also need to take elliptic curves over $\mathbf{Q}$ and reduce them mod n, where n is an integer. Both situations are somewhat subtle, as the following three examples show.

Example 2.7
Let E be given by
$$y^2 = x^3 - x + 1 \quad (\text{mod } 5^2).$$

Suppose we want to compute $(1, 1) + (21, 4)$. The slope of the line through the two points is $3/20$. The denominator is not zero mod 25, but it is also not invertible. Therefore the slope is neither infinite nor finite mod 25. If we compute the sum using the formulas for the group law, the x-coordinate of the sum is
$$\left(\frac{3}{20}\right)^2 - 1 - 21 \equiv \infty \quad (\text{mod } 25).$$

But $(1, 1) + (1, 24) = \infty$, so we cannot also have $(1, 1) + (21, 4) = \infty$.   □

Example 2.8
Let E be given by
$$y^2 = x^3 - x + 1 \quad (\text{mod } 35).$$

Suppose we want to compute $(1, 1) + (26, 24)$. The slope is $23/25$, which is infinite mod 5 but finite mod 7. Therefore, the formulas for the sum yield a point that is ∞ mod 5 but is finite mod 7. In a sense, the point is partially at ∞. We cannot express it in affine coordinates mod 35. One remedy is to use the Chinese Remainder Theorem to write

$$E(\mathbf{Z}_{35}) = E(\mathbf{Z}_5) \oplus E(\mathbf{Z}_7)$$

and then work mod 5 and mod 7 separately. This strategy works well in the present case, but it doesn't help in the previous example.   □

Example 2.9
Let E be given by
$$y^2 = x^3 + 3x + 1$$

over $\mathbf{Q}$. Suppose we want to compute

$$(1, 1) + \left(\frac{571}{361}, -\frac{16379}{6859}\right).$$

Since the points are distinct, we compute the slope of the line through them in the usual way. This allows us to find the sum. Now consider E mod 7. The two points are seen to be congruent mod 7, so the line through them mod 7 is the tangent line. Therefore, the formula we use to add the points mod 7 is different from the one used in $\mathbf{Q}$. Suppose we want to show that the reduction map from $E(\mathbf{Q})$ to $E(\mathbf{F}_7)$ is a homomorphism. At first, it would seem that this is obvious, since we just take the formulas for the group law over $\mathbf{Q}$ and reduce them mod 7. But the present example says that sometimes we are using different formulas over $\mathbf{Q}$ and mod 7. A careful analysis shows that this does not cause problems, but it should be clear that the reduction map is more subtle than one might guess.   □

The remedy for the above problems is to develop a theory of elliptic curves over rings. We follow [59]. The reader willing to believe Corollaries 2.31, 2.32, and 2.33 can safely skip the details in this section.

Let R be a ring (always assumed to be commutative with 1). A tuple of elements $(x_1, x_2, \dots)$ from R is said to be **primitive** if there exist elements $r_1, r_2, \dots \in R$ such that

$$r_1 x_1 + r_2 x_2 + \dots = 1.$$

When $R = \mathbf{Z}$, this means that $\gcd(x_1, x_2, \dots) = 1$. When $R = \mathbf{Z}_n$, primitivity means that $\gcd(n, x_1, x_2, \dots) = 1$. When R is a field, primitivity means that at least one of the x_i is nonzero. In general, primitivity means that the ideal generated by $x_1, x_2, \dots$ is R. We say that two primitive triples (x, y, z) and (x', y', z') are equivalent if there exists a unit $u \in R^\times$ such that

$$(x', y', z') = (ux, uy, uz)$$

(in fact, it follows easily from the existence of r, s, t with $rx' + sy' + tz' = 1$ that any u satisfying this equation must be a unit). Define 2-dimensional **projective space** over R to be

$$\mathbf{P}^2(R) = \{(x, y, z) \in R^3 \,|\, (x, y, z) \text{ is primitive}\} \text{ mod equivalence.}$$

The equivalence class of (x, y, z) is denoted by $(x : y : z)$.

If R is a field, $\mathbf{P}^2(R)$ is the same as that defined in Section 2.3. If $(x : y : z) \in \mathbf{P}^2(\mathbf{Q})$, we can multiply by a suitable rational number to clear denominators and remove common factors from the numerators and therefore obtain a triple of integers with gcd=1. Therefore, $\mathbf{P}^2(\mathbf{Q})$ and $\mathbf{P}^2(\mathbf{Z})$ will be regarded as equal. Similarly, if R is a ring with

$$\mathbf{Z} \subseteq R \subseteq \mathbf{Q},$$

then $\mathbf{P}^2(R) = \mathbf{P}^2(\mathbf{Z})$.

In order to work with elliptic curves over R, we need to impose two conditions on R.

1. $2 \in R^\times$

2. If (a_{ij}) is a primitive $m \times n$ matrix such that $(a_{11}, a_{12}, \dots, a_{mn})$ is primitive and such that all 2×2 subdeterminants vanish (that is, $a_{ij}a_{k\ell} - a_{i\ell}a_{kj} = 0$ for all i, j, k, ℓ), then some R-linear combination of the rows is a primitive n-tuple.

The first condition is needed since we'll be working with the Weierstrass equation. In fact, we should add the condition that $3 \in R^\times$ if we want to change an arbitrary elliptic curve into Weierstrass form. Note that $\mathbf{Z}$ does not satisfy the first condition. This can be remedied by working with

$$\mathbf{Z}_{(2)} = \{\frac{x}{2^k} \mid x \in \mathbf{Z},\, k \geq 0\}.$$

This is a ring. As pointed out above, $\mathbf{P}^2(\mathbf{Z}_{(2)})$ equals $\mathbf{P}^2(\mathbf{Z})$, so the introduction of $\mathbf{Z}_{(2)}$ is a minor technicality.

The second condition is perhaps best understood when R is a field. In this case, the primitivity of the matrix simply means that at least one entry is nonzero. The vanishing of the 2×2 subdeterminants says that the rows are proportional to each other. The conclusion is that some linear combination of the rows (in this case, some row itself) is a nonzero vector.

When $R = \mathbf{Z}$, the primitivity of the matrix means that the gcd of the elements in the matrix is 1. Since the rows are assumed to be proportional, there is a vector $\mathbf{v}$ and integers $a_1, \ldots, a_m$ such that the ith row is $a_i\mathbf{v}$. The m-tuple $(a_1, \ldots, a_m)$ must be primitive since the gcd of its entries divides the gcd of the entries of the matrix. Therefore, there is a linear combination of the a_i's that equals 1. This means that some linear combination of the rows of the matrix is $\mathbf{v}$. The vector $\mathbf{v}$ is primitive since the gcd of its entries divides the gcd of the entries of the matrix. Therefore, we have obtained a primitive vector as a linear combination of the rows of the matrix. This shows that $\mathbf{Z}$ satisfies the second condition. The same argument, slightly modified to handle powers of 2, shows that $\mathbf{Z}_{(2)}$ also satisfies the second condition.

In general, condition 2 says that projective modules over R of rank 1 are free (see [59]). In particular, this holds for local rings, for finite rings, and for $\mathbf{Z}_{(2)}$. These suffice for our purposes.

For the rest of this section, assume R is a ring satisfying 1 and 2. An **elliptic curve** E over R is given by a homogeneous equation

$$y^2 z = x^3 + Axz^2 + Bz^3$$

with $A, B \in R$ such that $4A^3 + 27B^2 \in R^\times$. Define

$$E(R) = \{(x : y : z) \in \mathbf{P}^2(R) \mid y^2 z = x^3 + Axz^2 + Bz^3\}.$$

The addition law is defined in essentially the same manner as in Section 2.2, but the formulas needed are significantly more complicated. To make a long story short (maybe not so short), the answer is the following.

GROUP LAW

Let $(x_i : y_i : z_i) \in E(R)$ for $i = 1, 2$. Consider the following three sets of *equations:*

I.

$$x'_3 = (x_1y_2 - x_2y_1)(y_1z_2 + y_2z_1) + (x_1z_2 - x_2z_1)y_1y_2$$
$$-A(x_1z_2 + x_2z_1)(x_1z_2 - x_2z_1) - 3B(x_1z_2 - x_2z_1)z_1z_2$$
$$y'_3 = -3x_1x_2(x_1y_2 - x_2y_1) - y_1y_2(y_1z_2 - y_2z_1) - A(x_1y_2 - x_2y_1)z_1z_2$$
$$+A(x_1z_2 + x_2z_1)(y_1z_2 - y_2z_1) + 3B(y_1z_2 - y_2z_1)z_1z_2$$
$$z'_3 = 3x_1x_2(x_1z_2 - x_2z_1) - (y_1z_2 + y_2z_1)(y_1z_2 - y_2z_1)$$
$$+A(x_1z_2 - x_2z_1)z_1z_2$$

II.

$$x''_3 = y_1y_2(x_1y_2 + x_2y_1) - Ax_1x_2(y_1z_2 + y_2z_1)$$
$$-A(x_1y_2 + x_2y_1)(x_1z_2 + x_2z_1) - 3B(x_1y_2 + x_2y_1)z_1z_2$$
$$-3B(x_1z_2 + x_2z_1)(y_1z_2 + y_2z_1) + A^2(y_1z_2 + y_2z_1)z_1z_2$$
$$y''_3 = y_1^2y_2^2 + 3Ax_1^2x_2^2 + 9Bx_1x_2(x_1z_2 + x_2z_1)$$
$$-A^2x_1z_2(x_1z_2 + 2x_2z_1) - A^2x_2z_1(2x_1z_2 + x_2z_1)$$
$$-3ABz_1z_2(x_1z_2 + x_2z_1) - (A^3 + 9B^2)z_1^2z_2^2$$
$$z''_3 = 3x_1x_2(x_1y_2 + x_2y_1) + y_1y_2(y_1z_2 + y_2z_1) + A(x_1y_2 + x_2y_1)z_1z_2$$
$$+A(x_1z_2 + x_2z_1)(y_1z_2 + y_2z_1) + 3B(y_1z_2 + y_2z_1)z_1z_2$$

III.

$$x'''_3 = (x_1y_2 + x_2y_1)(x_1y_2 - x_2y_1) + Ax_1x_2(x_1z_2 - x_2z_1)$$
$$+3B(x_1z_2 + x_2z_1)(x_1z_2 - x_2z_1) - A^2(x_1z_2 - x_2z_1)z_1z_2$$
$$y'''_3 = (x_1y_2 - x_2y_1)y_1y_2 - 3Ax_1x_2(y_1z_2 - y_2z_1)$$
$$+A(x_1y_2 + x_2y_1)(x_1z_2 - x_2z_1) + 3B(x_1y_2 - x_2y_1)z_1z_2$$
$$-3B(x_1z_2 + x_2z_1)(y_1z_2 - y_2z_1) + A^2(y_1z_2 - y_2z_1)z_1z_2$$
$$z'''_3 = -(x_1y_2 + x_2y_1)(y_1z_2 - y_2z_1) - (x_1z_2 - x_2z_1)y_1y_2$$
$$-A(x_1z_2 + x_2z_1)(x_1z_2 - x_2z_1) - 3B(x_1z_2 - x_2z_1)z_1z_2$$

Then the matrix

$$\begin{pmatrix} x'_3 & y'_3 & z'_3 \\ x''_3 & y''_3 & z''_3 \\ x'''_3 & y'''_3 & z'''_3 \end{pmatrix}$$

is primitive and all 2×2 subdeterminants vanish. Take a primitive R-linear combination (x_3, y_3, z_3) of the rows. Define

$$(x_1 : y_1 : z_1) + (x_2 : y_2 : z_2) = (x_3 : y_3 : z_3).$$

Also, define

$$-(x_1 : y_1 : z_1) = (x_1 : -y_1 : z_1).$$

Then $E(R)$ is an abelian group under this definition of point addition. The identity element is $(0 : 1 : 0)$.

For some of the details concerning this definition, see [59]. The equations are deduced (with a slight correction) from those in [13]. A similar set of equations is given in [57].

When R is a field, each of these equations can be shown to give the usual group law when the output is a point in $\mathbf{P}^2(R)$ (that is, not all three coordinates vanish). If two or three of the equations yield points in $\mathbf{P}^2(R)$, then these points are equal (since the 2×2 subdeterminants vanish). If R is a ring, then it is possible that each of the equations yields a nonprimitive output (for example, perhaps 5 divides the output of I, 7 divides the output of II, and 11 divides the output of III). If we are working with $\mathbf{Z}$ or $\mathbf{Z}_{(2)}$, this is no problem. Simply divide by the gcd of the entries in an output. But in an arbitrary ring, gcd's might not exist, so we must take a linear combination to obtain a primitive vector, and hence an element in $\mathbf{P}^2(R)$.

Example 2.10
Let $R = \mathbf{Z}_{25}$ and let E be given by

$$y^2 = x^3 - x + 1 \quad (\text{mod } 5^2).$$

Suppose we want to compute $(1, 1) + (21, 4)$, as in Example 2.7 above. Write the points in homogeneous coordinates as

$$(x_1 : y_1 : z_1) = (1 : 1 : 1), \quad (x_2 : y_2 : z_2) = (21 : 4 : 1).$$

Formulas I, II, III yield

$$(x_3', y_3', z_3') = (5, 23, 0)$$
$$(x_3'', y_3'', z_3'') = (5, 8, 0)$$
$$(x_3''', y_3''', z_3''') = (20, 12, 0),$$

respectively. Note that these are all the same point in $\mathbf{P}^2(\mathbf{Z}_{25})$ since

$$(5, 23, 0) = 6(5, 8, 0) = 4(20, 12, 0).$$

If we reduce the point $(5 : 8 : 0) \bmod 5$, we obtain $(0 : 3 : 0) = (0 : 1 : 0)$, which is the point ∞. The fact that the point is at infinity mod 5 but not mod 25 is what caused the difficulties in our calculations in Example 2.7. $\Box$

Example 2.11
Let E be an elliptic curve. Suppose we use the formulas to calculate

$$(0 : 1 : 0) + (0 : 1 : 0).$$

Formulas I, II, III yield

$$(0,0,0), \quad (0,1,0), \quad (0,0,0),$$

respectively. The first and third outputs do not yield points in projective space. The second says that

$$(0:1:0) + (0:1:0) = (0:1:0).$$

This if of course the rule $\infty + \infty = \infty$ from the usual group law on elliptic curves.   □

The present version of the group law allows us to work with elliptic curves over rings in theoretical settings. We give three examples.

COROLLARY 2.31

Let n_1 and n_2 be odd integers with $\gcd(n_1, n_2) = 1$. Let E be an elliptic curve defined over $\mathbf{Z}_{n_1 n_2}$. Then there is a group isomorphism

$$E(\mathbf{Z}_{n_1 n_2}) \simeq E(\mathbf{Z}_{n_1}) \oplus E(\mathbf{Z}_{n_2}).$$

PROOF   Suppose that E is given by $y^2 z = x^3 + A x z^2 + B z^3$ with $A, B \in \mathbf{Z}_{n_1 n_2}$ and $4A^3 + 27B^2 \in \mathbf{Z}_{n_1 n_2}^{\times}$. Then we can regard A and B as elements of $\mathbf{Z}_{n_i}$ and we have $4A^3 + 27B^2 \in \mathbf{Z}_{n_i}^{\times}$. Therefore, we can regard E as an elliptic curve over $\mathbf{Z}_{n_i}$, so the statement of the corollary makes sense.

The Chinese remainder theorem says that there is an isomorphism of rings

$$\mathbf{Z}_{n_1 n_2} \simeq \mathbf{Z}_{n_1} \oplus \mathbf{Z}_{n_2}$$

given by

$$x \bmod n_1 n_2 \longleftrightarrow (x \bmod n_1, \, x \bmod n_2).$$

This yields a bijection between triples in $\mathbf{Z}_{n_1 n_2}$ and pairs of triples, one in $\mathbf{Z}_{n_1}$ and one in $\mathbf{Z}_{n_2}$. It is not hard to see that primitive triples for $\mathbf{Z}_{n_1 n_2}$ correspond to pairs of primitive triples in $\mathbf{Z}_{n_1}$ and $\mathbf{Z}_{n_2}$. Moreover,

$$y^2 z \equiv x^3 + A x z^2 + B z^3 \pmod{n_1 n_2}$$

$$\Longleftrightarrow \begin{cases} y^2 z \equiv x^3 + A x z^2 + B z^3 \pmod{n_1} \\ y^2 z \equiv x^3 + A x z^2 + B z^3 \pmod{n_2} \end{cases}$$

Therefore, there is a bijection

$$\psi : E(\mathbf{Z}_{n_1 n_2}) \longrightarrow E(\mathbf{Z}_{n_1}) \oplus E(\mathbf{Z}_{n_2}).$$

It remains to show that ψ is a homomorphism. Let $P_1, P_2 \in E(\mathbf{Z}_{n_1 n_2})$ and let $P_3 = P_1 + P_2$. This means that there is a linear combination of the outputs of formulas I, II, III that is primitive and yields P_3. Reducing all of these calculations mod n_i (for $i = 1, 2$) yields exactly the same result, namely the primitive point P_3 (mod n_i) is the sum of P_1 (mod n_i) and P_2 (mod n_i). This means that $\psi(P_3) = \psi(P_1) + \psi(P_2)$, so ψ is a homomorphism.   ∎

COROLLARY 2.32

Let E be an elliptic curve over $\mathbf{Q}$ given by

$$y^2 = x^3 + Ax + B$$

with $A, B \in \mathbf{Z}$. Let n be a positive odd integer such that $\gcd(n, 4A^3 + 27B^2) = 1$. Represent the elements of $E(\mathbf{Q})$ as primitive triples $(x : y : z) \in \mathbf{P}^2(\mathbf{Z})$. The map

$$red_n : E(\mathbf{Q}) \longrightarrow E(\mathbf{Z}_n)$$
$$(x : y : z) \mapsto (x : y : z) \pmod{n}$$

is a group homomorphism.

PROOF   If $P_1, P_2 \in E(\mathbf{Q})$ and $P_1 + P_2 = P_3$, then P_3 is a primitive point that can be expressed as a linear combination of the outputs of formulas I, II, III. Reducing all of the calculations mod n yields the result.   ∎

Corollary 2.32 can be generalized as follows.

COROLLARY 2.33

Let R be a ring and let I be an ideal of R. Assume that both R and R/I satisfy conditions (1) and (2) on page 61. Let E be given by

$$y^2 z = x^3 + Axz^2 + Bz^3$$

with $A, B \in R$ and assume there exists $r \in R$ such that

$$(4A^3 + 27B^2)r - 1 \in I.$$

Then the map

$$red_I : E(R) \longrightarrow E(R/I)$$
$$(x : y : z) \mapsto (x : y : z) \mod I$$

is a group homomorphism.

PROOF   The proof is the same as for Corollary 2.32, with R in place of $\mathbf{Z}$ and mod I in place of mod n. The condition that $(4A^3 + 27B^2)r - 1 \in I$ for some r is the requirement that $4A^3 + 27B^2$ is a unit in R/I, which was required in the definition of an elliptic curve over the ring R/I.   ∎

Exercises

2.1 (a) Show that the constant term of a monic cubic polynomial is the negative of the product of the roots.

   (b) Use (a) to derive the formula for the sum of two distinct points P_1, P_2 in the case that the x-coordinates x_1 and x_2 are nonzero, as in Section 2.2. Note that when one of these coordinates is 0, you need to divide by zero to obtain the usual formula.

2.2 Let (x, y) be a point on the elliptic curve E given by $y^2 = x^3 + Ax + B$. Show that if $y = 0$ then $3x^2 + A \neq 0$. (*Hint:* What is the condition for a polynomial to have x as a multiple root?)

2.3 Show that three points on an elliptic curve add to ∞ if and only if they are collinear.

2.4 Show that the method of 2.1 actually computes kP. (*Hint:* Use induction on the length of the binary expansion of k. If $k = k_0 + 2k_1 + 4k_2 + \cdots + 2^\ell a_\ell$, assume the result holds for $k' = k_0 + 2k_1 + 4k_2 + \cdots + 2^{\ell-1} a_{\ell-1}$.)

2.5 If $P = (x, y) \neq \infty$ is on the curve described by (2.1), then $-P$ is the other finite point of intersection of the curve and the vertical line through P. Show that $-P = (x, -a_1 x - a_3 - y)$. (*Hint:* This involves solving a quadratic in y. Note that the sum of the roots of a monic quadratic polynomial equals the negative of the coefficient of the linear term.)

2.6 Let $\mathbf{R}$ be the real numbers. Show that the map $(x, y, z) \mapsto (x : y : z)$ gives a two-to-one map from the sphere $x^2 + y^2 + z^2 = 1$ in $\mathbf{R}^3$ to $\mathbf{P}_\mathbf{R}^2$. Since the sphere is compact, this shows that $\mathbf{P}_\mathbf{R}^2$ is compact under the topology inherited from the sphere (a set is open in $\mathbf{P}_\mathbf{R}^2$ if and only if its inverse image is open in the sphere).

2.7 (a) Show that two lines $a_1 x + b_1 y + c_1 z = 0$ and $a_2 x + b_2 y + c_2 z = 0$ in two-dimensional projective space have a point of intersection.

   (b) Show that there is exactly one line through two distinct given points in $\mathbf{P}_K^2$.

2.8 Suppose that the matrix

$$M = \begin{pmatrix} a_1 & b_1 \\ a_2 & b_2 \\ a_3 & b_3 \end{pmatrix}$$

has rank 2. Let (a, b, c) be a nonzero vector in the left nullspace of M, so $(a, b, c)M = 0$. Show that the parametric equations

$$x = a_1 u + b_1 v, \quad y = a_2 u + b_2 v, \quad z = a_3 u + b_3 v,$$

describe the line $ax + by + cz = 0$ in $\mathbf{P}^2_K$. (It is easy to see that the points $(x : y : z)$ lie on the line. The main point is that each point on the line corresponds to a pair (u, v).)

2.9 (a) Put the Legendre equation $y^2 = x(x-1)(x-\lambda)$ into Weierstrass form and use this to show that the j-invariant is

$$j = 2^8 \frac{(\lambda^2 - \lambda + 1)^3}{\lambda^2(\lambda - 1)^2}.$$

(b) Show that if $j \neq 0, 1728$ then there are six distinct values of λ giving this j, and that if λ is one such value then the full set is

$$\{\lambda, \frac{1}{\lambda}, 1 - \lambda, \frac{1}{1 - \lambda}, \frac{\lambda}{\lambda - 1}, \frac{\lambda - 1}{\lambda}\}.$$

(c) Show that if $j = 1728$ then $\lambda = -1, 2, 1/2$, and if $j = 0$ then $\lambda^2 - \lambda + 1 = 0$.

2.10 Consider the equation $u^2 - v^2 = 1$, and the point $(u_0, v_0) = (1, 0)$.

(a) Use the method of Section 2.5.4 to obtain the parameterization

$$u = \frac{m^2 + 1}{m^2 - 1}, \quad v = \frac{2m}{m^2 - 1}.$$

(b) Show that the projective curve $u^2 - v^2 = w^2$ has two points at infinity, $(1 : 1, 0)$ and $(1 : -1 : 0)$.

(c) The parameterization obtained in (a) can be written in projective coordinates as $(u : v : w) = (m^2 + 1 : 2m : m^2 - 1)$ (or $(m^2 + n^2 : 2mn : m^2 - n^2)$ in a homogeneous form). Show that the values $m = \pm 1$ correspond to the two points at infinity. Explain why this is to be expected from the graph (using real numbers) of $u^2 - v^2 = 1$. (*Hint:* Where does an asymptote intersect a hyperbola?)

2.11 Suppose $(u_0, v_0, w_0) = (u_0, 0, 0)$ lies in the intersection

$$au^2 + bv^2 = e, \quad cu^2 + dw^2 = f.$$

(a) Show that the procedure of Section 2.5.4 leads to an equation of the form "square $=$ degree 2 polynomial in m."

(b) Let $F = au^2 + bv^2 = e$ and $G = cu^2 + dw^2 = f$. Show that the Jacobian matrix $\begin{pmatrix} F_u & F_v & F_w \\ G_u & G_v & G_w \end{pmatrix}$ at $(u_0, 0, 0)$ has rank 1. Since the rank is less than 2, this means that the point is a singular point.

2.12 (a) Show that $(x, y) \mapsto (x, -y)$ is a group homomorphism from E to itself, for any elliptic curve in Weierstrass form.

(b) Show that $(x, y) \mapsto (\zeta x, -y)$, where ζ is a nontrivial cube root of 1, is an automorphism of the elliptic curve $y^2 = x^3 + B$.

(c) Show that $(x, y) \mapsto (-x, iy)$, where $i^2 = -1$, is an automorphism of the elliptic curve $y^2 = x^3 + Ax$.

2.13 Let K have characteristic 3 and let E be defined by $y^2 = x^3 + a_2 x^2 + a_4 x + a_6$. The j-invariant in this case is defined to be

$$j = \frac{a_2^6}{a_2^2 a_4^2 - a_2^3 a_6 - a_4^3}$$

(this formula is false if the characteristic is not 3).

(a) Show that either $a_2 \neq 0$ or $a_4 \neq 0$ (otherwise, the cubic has a triple root, which is not allowed).

(b) Show that if $a_2 \neq 0$, then the change of variables $x_1 = x - (a_4/a_2)$ yields an equation of the form $y_1^2 = x_1^3 + a_2' x_1^2 + a_6'$. This means that we may always assume that exactly one of a_2 and a_4 is 0.

(c) Show that if two elliptic curves $y^2 = x^3 + a_2 x^2 + a_6$ and $y^2 = x^3 + a_2' x^2 + a_6'$ have the same j-invariant, then there exists $\mu \in \overline{K}^\times$ such that $a_2' = \mu^2 a_2$ and $a_6' = \mu^6 a_6$.

(d) Show that if $y^2 = x^3 + a_4 x + a_6$ and $y^2 = x^3 + a_4' x^2 + a_6'$ are two elliptic curves (in characteristic 3), then there is a change of variables $x \mapsto x + c$; with $c \in K$, that changes one equation into the other. $y \mapsto ay$ $x \mapsto bx + c$

(e) Observe that if $a_2 = 0$ then $j = 0$ and if $a_4 = 0$ then $j = -a_2^3/a_6$. Show that every element of K appears as the j-invariant of a curve defined over K.

(f) Show that if two curves have the same j-invariant then there is a change of variables over $\overline{K}$ that changes one into the other.

2.14 Let $\alpha(x, y) = (p(x)/q(x), \, y \cdot s(x)/t(x))$ be an endomorphism of the elliptic curve E given by $y^2 = x^3 + Ax + B$, where p, q, s, t are polynomials such that p and q have no common root and s and t have no common root.

(a) Using the fact that (x, y) and $\alpha(x, y)$ lie on E, show that

$$\frac{(x^3 + Ax + B) \, s(x)^2}{t(x)^2} = \frac{u(x)}{q(x)^3}.$$

for some polynomial $u(x)$ such that q and u have no common root. (*Hint:* Show that a common root of u and q must also be a root of p.)

(b) Suppose $t(x_0) = 0$. Use the facts that $x^3 + Ax + B$ has no multiple roots and all roots of t^2 are multiple roots to show that $q(x_0) = 0$. This shows that if $q(x_0) \neq 0$ then $\alpha(x_0, y_0)$ is defined.

2.15 Consider the singular curve $y^2 = x^3 + ax^2$ with $a \neq 0$. Let $y = mx$ be a line through $(0,0)$. Show that the line always intersects the curve to order at least 2, and show that the order is 3 exactly when $m^2 = a$. This may be interpreted as saying that the lines $y = \pm\sqrt{a}x$ are the two tangents to the curve at $(0,0)$.

2.16 (a) Apply the method of Section 2.5.4 to the circle $u^2 + v^2 = 1$ and the point $(-1, 0)$ to obtain the parameterization

$$u = \frac{1 - t^2}{1 + t^2}, \quad v = \frac{2t}{1 + t^2}.$$

(b) Suppose x, y, z are integers such that $x^2 + y^2 = z^2$, $\gcd(x, y, z) = 1$, and x is even. Use (a) to show that there are integers m, n such that

$$x = 2mn, \quad y = m^2 - n^2, \quad z = m^2 + n^2.$$

Also, show that $\gcd(x, y, z) = 1$ implies that $\gcd(m, n) = 1$ and that $m \not\equiv n \pmod 2$.

2.17 Let $p(x)$ and $q(x)$ be polynomials with no common roots. Show that

$$\frac{d}{dx}\left(\frac{p(x)}{q(x)}\right) = 0$$

(that is, the identically 0 rational function) if and only if both $p'(x) = 0$ and $q'(x) = 0$. (If p or q is nonconstant, then this can happen only in positive characteristic.)

2.18 Let E be given by $y^2 = x^3 + Ax + B$ over a field K and let $d \in K^\times$. The **twist** of E by d is the elliptic curve $E^{(d)}$ given by $y^2 = x^3 + Ad^2x + Bd^3$.

(a) Show that $j(E^{(d)}) = j(E)$.

(b) Show that $E^{(d)}$ can be transformed into E over $K(\sqrt{d})$.

(c) Show that $E^{(d)}$ can be transformed over K to the form $dy_1^2 = x_1^3 + Ax_1 + B$.

2.19 Let $\alpha, \beta \in \mathbf{Z}$ be such that $\gcd(\alpha, \beta) = 1$. Assume that $\alpha \equiv -1 \pmod 4$ and $\beta \equiv 0 \pmod{32}$. Let E be given by $y^2 = x(x - \alpha)(x - \beta)$.

(a) Let p be prime. Show that the cubic polynomial $x(x - \alpha)(x - \beta)$ cannot have a triple root mod p.

(b) Show that the substitution

$$x = 4x_1, \quad y = 8y_1 + 4x_1$$

changes E into E_1, given by

$$y_1^2 + x_1 y_1 = x_1^3 + \frac{-\beta - \alpha - 1}{4} x_1^2 + \frac{\alpha\beta}{16} x_1.$$

(c) Show that the reduction mod 2 of the equation for E_1 is

$$y_1^2 + x_1 y_1 = x_1^3 + e x_1^2$$

for some $e \in \mathbf{F}_2$. This curve is singular at $(0, 0)$.

(d) Let γ be a constant and consider the line $y_1 = \gamma x_1$. Show that if $\gamma^2 + \gamma = e$, then the line intersects the curve in part (c) to order 3, and if $\gamma^2 + \gamma \neq e$ then this line intersects the curve to order 2.

(e) Show that there are two distinct values of $\gamma \in \overline{\mathbf{F}}_2$ such that $\gamma^2 + \gamma = e$. This implies that there are two distinct tangent lines to the curve E_1 mod 2 at $(0,0)$, as in Exercise 2.15.

We take the property of part (e) to be the definition of multiplicative reduction in characteristic 2. Therefore, parts (a) and (e) show that the curve E_1 has good or multiplicative reduction at all primes. A **semistable** elliptic curve over $\mathbf{Q}$ is one that has good or multiplicative reduction at all primes, possibly after a change of variables (over $\mathbf{Q}$) such as the one in part (b). Therefore, E is semistable. See Section 13.1 for a situation where this fact is used.

Chapter 3

Torsion Points

The torsion points, namely those whose orders are finite, play an important role in the study of elliptic curves. We'll see this in Chapter 4 for elliptic curves over finite fields, where all points are torsion points, and in Chapter 8, where we use 2-torsion points in a procedure known as descent. In the present chapter, we first consider the elementary cases of 2- and 3-torsion, then determine the general situation. Finally, we discuss the important Weil pairing on torsion points.

3.1 Torsion Points

Let E be an elliptic curve defined over a field K. Let n be a positive integer. We are interested in

$$E[n] = \{P \in E(\overline{K}) \mid nP = \infty\}$$

(recall that $\overline{K} = $ algebraic closure of K). We emphasize that $E[n]$ contains points with coordinates in $\overline{K}$, not just in K.

When the characteristic of K is not 2, E can be put in the form $y^2 = $ cubic, and it is easy to determine $E[2]$. Let

$$y^2 = (x - e_1)(x - e_2)(x - e_3),$$

with $e_1, e_2, e_3 \in \overline{K}$. A point P satisfies $2P = \infty$ if and only if the tangent line at P is vertical. It is easy to see that this means that $y = 0$, so

$$E[2] = \{\infty, (e_1, 0), (e_2, 0), (e_3, 0)\}.$$

As an abstract group, this is isomorphic to $\mathbf{Z}_2 \oplus \mathbf{Z}_2$.

The situation in characteristic 2 is more subtle. In Section 2.7 we showed that E can be assumed to have one of the following two forms:

$$(I) \quad y^2 + xy + x^3 + a_2 x^2 + a_6 = 0 \quad \text{or} \quad (II) \quad y^2 + a_3 y + x^3 + a_4 x + a_6 = 0.$$

In the first case, $a_6 \neq 0$ and in the second case, $a_3 \neq 0$ (otherwise the curves would be singular). If $P = (x, y)$ is a point of order 2, then the tangent at P must be vertical, which means that the partial derivative with respect to y must vanish. In case I, this means that $x = 0$. Substitute $x = 0$ into (I) to obtain $0 = y^2 + a_6 = (y + \sqrt{a_6})^2$. Therefore $(0, \sqrt{a_6})$ is the only point of order 2 (square roots are unique in characteristic 2), so

$$E[2] = \{\infty, (0, \sqrt{a_6})\}.$$

As an abstract group, this is isomorphic to $\mathbf{Z}_2$.

In case II, the partial derivative with respect to y is $a_3 \neq 0$. Therefore, there is no point of order 2, so

$$E[2] = \{\infty\}.$$

We summarize the preceding discussion as follows.

PROPOSITION 3.1

Let E be an elliptic curve over a field K. If the characteristic of K is not 2, then

$$E[2] \simeq \mathbf{Z}_2 \oplus \mathbf{Z}_2.$$

If the characteristic of K is 2, then

$$E[2] \simeq 0 \ \ or \ \mathbf{Z}_2.$$

Now let's look at $E[3]$. Assume first that the characteristic of K is not 2 or 3, so that E can be given by the equation $y^2 = x^3 + Ax + B$. A point P satisfies $3P = \infty$ if and only if $2P = -P$. This means that the x-coordinate of $2P$ equals the x-coordinate of P (the y-coordinates therefore differ in sign; of course, if they were equal, then $2P = P$, hence $P = \infty$). In equations, this becomes

$$m^2 - 2x = x, \quad \text{where } m = \frac{3x^2 + A}{2y}.$$

Using the fact that $y^2 = x^3 + Ax + B$, we find that

$$(3x^2 + A)^2 = 12x(x^3 + Ax + B).$$

This simplifies to

$$3x^4 + 6Ax^2 + 12Bx - A^2 = 0.$$

The discriminant of this polynomial is $-6912(4A^3 + 27B^2)^2$, which is nonzero. Therefore the polynomial has no multiple roots. There are 4 distinct values of x (in $\overline{K}$), and each x yields two values of y, so we have eight points of order 3. Since ∞ is also in $E[3]$, we see that $E[3]$ is a group of order 9 in which every element is 3-torsion. It follows that

$$E[3] \simeq \mathbf{Z}_3 \oplus \mathbf{Z}_3.$$

The case where K has characteristic 2 is Exercise 3.1.

Now let's look at characteristic 3. We may assume that E has the form $y^2 = x^3 + a_2 x^2 + a_4 x + a_6$. Again, we want the x-coordinate of $2P$ to equal the x-coordinate of P. We calculate the x-coordinate of $2P$ by the usual procedure and set it equal to the x-coordinate x of P. Some terms disappear because $3 = 0$. We obtain

$$\left(\frac{2a_2 x + a_4}{2y} \right)^2 - a_2 = 3x = 0.$$

This simplifies to (recall that $4 = 1$)

$$a_2 x^3 + a_2 a_6 - a_4^2 = 0.$$

Note that we cannot have $a_2 = a_4 = 0$ since then $x^3 + a_6 = (x + a_6^{1/3})^3$ has multiple roots, so at least one of a_2, a_4 is nonzero.

If $a_2 = 0$, then we have $-a_4^2 = 0$, which cannot happen, so there are no values of x. Therefore $E[3] = \{\infty\}$ in this case.

If $a_2 \neq 0$, then we obtain an equation of the form $a_2(x^3 + a) = 0$, which has a single triple root in characteristic 3. Therefore, there is one value of x, and two corresponding values of y. This yields 2 points of order 3. Since there is also the point ∞, we see that $E[3]$ has order 3, so $E[3] \simeq \mathbf{Z}_3$ as abstract groups.

The general situation is given by the following.

THEOREM 3.2

Let E be an elliptic curve over a field K and let n be a positive integer. If the characteristic of K does not divide n, or is 0, then

$$E[n] \simeq \mathbf{Z}_n \oplus \mathbf{Z}_n.$$

If the characteristic of K is $p > 0$ and $p|n$, write $n = p^r n'$ with $p \nmid n'$. Then

$$E[n] \simeq \mathbf{Z}_{n'} \oplus \mathbf{Z}_{n'} \quad or \quad \mathbf{Z}_n \oplus \mathbf{Z}_{n'}.$$

The theorem will be proved in Section 3.2.

An elliptic curve E in characteristic p is called **ordinary** if $E[p] \simeq \mathbf{Z}_p$. It is called **supersingular** if $E[p] \simeq 0$. Note that the terms "supersingular" and "singular" (as applied to bad points on elliptic curves) are unrelated. In the theory of complex multiplication (see Chapter 10), the "singular" j-invariants are those corresponding to elliptic curves with endomorphism rings larger than $\mathbf{Z}$, and the "supersingular" j-invariants are those corresponding to elliptic curves with the largest possible endomorphism rings, namely, orders in quaternion algebras.

Let n be a positive integer not divisible by the characteristic of K. Choose a **basis** $\{\beta_1, \beta_2\}$ for $E[n] \simeq \mathbf{Z}_n \oplus \mathbf{Z}_n$. This means that every element of $E[n]$ is expressible in the form $m_1\beta_1 + m_2\beta$ with integers m_1, m_2. Note that m_1, m_2 are uniquely determined mod n. Let $\alpha : E(\overline{K}) \to E(\overline{K})$ be a homomorphism. Then α maps $E[n]$ into $E[n]$. Therefore, there are $a, b, c, d \in \mathbf{Z}_n$ such that

$$\alpha(\beta_1) = a\beta_1 + c\beta_2, \quad \alpha(\beta_2) = b\beta_1 + d\beta_2.$$

Therefore each homomorphism $\alpha : E(\overline{K}) \to E(\overline{K})$ is represented by a 2×2 matrix

$$\alpha_n = \begin{pmatrix} a & b \\ c & d \end{pmatrix}.$$

Composition of homomorphisms corresponds to multiplication of the corresponding matrices.

In many cases, the homomorphism α will be taken to be an endomorphism, which means that it is given by rational functions (see Section 2.8). But α can also come from an automorphism of $\overline{K}$ that fixes K. This leads to the important subject of representations of Galois groups (that is, homomorphisms from Galois groups to groups of matrices).

Example 3.1
Let E be the elliptic curve defined over $\mathbf{R}$ by $y^2 = x^3 - 2$, and let $n = 2$. Then

$$E[2] = \{\infty, (2^{1/3}, 0), (\zeta 2^{1/3}, 0), (\zeta^2 2^{1/3}, 0)\},$$

where ζ is a nontrivial cube root of unity. Let

$$\beta_1 = (2^{1/3}, 0), \quad \beta_2 = (\zeta 2^{1/3}, 0).$$

Then $\{\beta_1, \beta_2\}$ is a basis for $E[2]$, and $\beta_3 = (\zeta^2 2^{1/3}, 0) = \beta_1 + \beta_2$.

Let $\alpha : E(\mathbf{C}) \to E(\mathbf{C})$ be complex conjugation: $\alpha(x, y) = (\overline{x}, \overline{y})$, where the bar denotes complex conjugation. It is easy to verify that α is a homomorphism. In fact, since all the coefficients of the formulas for the group law have real coefficients, we have $\overline{P_1} + \overline{P_2} = \overline{P_1 + P_2}$. This is the same as $\alpha(P_1) + \alpha(P_2) = \alpha(P_1 + P_2)$. We have

$$\alpha(\beta_1) = 1 \cdot \beta_1 + 0 \cdot \beta_2, \quad \alpha(\beta_2) = \beta_3 = 1 \cdot \beta_1 + 1 \cdot \beta_2.$$

Therefore we obtain the matrix $\alpha_2 = \begin{pmatrix} 1 & 1 \\ 0 & 1 \end{pmatrix}$. Note that $\alpha \circ \alpha$ is the identity, which corresponds to the fact that α_2^2 is the identity matrix mod 2.   ☐

3.2   Division Polynomials

The goal of this section is to prove Theorem 3.2. We'll also obtain a few other results that will be needed in proofs in Section 4.2.

In order to study the torsion subgroups, we need to describe the map on an elliptic curve given by multiplication by an integer. As in Section 2.8, this is an endomorphism of the elliptic curve and can be described by rational functions. We shall give formulas for these functions.

We start with variables A, B. Define the **division polynomials** $\psi_m \in \mathbf{Z}[x, y, A, B]$ by

$$\psi_0 = 0$$
$$\psi_1 = 1$$
$$\psi_2 = 2y$$
$$\psi_3 = 3x^4 + 6Ax^2 + 12Bx - A^2$$
$$\psi_4 = 4y(x^6 + 5Ax^4 + 20Bx^3 - 5A^2x^2 - 4ABx - 8B^2 - A^3)$$
$$\psi_{2m+1} = \psi_{m+2}\psi_m^3 - \psi_{m-1}\psi_{m+1}^3 \text{ for } m \geq 2$$
$$\psi_{2m} = (2y)^{-1}(\psi_m)(\psi_{m+2}\psi_{m-1}^2 - \psi_{m-2}\psi_{m+1}^2) \text{ for } m \geq 2.$$

LEMMA 3.3

ψ_n is a polynomial in $\mathbf{Z}[x, y^2, A, B]$ when n is odd, and ψ_n is a polynomial in $2y\mathbf{Z}[x, y^2, A, B]$ when n is even.

PROOF   The lemma is true for $n \leq 4$. Assume, by induction, that it holds for all $n < 2m$. We may assume that $2m > 4$, so $m > 2$. Then $2m > m + 2$, so all polynomials appearing in the definition of ψ_{2m} satisfy the induction assumptions. If m is even, then $\psi_m, \psi_{m+2}, \psi_{m-2}$ are in $2y\mathbf{Z}[x, y^2, A, B]$, from which it follows that ψ_{2m} is in $2y\mathbf{Z}[x, y^2, A, B]$. If m is odd, then ψ_{m-1} and ψ_{m+1} are in $2y\mathbf{Z}[x, y^2, A, B]$, so again we find that ψ_{2m} is in $2y\mathbf{Z}[x, y^2, A, B]$. Therefore, the lemma holds for $n = 2m$. Similarly, it holds for $n = 2m + 1$. ∎

Define polynomials

$$\phi_m = x\psi_m^2 - \psi_{m+1}\psi_{m-1}$$
$$\omega_m = (4y)^{-1}(\psi_{m+2}\psi_{m-1}^2 - \psi_{m-2}\psi_{m+1}^2).$$

LEMMA 3.4

$\phi_n \in \mathbf{Z}[x, y^2, A, B]$ for all n. If n is odd, then $\omega_n \in y\mathbf{Z}[x, y^2, A, B]$. If n is even, then $\omega_n \in \mathbf{Z}[x, y^2, A, B]$.

PROOF   If n is odd, then ψ_{n+1} and ψ_{n-1} are in $y\mathbf{Z}[x, y^2, A, B]$, so their product is in $\mathbf{Z}[x, y^2, A, B]$. Therefore, $\phi_n \in \mathbf{Z}[x, y^2, A, B]$. If n is even, the proof is similar.

The facts that $y^{-1}\omega_n \in \mathbf{Z}[x, y^2, A, B]$ for odd n and $\omega_n \in \frac{1}{2}\mathbf{Z}[x, y^2, A, B]$ for even n follow from Lemma 3.3, and these are all that we need for future applications. However, to get rid of the extra 2 in the denominator, we proceed as follows. Induction (treating separately the various possibilities for n mod 4) shows that

$$\psi_n \equiv (x^2 + A)^{(n^2-1)/4} \pmod 2 \quad \text{when } n \text{ is odd}$$

and

$$(2y)^{-1}\psi_n \equiv \left(\frac{n}{2}\right)(x^2 + A)^{(n^2-4)/4} \pmod 2 \quad \text{when } n \text{ is even.}$$

A straightforward calculation now yields the lemma.   ∎

We now consider an elliptic curve

$$E: \quad y^2 = x^3 + Ax + B, \quad 4A^3 + 27B^2 \neq 0.$$

We don't specify what ring or field the coefficients A, B are in, so we continue to treat them as variables. We regard the polynomials in $\mathbf{Z}[x, y^2, A, B]$ as polynomials in $\mathbf{Z}[x, A, B]$ by replacing y^2 with $x^3 + Ax + B$. Therefore, we write $\phi_n(x)$ and $\psi_n^2(x)$. Note that ψ_n is not necessarily a polynomial in x alone, while ψ_n^2 is always a polynomial in x.

LEMMA 3.5

$$\phi_n(x) = x^{n^2} + \text{lower degree terms}$$
$$\psi_n^2(x) = n^2 x^{n^2-1} + \text{lower degree terms}$$

PROOF   In fact, we claim that

$$\psi_n = \begin{cases} y(nx^{(n^2-4)/2} + \cdots) & \text{if } n \text{ is even} \\ nx^{(n^2-1)/2} + \cdots & \text{if } n \text{ is odd.} \end{cases}$$

This is proved by induction. For example, if $n = 2m + 1$ with m even, then the leading term of $\psi_{m+2}\psi_m^3$ is

$$(m+2)m^3 y^4 x^{\frac{(m+2)^2-4}{2} + \frac{3m^2-12}{2}}.$$

Changing y^4 to $(x^3 + Ax + B)^2$ yields

$$(m+2)m^3 x^{\frac{(2m+1)^2-1}{2}}.$$

Similarly, the leading term of $\psi_{m-1}\psi_{m+1}^3$ is

$$(m-1)(m+1)^3 x^{\frac{(2m+1)^2-1}{2}}.$$

Subtracting and using the recursion relation shows that the leading term of ψ_{2m+1} is as claimed in the lemma. The other cases are treated similarly. ∎

We can now state the main theorem.

THEOREM 3.6

Let $P = (x, y)$ be a point on the elliptic curve $y^2 = x^3 + Ax + B$ (over some field of characteristic not 2), and let n be a positive integer. Then

$$nP = \left(\frac{\phi_n(x)}{\psi_n^2(x)}, \frac{\omega_n(x, y)}{\psi_n(x, y)^3} \right).$$

The proof will be given in Section 9.5.

COROLLARY 3.7

Let E be an elliptic curve. The endomorphism of E given by multiplication by n has degree n^2.

PROOF   From Lemma 3.5, we have that the maximum of the degrees of the numerator and denominator of $\phi_n(x)/\psi_n^2(x)$ is n^2. Therefore, the degree of the endomorphism is n^2 if this rational function is reduced, that is, if $\phi_n(x)$ and $\psi_n^2(x)$ have no common roots. We'll show that this is the case. Suppose not. Let n be the smallest index for which they have a common root.

Suppose $n = 2m$ is even. A quick calculation shows that

$$\phi_2(x) = x^4 - 2Ax^2 - 8Bx + A^2.$$

Computing the x-coordinate of $2m(x, y)$ in two steps by multiplying by m and then by 2, and using the fact that

$$\psi_2^2 = 4y^2 = 4(x^3 + Ax + B),$$

we obtain

$$\frac{\phi_{2m}}{\psi_{2m}^2} = \frac{\phi_2(\phi_m/\psi_m^2)}{\psi_2^2(\phi_m/\psi_m^2)}$$

$$= \frac{\phi_m^4 - 2A\phi_m^2\psi_m^4 - 8B\phi_m\psi_m^6 + A^2\psi_m^8}{(4\psi_m^2)(\phi_m^3 + A\phi_m\psi_m^4 + B\psi_m^6)}$$

$$= \frac{U}{V},$$

where U and V are the numerator and denominator of the preceding expression. To show U and V have no common roots, we need the following.

LEMMA 3.8

Let $\Delta = 4A^3 + 27B^2$ and let

$$F(x, z) = x^4 - 2Ax^2z^2 - 8Bxz^3 + A^2z^4$$
$$G(x, z) = 4z(x^3 + Axz^2 + Bz^3)$$
$$f_1(x, z) = 12x^2z + 16Az^3$$
$$g_1(x, z) = 3x^3 - 5Axz^2 - 27Bz^3$$
$$f_2(x, z) = 4\Delta x^3 - 4a^2bx^2z + 4A(3A^3 + 22B^2)xz^2 + 12B(A^3 + 8B^2)z^3$$
$$g_2(x, z) = A^2Bx^3 + A(5A^3 + 32B^2)x^2z + 2B(13A^3 + 96B^2)xz^2$$
$$- 3A^2(A^3 + 8B^2)z^3.$$

Then

$$Ff_1 - Gg_1 = 4\Delta z^7 \text{ and } Ff_2 + Gg_2 = 4\Delta x^7.$$

PROOF   This is verified by a straightforward calculation. Where do these identities come from? The polynomials $F(x, 1)$ and $G(x, 1)$ have no common roots, so the extended Euclidean algorithm, applied to polynomials, finds polynomials $f_1(x), g_1(x)$ such that $F(x, 1)f_1(x) + G(x, 1)g_1(x) = 1$. Changing x to x/z, multiplying by z^7 (to make everything homogeneous), then multiplying by 4Δ to clear denominators yields the first identity. The second is obtained by reversing the roles of x and z.   ∎

The lemma implies that

$$U \cdot f_1(\phi_m, \psi_m^2) - V \cdot g_1(\phi_m, \psi_m^2) = 4\psi_m^{14}\Delta$$
$$U \cdot f_2(\phi_m, \psi_m^2) + V \cdot g_2(\phi_m, \psi_m^2) = 4\phi_m^7\Delta.$$

If U, V have a common root, then so do ϕ_m and ψ_m^2. Since $n = 2m$ is the first index for which there is a common root, this is impossible.

It remains to show that $U = \phi_{2m}$ and $V = \psi_{2m}^2$. Since $U/V = \phi_{2m}/\psi_{2m}^2$ and since U, V have no common root, it follows that ϕ_{2m} is a multiple of U and ψ_{2m}^2 is a multiple of V. A quick calculation using Lemma 3.5 shows that

$$U = x^{4m^2} + \text{lower degree terms.}$$

Lemma 3.5 and the fact that ϕ_{2m} is a multiple of U imply that $\phi_{2m} = U$. Therefore, $V = \psi_{2m}^2$. It follows that ϕ_{2m} and ψ_{2m}^2 have no common roots.

Now suppose that the smallest index n such that there is a common root is odd: $n = 2m + 1$. Let r be a common root of ϕ_n and ψ_n^2. Since

$$\phi_n = x\psi_n^2 - \psi_{n-1}\psi_{n+1},$$

and since $\psi_{n+1}\psi_{n-1}$ is a polynomial in x, we have $(\psi_{n+1}\psi_{n-1})(r) = 0$. But $\psi_{n\pm1}^2$ are polynomials in x and their product vanishes at r. Therefore $\psi_{n+\delta}^2(r) = 0$, where δ is either 1 or -1.

Since n is odd, both ψ_n and $\psi_{n+2\delta}$ are polynomials in x. Moreover,

$$(\psi_n \psi_{n+2\delta})^2 = \psi_n^2 \psi_{n+2\delta}^2$$

vanishes at r. Therefore $\psi_n \psi_{n+2\delta}$ vanishes at r. Since

$$\phi_{n+\delta} = x\psi_{n+\delta}^2 - \psi_n \psi_{n+2\delta},$$

we find that $\phi_{n+\delta}(r) = 0$. Therefore, $\phi_{n+\delta}$ and $\psi_{n+\delta}^2$ have a common root. Note that $n + \delta$ is even.

When considering the case that n is even, we showed that if ϕ_{2m} and ψ_{2m}^2 have a common root, then ϕ_m and ψ_m^2 have a common root. In the present case, we apply this to $2m = n + \delta$. Since n is assumed to be the smallest index for which there is a common root, we have

$$\frac{n+\delta}{2} \geq n.$$

This implies that $n = 1$. But clearly $\phi_1 = x$ and $\psi_1^2 = 1$ have no common roots, so we have a contradiction.

This proves that ϕ_n and ψ_n^2 have no common roots in all cases. Therefore, as pointed out at the beginning of the proof, the multiplication by n map has degree n^2. This completes the proof of Corollary 3.7. ∎

Recall from Section 2.8 that if $\alpha(x, y) = (R(x), yS(x))$ is an endomorphism of an elliptic curve E, then α is separable if $R'(x)$ is not identically 0. Assume n is not a multiple of the characteristic p of the field. From Theorem 3.6 we see that the multiplication by n map has

$$R(x) = \frac{x^{n^2} + \cdots}{n^2 x^{n^2-1} + \cdots}.$$

The numerator of the derivative is $n^2 x^{2n^2-2} + \cdots \neq 0$, so $R'(x) \neq 0$. Therefore, multiplication by n is separable. From Corollary 3.7 and Proposition 2.20, $E[n]$, the kernel of multiplication by n, has order n^2. The structure theorem for finite abelian groups (see Appendix B) says that $E[n]$ is isomorphic to

$$\mathbf{Z}_{n_1} \oplus \mathbf{Z}_{n_2} \oplus \cdots \oplus \mathbf{Z}_{n_k},$$

for some integers $n_1, n_2, \ldots, n_k$ with $n_i | n_{i+1}$ for all i. Let ℓ be a prime dividing n_1. Then $\ell | n_i$ for all i. This means that $E[\ell] \subseteq E[n]$ has order ℓ^k. Since we have just proved that $E[\ell]$ has order ℓ^2, we must have $k = 2$. Multiplication by n annihilates $E[n] \simeq \mathbf{Z}_{n_1} \oplus \mathbf{Z}_{n_2}$, so we must have $n_2 | n$. Since $n^2 = \#E[n] = n_1 n_2$, it follows that $n_1 = n_2 = n$. Therefore,

$$E[n] \simeq \mathbf{Z}_n \oplus \mathbf{Z}_n$$

when the characteristic p of the field does not divide n.

It remains to consider the case where $p|n$. We first determine the p-power torsion on E. By Proposition 2.27, multiplication by p is not separable. By Proposition 2.20, the kernel $E[p]$ of multiplication by p has order strictly less than the degree of this endomorphism, which is p^2 by Corollary 3.7. Since every element of $E[p]$ has order 1 or p, the order of $E[p]$ is a power of p, hence must be 1 or p. If $E[p]$ is trivial, then $E[p^k]$ must be trivial for all k. Now suppose $E[p]$ has order p. We claim that $E[p^k] \simeq \mathbf{Z}_{p^k}$ for all k. It is easy to see that $E[p^k]$ is cyclic. The hard part is to show that the order is p^k, rather than something smaller (for example, why can't we have $E[p^k] = E[p] \simeq \mathbf{Z}_p$ for all k?). Suppose there exists an element P of order p^j. By Theorem 2.21, multiplication by p is surjective, so there exists a point Q with $pQ = P$. Since

$$p^j Q = p^{j-1} P \neq \infty \quad \text{but} \quad p^{j+1} Q = p^j P = \infty,$$

Q has order p^{j+1}. By induction, there are points of order p^k for all k. Therefore, $E[p^k]$ is cyclic of order p^k.

We can now put everything together. Write $n = p^r n'$ with $r \geq 0$ and $p \nmid n'$. Then

$$E[n] \simeq E[n'] \oplus E[p^r].$$

We have $E[n'] \simeq \mathbf{Z}_{n'} \oplus \mathbf{Z}_{n'}$, since $p \nmid n'$. We have just showed that $E[p^r] \simeq 0$ or $\mathbf{Z}_{p^r}$. Recall that

$$\mathbf{Z}_{n'} \oplus \mathbf{Z}_{p^r} \simeq \mathbf{Z}_{n'p^r} \simeq \mathbf{Z}_n$$

(see Appendix A). Therefore, we obtain

$$E[n] \simeq \mathbf{Z}_{n'} \oplus \mathbf{Z}_{n'} \quad \text{or} \quad \mathbf{Z}_n \oplus \mathbf{Z}_{n'}.$$

This completes the proof of Theorem 3.2.   ∎

3.3   The Weil Pairing

The Weil pairing on the n-torsion on an elliptic curve is a major tool in the study of elliptic curves. For example, it will be used in Chapter 4 to prove Hasse's theorem on the number of points on an elliptic curve over a finite field. It will be used in Chapter 5 to attack the discrete logarithm problem for elliptic curves. In Chapter 6, it will be used in a cryptographic setting.

Let E be an elliptic curve over a field K and let n be an integer not divisible by the characteristic of K. Then $E[n] \simeq \mathbf{Z}_n \oplus \mathbf{Z}_n$. Let

$$\mu_n = \{x \in \overline{K} \mid x^n = 1\}$$

be the group of nth roots of unity in $\overline{K}$. Since the characteristic of K does not divide n, the equation $x^n = 1$ has no multiple roots, hence has n roots in

$\overline{K}$. Therefore, μ_n is a cyclic group of order n. Any generator ζ of μ_n is called a **primitive nth root of unity**. This is equivalent to saying that $\zeta^k = 1$ if and only if n divides k.

THEOREM 3.9

Let E be an elliptic curve defined over a field K and let n be a positive integer. Assume that the characteristic of K does not divide n. Then there is a pairing

$$e_n : E[n] \times E[n] \to \mu_n,$$

called the **Weil pairing**, *that satisfies the following properties:*

1. *e_n is bilinear in each variable. This means that*

$$e_n(S_1 + S_2, T) = e_n(S_1, T)e_n(S_2, T)$$

   and

$$e_n(S, T_1 + T_2) = e_n(S, T_1)e_n(S, T_2)$$

   for all $S, S_1, S_2, T, T_1, T_2 \in E[n]$.

2. *e_n is nondegenerate in each variable. This means that if $e_n(S, T) = 1$ for all $T \in E[n]$ then $S = \infty$ and also that if $e_n(S, T) = 1$ for all $S \in E[n]$ then $T = \infty$.*

3. *$e_n(T, T) = 1$ for all $T \in E[n]$.*

4. *$e_n(T, S) = e_n(S, T)^{-1}$ for all $S, T \in E[n]$.*

5. *$e_n(\sigma S, \sigma T) = \sigma(e_n(S, T))$ for all automorphisms σ of $\overline{K}$ such that σ is the identity map on the coefficients of E (if E is in Weierstrass form, this means that $\sigma(A) = A$ and $\sigma(B) = B$).*

6. *$e_n(\alpha(S), \alpha(T)) = e_n(S, T)^{\deg(\alpha)}$ for all separable endomorphisms α of E. If the coefficients of E lie in a finite field $\mathbf{F}_q$, then the statement also holds when α is the Frobenius endomorphism ϕ_q. (Actually, the statement holds for all endomorphisms α, separable or not. See [28].)*

The proof of the theorem will be given in Chapter 11. In the present section, we'll derive some consequences.

COROLLARY 3.10

Let $\{T_1, T_2\}$ be a basis of $E[n]$. Then $e_n(T_1, T_2)$ is a primitive nth root of unity.

PROOF   Suppose $e_n(T_1, T_2) = \zeta$ with $\zeta^d = 1$. Then $e_n(T_1, dT_2) = 1$. Also, $e_n(T_2, dT_2) = e_n(T_2, T_2)^d = 1$ (by (1) and (3)). Let $S \in E[n]$. Then

$S = aT_1 + bT_2$ for some integers a, b. Therefore,

$$e_n(S, dT_2) = e_n(T_1, dT_2)^a e_n(T_2, dT_2)^b = 1.$$

Since this holds for all S, (2) implies that $dT_2 = \infty$. Since $dT_2 = \infty$ if and only if $n|d$, it follows that ζ is a primitive nth root of unity.

COROLLARY 3.11
  If $E[n] \subseteq E(K)$, then $\mu_n \subset K$.

REMARK 3.12   Recall that points in $E[n]$ are allowed to have coordinates in $\overline{K}$. The hypothesis of the corollary is that these points all have coordinates in K.   ∎

PROOF   Let σ be any automorphism of $\overline{K}$ such that σ is the identity on K. Let T_1, T_2 be a basis of $E[n]$. Since T_1, T_2 are assumed to have coordinates in K, we have $\sigma T_1 = T_1$ and $\sigma T_2 = T_2$. By (5),

$$\zeta = e_n(T_1, T_2) = e_n(\sigma T_1, \sigma T_2) = \sigma(e_n(T_1, T_2)) = \sigma(\zeta).$$

The fundamental theorem of Galois theory says that if an element $x \in \overline{K}$ is fixed by all such automorphisms σ, then $x \in K$. Therefore, $\zeta \in K$. Since ζ is a primitive nth root of unity by Corollary 3.10, it follows that $\mu_n \subset K$. (*Technical point:* The fundamental theorem of Galois theory only implies that ζ lies in a purely inseparable extension of K. But an nth root of unity generates a separable extension of K when the characteristic does not divide n, so we conclude that $\zeta \in K$.)   ∎

COROLLARY 3.13
  Let E be an elliptic curve defined over $\mathbf{Q}$. Then $E[n] \not\subseteq E(\mathbf{Q})$ for $n \geq 3$.

PROOF   If $E[n] \subseteq E(\mathbf{Q})$, then $\mu_n \subset \mathbf{Q}$, which is not the case when $n \geq 3$.
∎

REMARK 3.14   When $n = 2$, it is possible to have $E[2] \subseteq E(\mathbf{Q})$. For example, if E is given by $y^2 = x(x-1)(x+1)$, then

$$E[2] = \{\infty, (0,0), (1,0), (-1,0)\}.$$

If $n = 3, 4, 5, 6, 7, 8, 9, 10, 12$, there are elliptic curves E defined over $\mathbf{Q}$ that have points of order n with rational coordinates. However, the corollary says that it is not possible for all points of order n to have rational coordinates for these n. The torsion subgroups of elliptic curves over $\mathbf{Q}$ will be discussed in Chapter 8.   ∎

We now use the Weil pairing to deduce two propositions that will be used in the proof of Hasse's theorem in Chapter 4. Recall that if α is an endomorphism of E, then we obtain a matrix $\alpha_n = \begin{pmatrix} a & b \\ c & d \end{pmatrix}$ with entries in $\mathbf{Z}_n$, describing the action of α on a basis $\{T_1, T_2\}$ of $E[n]$.

PROPOSITION 3.15

Let α be an endomorphism of an elliptic curve E defined over a field K. Let n be a positive integer not divisible by the characteristic of K. Then $\det(\alpha_n) \equiv \deg(\alpha) \pmod{n}$.

PROOF   By Corollary 3.10, $\zeta = e_n(T_1, T_2)$ is a primitive nth root of unity. By part (6) of Theorem 3.9, we have

$$
\begin{aligned}
\zeta^{\deg(\alpha)} &= e_n(\alpha(T_1), \alpha(T_2)) = e_n(aT_1 + cT_2, bT_1 + dT_2) \\
&= e_n(T_1, T_1)^{ab} e_n(T_1, T_2)^{ad} e_n(T_2, T_1)^{cb} e_n(T_2, T_2)^{cd} \\
&= \zeta^{ad-bc},
\end{aligned}
$$

by the properties of the Weil pairing. Since ζ is a primitive nth root of unity, $\deg(\alpha) \equiv ad - bc \pmod{n}$.   ∎

As we'll see in the proof of the next result, Proposition 3.15 allows us to reduce questions about the degree to calculations with matrices. Both Proposition 3.15 and Proposition 3.16 hold for all endomorphisms, since part (6) of Theorem 3.9 holds in general. However, we prove part (6) only for separable endomorphisms and for the Frobenius map, which is sufficient for our purposes. We'll state Proposition 3.16 in general, and the proof is sufficient for separable endomorphisms and for all endomorphisms of the form $r + s\phi_q$ with arbitrary integers r, s.

Let α and β be endomorphisms of E and let a, b be integers. The endomorphism $a\alpha + b\beta$ is defined by

$$(a\alpha + b\beta)(P) = a\alpha(P) + b\beta(P).$$

Here $a\alpha(P)$ means multiplication on E of $\alpha(P)$ by the integer a. The result is then added on E to $b\beta(P)$. This process can all be described by rational functions, since this is true for each of the individual steps. Therefore $a\alpha + b\beta$ is an endomorphism.

PROPOSITION 3.16

$$\deg(a\alpha + b\beta) = a^2 \deg\alpha + b^2 \deg\beta + ab(\deg(\alpha + \beta) - \deg\alpha - \deg\beta).$$

PROOF   Let n be any integer not divisible by the characteristic of K. Represent α and β by matrices α_n and β_n (with respect to some basis of $E[n]$). Then $a\alpha_n + b\beta_n$ gives the action of $a\alpha + b\beta$ on $E[n]$. A straightforward calculation yields

$$\det(a\alpha_n + b\beta_n) = a^2 \det \alpha_n + b^2 \det \beta_n + ab(\det(\alpha_n + \beta_n) - \det \alpha_n - \det \beta_n)$$

for any matrices α_n and β_n (see Exercise 3.2). Therefore

$$\deg(a\alpha + b\beta) \equiv$$
$$a^2 \deg \alpha + b^2 \deg \beta + ab(\deg(\alpha + \beta) - \deg \alpha - \deg \beta) \pmod{n}.$$

Since this holds for infinitely many n, it must be an equality.   ∎

Exercises

3.1 Let E be an elliptic curve in characteristic 2. Show that $E[3] \simeq \mathbf{Z}_3 \oplus \mathbf{Z}_3$. (*Hint:* Use the formulas at the end of Section 2.7.)

3.2 Let M and N be 2×2 matrices with $N = \begin{pmatrix} w & x \\ y & z \end{pmatrix}$. Define $\tilde{N} = \begin{pmatrix} z & -x \\ -y & w \end{pmatrix}$ (this is the adjoint matrix).

(a) Show that $\mathrm{Trace}(M\tilde{N}) = \det(M + N) - \det(M) - \det(N)$.

(b) Use (a) to show that

$$\det(aM + bN) - a^2 \det M - b^2 \det N$$
$$= ab(\det(M + N) - \det M - \det N)$$

for all scalars a, b. This is the relation used in the proof of Proposition 3.16.

3.3 Let E be an elliptic curve over a field K and let P be a point of order n (where n is not divisible by the characteristic of the field K). Let $Q \in E[n]$. Show that there exists an integer k such that $Q = kP$ if and only if $e_n(P, Q) = 1$.

3.4 Write the equation of the elliptic curve E as

$$F(x, y, z) = y^2 z - x^3 - Axz^2 - Bz^3 = 0.$$

Show that a point P on E is in $E[3]$ if and only if

$$\det \begin{pmatrix} F_{xx} & F_{xy} & F_{xz} \\ F_{yx} & F_{yy} & F_{yz} \\ F_{zx} & F_{zy} & F_{zz} \end{pmatrix} = 0$$

at the point P, where F_{ab} denotes the 2nd partial derivative with respect to a, b. The determinant is called the *Hessian*. For a curve in $\mathbf{P}^2$ defined by an equation $F = 0$, a point where the Hessian is zero is called a *flex* of the curve.

Chapter 4

Elliptic Curves over Finite Fields

Let $\mathbf{F}$ be a finite field and let E be an elliptic curve defined over $\mathbf{F}$. Since there are only finitely many pairs (x, y) with $x, y \in \mathbf{F}$, the group $E(\mathbf{F})$ is finite. Various properties of this group, for example, its order, turn out to be important in many contexts. In this chapter, we present the basic theory of elliptic curves over finite fields. Not only are the results interesting in their own right, but also they are the starting points for the cryptographic applications discussed in Chapter 6.

4.1 Examples

First, let's consider some examples.

Example 4.1

Let E be the curve $y^2 = x^3 + x + 1$ over $\mathbf{F}_5$. To count points on E, we make a list of the possible values of x, then of $x^3 + x + 1 \pmod 5$, then of the square roots y of $x^3 + x + 1 \pmod 5$. This yields the points on E.

x	$x^3 + x + 1$	y	Points
0	1	± 1	$(0, 1), (0, 4)$
1	3	$-$	$-$
2	1	± 1	$(2, 1), (2, 4)$
3	1	± 1	$(3, 1), (3, 4)$
4	4	± 2	$(4, 2), (4, 3)$
∞	∞	∞	

Therefore, $E(\mathbf{F}_5)$ has order 9.

Let's compute $(3,1) + (2,4)$ on E. The slope of the line through the two points is

$$\frac{4-1}{2-3} \equiv 2 \pmod 5.$$

The line is therefore $y = 2(x-3)+1 \equiv 2x$. Substituting this into $y^2 = x^3+x+1$ and rearranging yields

$$0 = x^3 - 4x^2 + 3x + 1.$$

The sum of the roots is 4, and we know the roots 3 and 2. Therefore the remaining root is $x = 4$. Since $y = 2x$, we have $y \equiv 3$. Reflecting across the x-axis yields the sum:

$$(3,1) + (2,4) = (4,2).$$

(Of course, we could have used the formulas of Section 2.2 directly.) A little calculation shows that $E(\mathbf{F}_5)$ is cyclic, generated by $(0,1)$ (Exercise 4.1).   ▯

Example 4.2
Let E be the elliptic curve $y^2 = x^3 + 2$ over $\mathbf{F}_7$. Then

$$E(\mathbf{F}_7) = \{\infty, \ (0,3), \ (0,4), \ (3,1), \ (3,6), \ (5,1), \ (5,6), \ (6,1), \ (6,6)\}.$$

An easy calculation shows that all of these points P satisfy $3P = \infty$, so the group is isomorphic to $\mathbf{Z}_3 \oplus \mathbf{Z}_3$.   ▯

Example 4.3
Let's consider the elliptic curve E given by $y^2 + xy = x^3 + 1$ defined over $\mathbf{F}_2$. We can find the points as before and obtain

$$E(\mathbf{F}_2) = \{\infty, \ (0,1), \ (1,0), \ (1,1)\}.$$

This is a cyclic group of order 4. The points $(1,0), (1,1)$ have order 4 and the point $(0,1)$ has order 2.

Now let's look at $E(\mathbf{F}_4)$. Recall that $\mathbf{F}_4$ is the finite field with 4 elements. We can write it as $\mathbf{F}_4 = \{0, 1, \omega, \omega^2\}$, with the relation $\omega^2 + \omega + 1 = 0$ (which implies, after multiplying by $\omega + 1$, that $\omega^3 = 1$). Let's list the elements of $E(\mathbf{F}_4)$.

$$x = 0 \Rightarrow y^2 = 1 \Rightarrow y = 1$$
$$x = 1 \Rightarrow y^2 + y = 0 \Rightarrow y = 0, 1$$
$$x = \omega \Rightarrow y^2 + \omega y = 0 \Rightarrow y = 0, \omega$$
$$x = \omega^2 \Rightarrow y^2 + \omega^2 y = 0 \Rightarrow y = 0, \omega^2$$
$$x = \infty \Rightarrow y = \infty.$$

Therefore

$$E(\mathbf{F}_4) = \left\{\infty, \ (0,1), \ (1,0), \ (1,1), \ (\omega,0), \ (\omega,\omega), \ (\omega^2,0), \ (\omega^2,\omega^2)\right\}.$$

Since we are in characteristic 2, there is at most one point of order 2 (see Proposition 3.1). In fact, $(0, 1)$ has order 2. Therefore, $E(\mathbf{F}_4)$ is cyclic of order 8. Any one of the four points containing ω or ω^2 is a generator. This may be verified by direct calculation, or by observing that they do not lie in the order 4 subgroup $E(\mathbf{F}_2)$. Let $\phi_2(x, y) = (x^2, y^2)$ be the Frobenius map. It is easy to see that ϕ_2 permutes the elements of $E(\mathbf{F}_4)$, and

$$E(\mathbf{F}_2) = \{(x, y) \in E(\mathbf{F}_4) \mid \phi_2(x, y) = (x, y)\}.$$

In general, for any elliptic curve E defined over $\mathbf{F}_q$ and any extension $\mathbf{F}$ of $\mathbf{F}_q$, the Frobenius map ϕ_q permutes the elements of $E(\mathbf{F})$ and is the identity on the subgroup $E(\mathbf{F}_q)$. See Lemma 4.5. $\quad\Box$

Two main restrictions on the groups $E(\mathbf{F}_q)$ are given in the next two theorems.

THEOREM 4.1

Let E be an elliptic curve over the finite field $\mathbf{F}_q$. Then

$$E(\mathbf{F}_q) \simeq \mathbf{Z}_n \quad or \quad \mathbf{Z}_{n_1} \oplus \mathbf{Z}_{n_2}$$

for some integer $n \geq 1$, or for some integers $n_1, n_2 \geq 1$ with n_1 dividing n_2.

PROOF A basic result in group theory (see Appendix B) says that a finite abelian group is isomorphic to a direct sum of cyclic groups

$$\mathbf{Z}_{n_1} \oplus \mathbf{Z}_{n_2} \oplus \cdots \oplus \mathbf{Z}_{n_r},$$

with $n_i | n_{i+1}$ for $i \geq 1$. Since, for each i, the group $\mathbf{Z}_{n_i}$ has n_1 elements of order dividing n_1, we find that $E(\mathbf{F}_q)$ has n_1^r elements of order dividing n_1. By Theorem 3.2, there are at most n_1^2 such points (even if we allow coordinates in the algebraic closure of $\mathbf{F}_q$). Therefore $r \leq 2$. This is the desired result (the group is trivial if $r = 0$; this case is covered by $n = 1$ in the theorem). $\blacksquare$

THEOREM 4.2 (Hasse)

Let E be an elliptic curve over the finite field $\mathbf{F}_q$. Then the order of $E(\mathbf{F}_q)$ satisfies

$$|q + 1 - \#E(\mathbf{F}_q)| \leq 2\sqrt{q}.$$

The proof will be given in Section 4.2.

A natural question is what groups can actually occur as groups $E(\mathbf{F}_q)$. The answer is given in the following two results, which are proved in [107] and [77], respectively.

THEOREM 4.3

Let $q = p^n$ be a power of a prime p and let $N = q+1-a$. There is an elliptic curve E defined over $\mathbf{F}_q$ such that $\#E(\mathbf{F}_q) = N$ if and only if $|a| \leq 2\sqrt{q}$ and a satisfies one of the following:

1. $\gcd(a,p) = 1$

2. n *is even and* $a = \pm 2\sqrt{q}$

3. n *is even,* $p \not\equiv 1 \pmod{3}$, *and* $a = \pm\sqrt{q}$

4. n *is odd,* $p = 2$ *or* 3, *and* $a = \pm p^{(n+1)/2}$

5. n *is even,* $p \not\equiv 1 \pmod{4}$ *and* $a = 0$

6. n *is odd and* $a = 0$.

THEOREM 4.4

Let N be an integer that occurs as the order of an elliptic curve over a finite field $\mathbf{F}_q$, as in Theorem 4.3. Write $N = p^e n_1 n_2$ with $p \nmid n_1 n_2$ and $n_1 | n_2$ (possibly $n_1 = 1$). There is an elliptic curve E over $\mathbf{F}_q$ such that

$$E(\mathbf{F}_q) \simeq \mathbf{Z}_{p^e} \oplus \mathbf{Z}_{n_1} \oplus \mathbf{Z}_{n_2}$$

if and only if

1. $n_1 | q - 1$ *in cases (1), (3), (4), (5), (6) of Theorem 4.3*

2. $n_1 = n_2$ *in case (2) of Theorem 4.3.*

These are the only groups that occur as groups $E(\mathbf{F}_q)$.

4.2   The Frobenius Endomorphism

Let $\mathbf{F}_q$ be a finite field with algebraic closure $\overline{\mathbf{F}}_q$ and let

$$\phi_q : \overline{\mathbf{F}}_q \longrightarrow \overline{\mathbf{F}}_q,$$
$$x \mapsto x^q$$

be the Frobenius map for $\mathbf{F}_q$ (see Appendix C for a review of finite fields). Let E be an elliptic curve defined over $\mathbf{F}_q$. Then ϕ_q acts on the coordinates of points in $E(\overline{\mathbf{F}}_q)$:

$$\phi(x,y) = (x^q, y^q), \quad \phi_q(\infty) = \infty.$$

LEMMA 4.5

Let E be defined over $\mathbf{F}_q$, and let $(x, y) \in E(\overline{\mathbf{F}}_q)$.

  1. $\phi_q(x, y) \in E(\overline{\mathbf{F}}_q)$

  2. $(x, y) \in E(\mathbf{F}_q)$ if and only if $\phi_q(x, y) = (x, y)$.

PROOF   One fact we need is that $(a + b)^q = a^q + b^q$ when q is a power of the characteristic of the field. We also need that $a^q = a$ for all $a \in \mathbf{F}_q$. See Appendix C.

Since the proof is the same for the Weierstrass and the generalized Weierstrass equations, we work with the general form. We have

$$y^2 + a_1 xy + a_3 y = x^3 + a_2 x^2 + a_4 x + a_6,$$

with $a_i \in \mathbf{F}_q$. Raise the equation to the qth power to obtain

$$(y^q)^2 + a_1(x^q y^q) + a_3(y^q) = (x^q)^3 + a_2(x^q)^2 + a_4(x^q) + a_6.$$

This means that (x^q, y^q) lies on E, which proves (1).

For (2), again recall that $x \in \mathbf{F}_q$ if and only if $\phi_q(x) = x$ (see Appendix C), and similarly for y. Therefore

$$(x, y) \in E(\mathbf{F}_q) \Leftrightarrow x, y \in \mathbf{F}_q$$
$$\Leftrightarrow \phi_q(x) = x \text{ and } \phi_q(y) = y$$
$$\Leftrightarrow \phi_q(x, y) = (x, y).$$

∎

LEMMA 4.6

Let E be an elliptic curve defined over $\mathbf{F}_q$. Then ϕ_q is an endomorphism of E of degree q, and ϕ_q is not separable.

This is the same as Lemma 2.19.

Note that the kernel of the endomorphism ϕ_q is trivial. This is related to the fact that ϕ_q is not separable. See Proposition 2.20.

The following result is the key to counting points on elliptic curves over finite fields. Since ϕ_q is an endomorphism of E, so are $\phi_q^2 = \phi_q \circ \phi_q$ and also $\phi_q^n = \phi_q \circ \phi_q \circ \cdots \circ \phi_q$ for every $n \geq 1$. Since multiplication by -1 is also an endomorphism, the sum $\phi_q^n - 1$ is an endomorphism of E.

PROPOSITION 4.7

Let E be defined over $\mathbf{F}_q$ and let $n \geq 1$.

  1. $Ker(\phi_q^n - 1) = E(\mathbf{F}_{q^n})$.

2. $\phi_q^n - 1$ is a separable endomorphism, so $\#E(\mathbf{F}_{q^n}) = \deg(\phi_q^n - 1)$.

PROOF   Since ϕ_q^n is the Frobenius map for the field $\mathbf{F}_{q^n}$, part (1) is just a restatement of Lemma 4.5. The fact that $\phi_q^n - 1$ is separable was proved in Proposition 2.28. Therefore (2) follows from Proposition 2.20.   ∎

Proof of Hasse's theorem:
We can now prove Hasse's theorem (Theorem 4.2). Let

$$a = q + 1 - \#E(\mathbf{F}_q) = q + 1 - \deg(\phi_q - 1). \tag{4.1}$$

We want to show that $|a| \leq 2\sqrt{q}$. We need the following.

LEMMA 4.8
Let r, s be integers with $\gcd(s, q) = 1$. Then $\deg(r\phi_q - s) = r^2 q + s^2 - rsa$.

PROOF   Proposition 3.16 implies that

$$\deg(r\phi_q - s) = r^2 \deg(\phi_q) + s^2 \deg(-1) + rs(\deg(\phi_q - 1) - \deg(\phi_q) - \deg(-1)).$$

Since $\deg(\phi_q) = q$ and $\deg(-1) = 1$, the result follows from (4.1).   ∎

REMARK 4.9   The assumption that $\gcd(s, q) = 1$ is not needed. We include it since we have proved Proposition 3.16 not in general, but only when the endomorphisms are separable or ϕ_q.   ∎

We can now finish the proof of Hasse's theorem. Since $\deg(r\phi_q - s) \geq 0$, the lemma implies that

$$q\left(\frac{r}{s}\right)^2 - a\left(\frac{r}{s}\right) + 1 \geq 0$$

for all r, s with $\gcd(s, q) = 1$. The set of rational numbers r/s such that $\gcd(s, q) = 1$ is dense in $\mathbf{R}$. (*Proof:* Take s to be a power of 2 or a power of 3, one of which must be relatively prime with q. The rationals of the form $r/2^m$ and those of the form $r/3^m$ are easily seen to be dense in $\mathbf{R}$.) Therefore,

$$qx^2 - ax + 1 \geq 0$$

for all real numbers x. Therefore the discriminant of the polynomial is negative or 0, which means that $a^2 - 4q \leq 0$, hence $|a| \leq 2\sqrt{q}$. This completes the proof of Hasse's theorem.   ∎

There are several major ingredients of the above proof. One is that we can identify $E(\mathbf{F}_q)$ as the kernel of $\phi_q - 1$. Another is that $\phi_q - 1$ is separable, so the order of the kernel is the degree of $\phi_q - 1$. A third major ingredient is the Weil pairing, especially part (6) of Theorem 3.9, and its consequence, Proposition 3.16.

Proposition 4.7 has another very useful consequence.

THEOREM 4.10

Let E be an elliptic curve defined over $\mathbf{F}_q$. Let a be as in Equation 4.1. Then

$$\phi_q^2 - a\phi_q + q = 0$$

as endomorphisms of E, and a is the unique integer k such that

$$\phi_q^2 - k\phi_q + q = 0.$$

In other words, if $(x,y) \in E(\overline{\mathbf{F}}_q)$, then

$$\left(x^{q^2}, y^{q^2}\right) - a\left(x^q, y^q\right) + q(x,y) = \infty,$$

and a is the unique integer such that this relation holds for all $(x,y) \in E(\overline{\mathbf{F}}_q)$. Moreover, a is the unique integer satisfying

$$a \equiv \mathrm{Trace}((\phi_q)_m) \quad \mod m$$

for all m with $\gcd(m, q) = 1$.

PROOF   If $\phi_q^2 - a\phi_q + q$ is not the zero endomorphism, then its kernel is finite (Proposition 2.20). We'll show that the kernel is infinite, hence the endomorphism is 0.

Let $m \geq 1$ be an integer with $\gcd(m, q) = 1$. Recall that ϕ_q induces a matrix $(\phi_q)_m$ that describes the action of ϕ_q on $E[m]$. Let

$$(\phi_q)_m = \begin{pmatrix} s & t \\ u & v \end{pmatrix}.$$

Since $\phi_q - 1$ is separable by Proposition 2.28, Propositions 2.20 and 3.15 imply that

$$\#\mathrm{Ker}(\phi_q - 1) = \deg(\phi_q - 1) \equiv \det((\phi_q)_m - I)$$
$$= sv - tu - (s + v) + 1 \quad (\mathrm{mod}\ m).$$

By Proposition 3.15, $sv - tu = \det((\phi_q)_m) \equiv q \pmod{m}$. By (4.1), $\#\mathrm{Ker}(\phi_q - 1) = q + 1 - a$. Therefore,

$$\mathrm{Trace}((\phi_q)_m) = s + v \equiv a \quad (\mathrm{mod}\ m).$$

By the Cayley-Hamilton theorem of linear algebra, or by a straightforward calculation (substituting the matrix into the polynomial), we have

$$(\phi_q)_m^2 - a(\phi_q)_m + qI \equiv 0 \pmod{m},$$

where I is the 2×2 identity matrix. (Note that $X^2 - aX + q$ is the characteristic polynomial of $(\phi_q)_m$.) This means that the endomorphism $\phi_q^2 - a\phi + q$ is identically zero on $E[m]$. Since there are infinitely many choices for m, the kernel of $\phi_q^2 - a\phi + q$ is infinite, so the endomorphism is 0.

Suppose $a_1 \neq a$ satisfies $\phi_q^2 - a_1\phi + q = 0$. Then

$$(a - a_1)\phi_q = (\phi_q^2 - a_1\phi + q) - (\phi_q^2 - a\phi + q) = 0.$$

By Theorem 2.21, $\phi_q : E(\overline{\mathbf{F}}_q) \to E(\overline{\mathbf{F}}_q)$ is surjective. Therefore, $(a - a_1)$ annihilates $E(\overline{\mathbf{F}}_q)$. In particular, $(a - a_1)$ annihilates $E[m]$ for every $m \geq 1$. Since there are points in $E[m]$ of order m when $\gcd(m, q) = 1$, we find that $a - a_1 \equiv 0 \pmod{m}$ for such m. Therefore $a - a_1 = 0$, so a is unique. ∎

We single out the following result, which was proved during the proof of Theorem 4.10, since it will be used in Chapter 13.

PROPOSITION 4.11
Let E be an elliptic curve over $\mathbf{F}_q$ and let $(\phi_q)_m$ denote the matrix giving the action of the Frobenius ϕ_q on $E[m]$. Let $a = q + 1 - \#E(\mathbf{F}_q)$. Then

$$\text{Trace}((\phi_q)_m) \equiv a \pmod{m}, \qquad \det((\phi_q)_m) \equiv q \pmod{m}.$$

The polynomial $X^2 - aX + q$ is often called the **characteristic polynomial of Frobenius**.

4.3   Determining the Group Order

Hasse's theorem gives bounds for the group of points on an elliptic curve over a finite field. In this section and in Section 4.5, we'll discuss some methods for actually determining the order of the group.

4.3.1   Subfield Curves

Sometimes we have an elliptic curve E defined over a small finite field $\mathbf{F}_q$ and we want to know the order of $E(\mathbf{F}_{q^n})$ for some n. We can determine the order of $E(\mathbf{F}_{q^n})$ when $n = 1$ by listing the points or by some other elementary procedure. The amazing fact is that this allows us to determine the order for all n.

THEOREM 4.12

Let $\#E(\mathbf{F}_q) = q + 1 - a$. Write $X^2 - aX + q = (X - \alpha)(X - \beta)$. Then

$$\#E(\mathbf{F}_{q^n}) = q^n + 1 - (\alpha^n + \beta^n)$$

for all $n \geq 1$.

PROOF   First, we need the fact that $\alpha^n + \beta^n$ is an integer. This could be proved by remarking that it is an algebraic integer and is also a rational number. However, it can also be proved by more elementary means.

LEMMA 4.13

Let $s_n = \alpha^n + \beta^n$. Then $s_0 = 2$, $s_1 = a$, and $s_{n+1} = as_n - qs_{n-1}$ for all $n \geq 1$.

PROOF   Multiply the relation $\alpha^2 - a\alpha + q = 0$ by α^{n-1} to obtain $\alpha^{n+1} = a\alpha^n - q\alpha^{n-1}$. There is a similar relation for β. Add the two relations to obtain the lemma.   ∎

It follows immediately from the lemma that $\alpha^n + \beta^n$ is an integer for all $n \geq 0$.

Let

$$f(X) = (X^n - \alpha^n)(X^n - \beta^n) = X^{2n} - (\alpha^n + \beta^n)X^n + q^n.$$

Then $X^2 - aX + q = (X - \alpha)(X - \beta)$ divides $f(X)$. It follows immediately from the standard algorithm for dividing polynomials that the quotient is a polynomial $Q(X)$ with integer coefficients (the main points are that the leading coefficient of $X^2 - aX + q$ is 1 and that this polynomial and $f(X)$ have integer coefficients). Therefore

$$(\phi_q^n)^2 - (\alpha^n + \beta^n)\phi_q^n + q^n = f(\phi_q) = Q(\phi_q)(\phi_q^2 - a\phi_q + q) = 0,$$

as endomorphisms of E, by Theorem 4.10. Note that $\phi_q^n = \phi_{q^n}$. By Theorem 4.10, there is only one integer k such that $\phi_{q^n}^2 - k\phi_{q^n} + q^n = 0$, and such a k is determined by $k = q^n + 1 - \#E(\mathbf{F}_{q^n})$. Therefore,

$$\alpha^n + \beta^n = q^n + 1 - \#E(\mathbf{F}_{q^n}).$$

This completes the proof of Theorem 4.12.   ∎

Example 4.4

In Example 4.3, we showed that the elliptic curve E given by $y^2 + xy = x^3 + 1$ over $\mathbf{F}_2$ satisfies $\#E(\mathbf{F}_2) = 4$. Therefore, $a = 2 + 1 - 4 = -1$, and we obtain the polynomial

$$X^2 + X + 2 = \left(X - \frac{-1 + \sqrt{-7}}{2}\right)\left(X - \frac{-1 - \sqrt{-7}}{2}\right).$$

Theorem 4.12 says that

$$\#E(\mathbf{F}_4) = 4 + 1 - \left(\frac{-1 + \sqrt{-7}}{2}\right)^2 - \left(\frac{-1 - \sqrt{-7}}{2}\right)^2.$$

Rather than computing the last expression directly, we can use the recurrence in Lemma 4.13:

$$s_2 = as_1 - 2s_0 = -(-1) - 2(2) = -3.$$

It follows that $\#E(\mathbf{F}_4) = 4 + 1 - (-3) = 8$, which is what we calculated by listing points.

Similarly, using the recurrence or using sufficiently high precision floating point arithmetic yields

$$\left(\frac{-1 + \sqrt{-7}}{2}\right)^{101} + \left(\frac{-1 + \sqrt{-7}}{2}\right)^{101} = 2969292210605269.$$

Therefore,

$$\#E(\mathbf{F}_{2^{101}}) = 2^{101} + 1 - 2969292210605269$$
$$= 2535301200456455833701195805484.$$

☐

The advantage of Theorem 4.12 is that it allows us to determine the group order for certain curves very quickly. The disadvantage is that it requires the curve to be defined over a small finite field.

4.3.2   Legendre Symbols

To make a list of points on $y^2 = x^3 + Ax + B$ over a finite field, we tried each possible value of x, then found the square roots y of $x^3 + Ax + B$, if they existed. This procedure is the basis for a simple point counting algorithm.

Recall the **Legendre symbol** $\left(\frac{x}{p}\right)$ for an odd prime p, which is defined as follows:

$$\left(\frac{x}{p}\right) = \begin{cases} +1 & \text{if } t^2 \equiv x \pmod{p} \text{ has a solution } t \not\equiv 0 \pmod{p}, \\ -1 & \text{if } t^2 \equiv x \pmod{p} \text{ has no solution } t \\ 0 & \text{if } x \equiv 0 \pmod{p}. \end{cases}$$

This can be generalized to any finite field $\mathbf{F}_q$ with q odd by defining, for $x \in \mathbf{F}_q$,

$$\left(\frac{x}{\mathbf{F}_q}\right) = \begin{cases} +1 \text{ if } t^2 = x \text{ has a solution } t \in \mathbf{F}_q^\times, \\ -1 \text{ if } t^2 = x \text{ has no solution } t \in \mathbf{F}_q, \\ 0 \text{ if } x = 0. \end{cases}$$

THEOREM 4.14

Let E be an elliptic curve defined by $y^2 = x^3 + Ax + B$ over $\mathbf{F}_q$. Then

$$\#E(\mathbf{F}_q) = q + 1 + \sum_{x \in \mathbf{F}_q} \left(\frac{x^3 + Ax + B}{\mathbf{F}_q}\right).$$

PROOF For a given x_0, there are two points (x, y) with x-coordinate x_0 if $x_0^3 + Ax_0 + B$ is a nonzero square in $\mathbf{F}_q$, one such point if it is zero, and no points if it is not a square. Therefore, the number of points with x-coordinate x_0 equals $1 + \left(\frac{x_0^3 + Ax_0 + B}{\mathbf{F}_q}\right)$. Summing over all $x_0 \in \mathbf{F}_q$, and including 1 for the point ∞, yields

$$\#E(\mathbf{F}_q) = 1 + \sum_{x \in \mathbf{F}_q} \left(1 + \left(\frac{x^3 + Ax + B}{\mathbf{F}_q}\right)\right).$$

Collecting the term 1 from each of the q summands yields the desired formula. ∎

COROLLARY 4.15

Let $x^3 + Ax + B$ be a polynomial with $A, B \in \mathbf{F}_q$, where q is odd. Then

$$\left|\sum_{x \in \mathbf{F}_q} \left(\frac{x^3 + Ax + B}{\mathbf{F}_q}\right)\right| \le 2\sqrt{q}.$$

PROOF When $x^3 + Ax + B$ has no repeated roots, $y^2 = x^3 + Ax + B$ gives an elliptic curve, so Theorem 4.14 says that

$$q + 1 - \#E(\mathbf{F}_q) = -\sum_{x \in \mathbf{F}_q} \left(\frac{x^3 + Ax + B}{\mathbf{F}_q}\right).$$

The result now follows from Hasse's theorem.

The case where $x^3 + Ax + B$ has repeated roots follows from Exercise 4.3. ∎

Example 4.5

Let E be the curve $y^2 = x^3 + x + 1$ over $\mathbf{F}_5$, as in Example 4.1. The nonzero squares mod 5 are 1 and 4. Therefore

$$\#E(\mathbf{F}_5) = 5 + 1 + \sum_{x=0}^{4} \left(\frac{x^3 + x + 1}{5} \right)$$

$$= 6 + \left(\frac{1}{5} \right) + \left(\frac{3}{5} \right) + \left(\frac{1}{5} \right) + \left(\frac{1}{5} \right) + \left(\frac{4}{5} \right)$$

$$= 6 + 1 - 1 + 1 + 1 + 1 = 9.$$

▯

When using Theorem 4.14, it is possible to compute each individual generalized Legendre symbol quickly (see Exercise 4.4), but it is more efficient to square all the elements of $\mathbf{F}_q^\times$ and store the list of squares. For simplicity, consider the case of $\mathbf{F}_p$. Make a vector with p entries, one for each element of $\mathbf{F}_p$. Initially, all entries in the vector are set equal to -1. For each j with $1 \leq j \leq (p-1)/2$, square j and reduce to get $k \mod p$. Change the kth entry in the vector to $+1$. Finally, change the 0th entry in the vector to 0. The resulting vector will be a list of the values of the Legendre symbol.

Theorem 4.14, which is sometimes known as the Lang-Trotter method, works quickly for small values of q, perhaps $q < 100$, but is slow for larger q, and is impossible to use when q is around 10^{100} or larger.

4.3.3   Orders of Points

Let $P \in E(\mathbf{F}_q)$. The **order** of P is the smallest positive integer k such that $kP = \infty$. A fundamental result from group theory (a corollary of Lagrange's theorem) is that the order of a point always divides the order of the group $E(\mathbf{F}_q)$. Also, for an integer n, we have $nP = \infty$ if and only if the order of P divides n. By Hasse's theorem, $\#E(\mathbf{F}_q)$ lies in an interval of length $4\sqrt{q}$. Therefore, if we can find a point of order greater than $4\sqrt{q}$, there can be only one multiple of this order in the correct interval, and it must be $\#E(\mathbf{F}_q)$. Even if the order of the point is smaller than $4\sqrt{q}$, we obtain a small list of possibilities for $\#E(\mathbf{F}_q)$. Using a few more points often shortens the list enough that there is a unique possibility for $\#E(\mathbf{F}_q)$.

How do we find the order of a point? If we know the order of the full group of points, then we can look at factors of this order. But, at present, the order of the group is what we're trying to find. In Section 4.3.4, we'll discuss a method (Baby Step, Giant Step) for finding the order of a point.

Example 4.6

Let E be the curve $y^2 = x^3 + 7x + 1$ over $\mathbf{F}_{101}$. It is possible to show that the point $(0,1)$ has order 116, so $N_{101} = \#E(\mathbf{F}_{101})$ is a multiple of 116. Hasse's theorem says that

$$101 + 1 - 2\sqrt{101} \leq N_{101} \leq 101 + 1 + 2\sqrt{101},$$

which means that $82 \leq N_{101} \leq 122$. The only multiple of 116 in this range is 116, so $N_{101} = 116$. As a corollary, we find that the group of points is cyclic of order 116, generated by $(0,1)$.   ☐

Example 4.7

Let E be the elliptic curve $y^2 = x^3 - 10x + 21$ over $\mathbf{F}_{557}$. The point $(2,3)$ can be shown to have order 189. Hasse's theorem implies that $511 \leq N_{557} \leq 605$. The only multiple of 189 in this range is $3 \cdot 189 = 567$. Therefore $N_{557} = 567$.
☐

Example 4.8

Let E be the elliptic curve $y^2 = x^3 + 7x + 12$ over $\mathbf{F}_{103}$. The point $(-1,2)$ has order 13 and the point $(19,0)$ has order 2. Therefore the order N_{103} of $E(\mathbf{F}_{103})$ is a multiple of 26. Hasse's theorem implies that $84 \leq N_{103} \leq 124$. The only multiple of 26 in that range is 104, so $N_{103} = 104$.   ☐

Example 4.9

Let E be the elliptic curve $y^2 = x^3 + 2$ over $\mathbf{F}_7$, as in Example 4.2. The group of points $E(\mathbf{F}_7)$ is isomorphic to $\mathbf{Z}_3 \oplus \mathbf{Z}_3$. Every point, except ∞, has order 3, so the best we can conclude with the present method is that the order N_7 of the group is a multiple of 3. Hasse's theorem says that $3 \leq N_7 \leq 13$, so the order is 3, 6, 9, or 12. Of course, if we find two independent points of order 3 (that is, one is not a multiple of the other), then they generate a subgroup of order 9. This means that the order of the full group is a multiple of 9, hence is 9.   ☐

The situation of the last example, where $E(\mathbf{F}_q) \simeq \mathbf{Z}_n \oplus \mathbf{Z}_n$, makes it more difficult to find the order of the group of points, but is fairly rare, as the next result shows.

PROPOSITION 4.16

Let E be an elliptic curve over $\mathbf{F}_q$ and suppose

$$E(\mathbf{F}_q) \simeq \mathbf{Z}_n \oplus \mathbf{Z}_n$$

for some integer n. Then either $q = n^2 + 1$ or $q = n^2 \pm n + 1$ or $q = (n \pm 1)^2$.

PROOF   By Hasse's theorem, $n^2 = q + 1 - a$, with $|a| \leq 2\sqrt{q}$. To prove the proposition, we use the following lemma, which puts a severe restriction on a.

LEMMA 4.17
$a \equiv 2 \pmod{n}$.

PROOF   Let p be the characteristic of $\mathbf{F}_q$. Then $p \nmid n$; otherwise, there would be p^2 points in $E[p]$, which is impossible in characteristic p by Theorem 3.2.

Since $E[n] \subseteq E(\mathbf{F}_q)$, Corollary 3.11 implies that the nth roots of unity are in $\mathbf{F}_q$, so $q - 1$ must be a multiple of n (see Appendix C). Therefore, $a = q + 1 - n^2 \equiv 2 \pmod{n}$.   ∎

Write $a = 2 + kn$ for some integer k. Then

$$n^2 = q + 1 - a = q - 1 - kn, \quad \text{so} \quad q = n^2 + kn + 1.$$

By Hasse's theorem,

$$|2 + kn| \leq 2\sqrt{q}.$$

Squaring this last inequality yields

$$4 + 4kn + k^2 n^2 \leq 4q = 4(n^2 + kn + 1).$$

Therefore, $|k| \leq 2$. The possibilities $k = 0, \pm 1, \pm 2$ give the values of q listed in the proposition. This completes the proof of Proposition 4.16.   ∎

Most values of q are not of the form given in the proposition, and even for such q most elliptic curves do not have $E(\mathbf{F}_q) \simeq \mathbf{Z}_n \oplus \mathbf{Z}_n$ (only a small fraction have order n^2), so we can regard $\mathbf{Z}_n \oplus \mathbf{Z}_n$ as rare.

More generally, most q are such that all elliptic curves over $\mathbf{F}_q$ have points of order greater than $4\sqrt{q}$ (Exercise 4.6). Therefore, with a little luck, we can usually find points with orders that allow us to determine $\#E(\mathbf{F}_q)$.

Suppose $E(\mathbf{F}_q) \simeq \mathbf{Z}_{n_1} \oplus \mathbf{Z}_{n_2}$ with $n_1 | n_2$. Then the order of every element divides n_2. If we choose some random points and compute their orders, what is the chance that the least common multiple of these orders is n_2? Let P_1, P_2 be points of orders n_1, n_2 such that every $P \in E(\mathbf{F}_q)$ is uniquely expressible in the form $P = a_1 P_1 + a_2 P_2$ with $0 \leq a_i < n_i$. Let p be a prime dividing n_2. If we take a random point P, then the probability is $1 - 1/p$ that $p \nmid a_2$. If $p \nmid a_2$, then the order of P contains the highest power of p possible. If p is large, then this means that it is very likely that the order of one randomly chosen point will contribute the correct power of p to the least common multiple of the orders of the points. If p is small, say $p = 2$, then the probability is at least $1/2$. This means that if we choose several randomly chosen points, the least common multiples of their orders should still have the correct power of p. The conclusion is that if we choose several random points and compute the least common multiple of their orders, it is very likely that we will obtain n_2, which is as large as possible.

4.3.4   Baby Step, Giant Step

Let $P \in E(\mathbf{F}_q)$. We want to find the order of P. First, we want to find an integer k such that $kP = \infty$. Let $\#E(\mathbf{F}_q) = N$. By Lagrange's theorem, $NP = \infty$. Of course, we might not know N yet, but we know that $q+1-2\sqrt{q} \leq N \leq q+1+2\sqrt{q}$. We could try all values of N in this range and see which ones satisfy $NP = \infty$. This takes around $4\sqrt{q}$ steps. However, it is possible to speed this up to around $4q^{1/4}$ steps by the following algorithm.

1. Compute $Q = (q+1)P$.

2. Choose an integer m with $m > q^{1/4}$. Compute and store the points jP for $j = 0, 1, 2, \ldots, m$.

3. Compute the points

$$Q + k(2mP) \text{ for } k = -m, -(m-1), \ldots, m$$

   until there is a match $Q + k(2mP) = \pm jP$ with a point (or its negative) on the stored list.

4. Conclude that $(q + 1 + 2mk \mp j)P = \infty$. Let $M = q + 1 + 2mk \mp j$.

5. Factor M. Let $p_1, \ldots, p_r$ be the distinct prime factors of M.

6. Compute $(M/p_i)P$ for $i = 1, \ldots, r$. If $(M/p_i)P = \infty$ for some i, replace M with M/p_i and go back to step (5). If $(M/p_i)P \neq \infty$ for all i then M is the order of the point M.

7. If we are looking for the $\#E(\mathbf{F}_q)$, then repeat steps (1)-(6) with randomly chosen points in $E(\mathbf{F}_q)$ until the greatest common multiple of the orders divides only one integer N with $q + 1 - 2\sqrt{q} \leq N \leq q + 1 + 2\sqrt{q}$. Then $N = \#E(\mathbf{F}_q)$.

There are two points that must be addressed.

I. Assuming that there is a match, this method clearly produces an integer that annihilates P. But why is there a match?

LEMMA 4.18
Let a be an integer with $|a| \leq 2m^2$. There exist integers a_0 and a_1 with $-m < a_0 \leq m$ and $-m \leq a_1 \leq m$ such that

$$a = a_0 + 2ma_1.$$

PROOF   Let $a_0 \equiv a \pmod{2m}$, with $-m < a_0 \leq m$ and $a_1 = (a - a_0)/2m$. Then

$$|a_1| \leq (2m^2 + m)/2m < m + 1.$$

∎

Let $a = a_0 + 2ma_1$ be as in the lemma and let $k = -a_1$. Then

$$Q + k(2mP) = (q + 1 - 2ma_1)P$$
$$= (q + 1 - a + a_0)P = NP + a_0P$$
$$= a_0P = \pm jP,$$

where $j = |a_0|$. Therefore, there is a match.

II. Why does step (6) yield the order of P?

LEMMA 4.19

Let G be an additive group (with identity element 0) and let $g \in G$. Suppose $Mg = 0$ for some positive integer M. Let $p_1, \ldots, p_r$ be the distinct primes dividing M. If $(M/p_i)g \neq 0$ for all i, then M is the order of g.

PROOF   Let k be the order of g. Then $k|M$. Suppose $k \neq M$. Let p_i be a prime dividing M/k. Then $p_i k|M$, so $k|(M/p_i)$. Therefore, $(M/p_i)g = 0$, contrary to assumption. Therefore $k = M$.   ∎

Therefore, step (6) finds the order of P.

REMARK 4.20   (1) To save storage space, it might be more efficient to store only the x coordinates of the points jP (along with the corresponding integer j), since looking for a match with $\pm jP$ only requires the x-coordinate (assuming we are working with a Weierstrass equation). When a match is found, the two possible y-coordinates can be recomputed.

(2) Computing $Q + k(2mP)$ can be done by computing Q and $2mP$ once for all. To get from $Q + k(2mP)$ to $Q + (k+1)(2mP)$, simply add $2mP$ rather than recomputing everything. Similarly, once jP has been computed, add P to get $(j+1)P$.

(3) We are assuming that we can factor M. If not, we can at least find all the small prime factors p_i and check that $(M/p_i)P \neq \infty$ for these. Then M will be a good candidate for the order of P.

(4) Why is the method called "Baby Step, Giant Step"? The **baby steps** are from a point jP to $(j+1)P$. The **giant steps** are from a point $k(2mP)$ to $(k+1)(2mP)$, since we take the "bigger" step $2mP$.   ∎

Example 4.10

Let E be the elliptic curve $y^2 = x^3 - 10x + 21$ over $\mathbf{F}_{557}$, as in Example 4.7. Let $P = (2, 3)$. We follow the procedure above.

1. $Q = 558P = (418, 33)$

2. Let $m = 5$, which is greater than $557^{1/4}$. The list of jP is

$$\infty, \ (2, 3), \ (58, 164), \ (44, 294), \ (56, 339), \ (132, 364).$$

3. When $k = 1$, we have $Q + k(2mP) = (2, 3)$, which matches the point on our list for $j = 1$.

4. We have $(q + 1 + 2mk - j)P = 567P = \infty$.

5. Factor $567 = 3^4 \cdot 7$. Compute $(567/3)P = 189P = \infty$. We now have 189 as a candidate for the order of P.

6. Factor $189 = 3^3 7$. Compute $(189/3)P = (38, 535) \neq \infty$ and $(189/7)P = (136, 360) \neq \infty$. Therefore 189 is the order of P.

As pointed out in Example 4.7, this suffices to determine that $\#E(\mathbf{F}_{557}) = 567$. ☐

4.4 A Family of Curves

In this section we give an explicit formula for the number of points in $E(\mathbf{F}_p)$, where E is the elliptic curve

$$y^2 = x^3 - kx,$$

and $k \not\equiv 0 \pmod{p}$. Counting the points on this curve mod a prime p has a long history, going back at least to Gauss.

THEOREM 4.21

Let p be an odd prime and let $k \not\equiv 0 \pmod{p}$. Let $N_p = \#E(\mathbf{F}_p)$, where E is the elliptic curve

$$y^2 = x^3 - kx.$$

1. If $p \equiv 3 \pmod{4}$, then $N_p = p + 1$.

2. If $p \equiv 1 \pmod{4}$, write $p = a^2 + b^2$, where a, b are integers with b even and $a + b \equiv 1 \pmod{4}$. Then

$$N_p = \begin{cases} p + 1 - 2a & \text{if } k \text{ is a fourth power mod } p \\ p + 1 + 2a & \text{if } k \text{ is a square mod } p \text{ but not a 4th power mod } p \\ p + 1 \pm 2b & \text{if } k \text{ is not a square mod } p \end{cases}$$

The proof of the theorem will take the rest of this section.

The integer a is uniquely determined by the conditions in the theorem, and b is uniquely determined up to sign. When k is not a square mod p, the proof below does not determine the sign of b. This is a much more delicate problem and we omit it.

Example 4.11

Let $p = 61 = (-5)^2 + 6^2$, where we chose the negative sign on 5 so that $-5 + 6 \equiv 1 \pmod 4$. Since $k = 1$ is a fourth power, the number of points on $y^2 = x^3 - x$ is $p + 1 - 2(-5) = 72$. □

It is well known that every prime $p \equiv 1 \pmod 4$ is a sum of two squares (this follows from Proposition 4.25 below). The next lemma shows that a and b are uniquely determined up to order and sign.

LEMMA 4.22

Suppose p is prime and a, b, c, d are integers such that $a^2 + b^2 = p = c^2 + d^2$. Then $a = \pm c$ and $b = \pm d$, or $a = \pm d$ and $b = \pm c$.

PROOF   We have $(a/b)^2 + 1 \equiv 0 \equiv (c/d)^2 + 1 \pmod p$, so $a/b \equiv \pm(c/d)$. By changing the sign of c if necessary, we may assume that $a/b \equiv c/d \pmod p$, hence $ad - bc \equiv 0 \pmod p$. A quick calculation shows that

$$p^2 = (ac + bd)^2 + (bc - ad)^2. \tag{4.2}$$

Suppose $ad = bc$. Then (4.2) implies that $ac + bd = \pm p$, so

$$\pm ap = a^2 c + abd = a^2 c + b^2 c = pc.$$

Hence, $\pm a = c$. It follows that $b = \pm d$.

Now suppose $ad \neq bc$. Since $ad - bc \equiv 0 \pmod p$, we have $(ad - bc)^2 \geq p^2$. Since $(ac + bd)^2 \geq 0$, it follows from (4.2) that $ad - bc = \pm p$ and $ac + bd = 0$. Therefore,

$$\pm cp = acd - bc^2 = -bd^2 - bc^2 = -bp,$$

so $c = \pm b$. This implies that $d = \pm a$.   ∎

If we require that a is odd and b is even, then a and b are uniquely determined up to sign. Suppose $b \equiv 2 \pmod 4$. Then $a + b \equiv 1 \pmod 4$ for a unique choice of the sign of a. Similarly, if $b \equiv 0 \pmod 4$, there is a unique choice of the sign of a that makes $a + b \equiv 1 \pmod 4$. Therefore, the integer a in the lemma is uniquely determined by p if we require that a is odd and $a + b \equiv 1 \pmod 4$.

The main part of the proof of Theorem 4.21 involves the case $p \equiv 1 \pmod 4$, so let's treat the case $p \equiv 3 \pmod 4$ first. The main point is that -1 is not a square mod p (*Proof:* if $x^2 \equiv -1$, then $1 \equiv x^{p-1} \equiv (x^2)^{(p-1)/2} \equiv (-1)^{(p-1)/2} = (-1)^{\text{odd}} = -1$, contradiction). Moreover, a nonsquare times a nonsquare is a square mod p. Therefore $x^3 - kx$ is a nonzero square mod p if and only if $(-x)^3 - k(-x) = -(x^3 - kx)$ is not a square mod p. Let's count points on E. Whenever $x^3 - kx = 0$, we obtain one point $(x,0)$. For the remaining values of x, we pair up x and $-x$. One of these gives two points (the one that makes $x^3 - kx$ a square) and the other gives no points. Therefore, each pair $x, -x$ gives two points. Therefore, we obtain a total of p points. The point ∞ gives one more, so we have $p + 1$ points.

Now assume $p \equiv 1 \pmod 4$. The proof, which takes the rest of this section, involves several steps and counts the points in terms of Jacobi sums. Rather than count the points on E directly, we make the transformation (see Theorem 2.17)

$$x = \frac{2(v+1)}{u^2}, \quad y = \frac{4(v+1)}{u^3},$$

which changes E into the curve C given by

$$v^2 = (k/4)u^4 + 1.$$

The inverse transformation is

$$u = \frac{2x}{y}, \quad v = -1 + \frac{2x^3}{y^2}.$$

We'll count the points on C mod p.

First, there are a few special points for the transformation from E to C. The point ∞ on E corresponds to $(0,1)$ on C. The point $(0,0)$ on E corresponds to $(0,-1)$ on C (see Theorem 2.17). If k is a square mod p, then the two 2-torsion points $(\pm\sqrt{k}, 0)$ correspond to the point at infinity on C. Therefore,

$$\#E(\mathbf{F}_p) = \#\{(u,v) \in \mathbf{F}_p \times \mathbf{F}_p \,|\, v^2 = (k/4)u^4 + 1\} + \delta,$$

where

$$\delta = \begin{cases} 2 & \text{if } k \text{ is a square mod } p \\ 0 & \text{if not.} \end{cases}$$

Let g be a primitive root mod p, which means that

$$\mathbf{F}_p^\times = \{g^j \,|\, 0 \le j < p - 1\}.$$

Let $i = \sqrt{-1} \in \mathbf{C}$. Define

$$\chi_2(g^j) = (-1)^j \quad \text{and} \quad \chi_4(g^j) = i^j.$$

Then χ_2 and χ_4 can be regarded as homomorphisms from $\mathbf{F}_p^\times$ to $\{\pm 1, \pm i\}$. Note that $\chi_4^2 = \chi_2$. The following lemma gets us started.

LEMMA 4.23

Let $p \equiv 1 \pmod 4$ be prime and let $x \in \mathbf{F}_p^\times$. Then

$$\#\{u \in \mathbf{F}_p^\times \,|\, u^2 = x\} = \sum_{\ell=0}^{1} \chi_2(x)^\ell,$$

and

$$\#\{u \in \mathbf{F}_p^\times \,|\, u^4 = x\} = \sum_{\ell=0}^{3} \chi_4(x)^\ell.$$

PROOF   Since $p \equiv 1 \pmod 4$, there are 4 fourth roots of 1 in $\mathbf{F}_p^\times$. Therefore, if there is a solution to $u^4 \equiv x$, there are 4 solutions. Write $x \equiv g^j$ $\pmod p$. Then x is a fourth power mod p if and only if $j \equiv 0 \pmod 4$. We have

$$\sum_{\ell=0}^{3} \chi_4(x)^\ell = \sum_{\ell=0}^{3} i^{j\ell} = \begin{cases} 4 \text{ if } j \equiv 0 \pmod 4 \\ 0 \text{ if } j \not\equiv 0 \pmod 4, \end{cases}$$

which is exactly the number of u with $u^4 \equiv x$. This proves the second half of the lemma. The proof of the first half is similar.   ∎

If, instead, we sum over the elements of $\mathbf{F}_p^\times$, we have the following result.

LEMMA 4.24

Let $p \equiv 1 \pmod 4$ be prime. Then

$$\sum_{b \in \mathbf{F}_p^\times} \chi_4(b)^\ell = \begin{cases} p - 1 & \text{if} \quad \ell \equiv 0 \pmod 4 \\ 0 & \text{if} \quad \ell \not\equiv 0 \pmod 4. \end{cases}$$

PROOF   If $\ell \equiv 0 \pmod 4$, all the terms in the sum are 1, so the sum is $p - 1$. If $\ell \not\equiv 0 \pmod 4$, then $\chi_4(g)^\ell \neq 1$. Multiplying by g permutes the elements of $\mathbf{F}_p^\times$, so

$$\chi_4(g)^\ell \sum_{b \in \mathbf{F}_p^\times} \chi_4(b)^\ell = \sum_{b \in \mathbf{F}_p^\times} \chi_4(gb)^\ell = \sum_{c \in \mathbf{F}_p^\times} \chi_4(c)^\ell,$$

which is the original sum. Since $\chi_4(g)^\ell \neq 1$, the sum must be 0.   ∎

Define the **Jacobi sums** by

$$J(\chi_2^j, \chi_4^\ell) = \sum_{\substack{a \in \mathbf{F}_p^\times \\ a \neq 1}} \chi_2(a)^j \chi_4(1 - a)^\ell.$$

PROPOSITION 4.25

$J(\chi_2, \chi_4^2) = -1$ and $|J(\chi_2, \chi_4)|^2 = p$.

PROOF   The first equality is proved as follows.

$$J(\chi_2, \chi_4^2) = \sum_{\substack{a \in \mathbf{F}_p^\times \\ a \neq 1}} \chi_2(a)\chi_4(1-a)^2 = \sum_{a \neq 0,1} \chi_2(a)\chi_2(1-a),$$

since $\chi_4^2 = \chi_2$. Since $\chi_2(a) = \pm 1$, we have $\chi_2(a) = \chi_2(a)^{-1}$ so the sum equals

$$\sum_{a \neq 0,1} \chi_2(a)^{-1}\chi_2(1-a) = \sum_{a \neq 0,1} \chi_2\left(\frac{1-a}{a}\right).$$

The map $x \mapsto 1 - \frac{1}{x}$ gives a permutation of the set of $x \in \mathbf{F}_p$, $x \neq 0, 1$. Therefore, letting $c = 1 - 1/a$, we obtain

$$\sum_{a \neq 0,1} \chi_2\left(\frac{1}{a} - 1\right) = \sum_{c \neq 0,1} \chi_2(-c) = -\chi_2(-1),$$

by Lemma 4.24. Since $g^{(p-1)/2} \equiv -1 \pmod{p}$ (both have order 2 in the cyclic group $\mathbf{F}_p^\times$), we have

$$1 = (\pm 1)^2 = \chi_2(g^{(p-1)/4})^2 = \chi_2(g^{(p-1)/2}) = \chi_2(-1).$$

This yields the first equality of the proposition.

To prove the second equality, multiply the Jacobi sum by its complex conjugate to obtain

$$|J(\chi_2, \chi_4)|^2 = \sum_{a \neq 0,1} \chi_2(a)\chi_4(1-a) \overline{\sum_{b \neq 0,1} \chi_2(b)\chi_4(1-b)}$$

$$= \sum_{a \neq 0,1} \sum_{b \neq 0,1} \chi_2\left(\frac{a}{b}\right) \chi_4\left(\frac{1-a}{1-b}\right).$$

We have used the fact that $\overline{\chi_4(x)} = \chi_4(x)^{-1}$. We now need the following.

LEMMA 4.26

Let $S = \{(x, y) \mid x, y \in \mathbf{F}_p^\times;\ x, y \neq 1;\ x \neq y\}$. The map

$$\sigma : (x, y) \mapsto \left(\frac{x}{y}, \frac{1-x}{1-y}\right)$$

is a permutation of S.

PROOF   Let $c = x/y$ and $d = (1-x)/(1-y)$. Then $x \neq 0$ yields $c \neq 0$ and $x \neq 1$ yields $d \neq 0$. The assumption that $x \neq y$ yields $c, d \neq 1$ and $c \neq d$. Therefore, $(c, d) \in S$.

To show that σ is surjective, let $c, d \in S$. Let

$$x = c\frac{d-1}{d-c}, \quad y = \frac{d-1}{d-c}.$$

It is easily verified that $(c, d) \in S$ implies $(x, y) \in S$ and that $\sigma(x, y) = (c, d)$.
∎

Returning to the proof of the proposition, we find that

$$|J(\chi_2, \chi_4)|^2 = \sum_{a=b} \chi_2\left(\frac{a}{b}\right) \chi_4\left(\frac{1-a}{1-b}\right) + \sum_{(a,b)\in S} \chi_2\left(\frac{a}{b}\right) \chi_4\left(\frac{1-a}{1-b}\right)$$

$$= (p-2) + \sum_{(c,d)\in S} \chi_2(c)\chi_4(d)$$

$$= (p-2) + \sum_{d\neq 0,1} \chi_4(d) \left(\sum_{c\in \mathbf{F}_p^\times} \chi_2(c) - \chi_2(1) - \chi_2(d)\right)$$

$$= (p-2) + \sum_{d\neq 0,1} \chi_4(d)(0 - 1 - \chi_4(d)^2)$$

$$= (p-2) - \sum_{d\neq 0,1} \chi_4(d) - \sum_{d\neq 0,1} \chi_4(d)^3$$

$$= (p-2) + \chi_4(1) + \chi_4(1)^3 = p.$$

This completes the proof of the second equality of Proposition 4.25.   ∎

We now show that the number of points on $v^2 = (k/4)u^4 + 1$ can be expressed in terms of Jacobi sums. By separating out the terms with $u = 0$ and the terms with $v = 0$, we obtain that the number of points is

$$\#\{v \mid v^2 = 1\} + \#\{u \mid u^4 = -4/k\}$$

$$+ \sum_{\substack{a+b=1 \\ a,b\neq 0}} \#\{v \mid v^2 = a\} \, \#\{u \mid u^4 = -4b/k\}$$

$$= \sum_{j=0}^{1} \chi_2(1)^j + \sum_{\ell=0}^{3} \chi_4(-4/k)^\ell + \sum_{\substack{a+b=1 \\ a,b\neq 0}} \sum_{j=0}^{1} \chi_2(a)^j \sum_{\ell=0}^{3} \chi_4(-4b/k)^\ell$$

$$= \sum_{j=0}^{1} \chi_2(1)^j + \sum_{\ell=0}^{3} \chi_4(-4/k)^\ell + \sum_{b\neq 0,1} \sum_{\ell=0}^{3} \chi_4(-4b/k)^\ell$$

$$+ \sum_{a\neq 0,1} \sum_{j=0}^{1} \chi_2(a)^j - (p-2)$$

$$+\chi_4(-4/k)^2 J(\chi_2,\chi_4^2) + \chi_4(-4/k) J(\chi_2,\chi_4) + \chi_4(-4/k)^3 J(\chi_2,\chi_4^3)$$

(Separate out the terms with $j = 0$ and $\ell = 0$. These yield the sums over ℓ and over j, respectively. The terms with $j = \ell = 0$, which sum to $p - 2$, are counted twice, so subtract $p - 2$. The terms with $j, \ell \neq 0$ contribute to the Jacobi sums.)

$$= \sum_{j=0}^{1} \sum_{a\neq 0} \chi_2(a)^j + \sum_{\ell=0}^{3} \sum_{b\neq 0} \chi_4(-4b/k)^\ell - (p-2)$$

$$-\chi_2(-4/k) + \chi_4(-4/k) J(\chi_2,\chi_4) + \chi_4(-4/k)^3 J(\chi_2,\chi_4^3)$$

$$= (p-1) + (p-1) - (p-2)$$
$$-\chi_2(-4/k) + \chi_4(-4/k) J(\chi_2,\chi_4) + \chi_4(-4/k)^3 J(\chi_2,\chi_4^3)$$
(by Lemma 4.24)

$$= p + 1 - \delta + \chi_4(-4/k) J(\chi_2,\chi_4) + \chi_4(-4/k)^3 J(\chi_2,\chi_4^3).$$

For the last equality, we used the fact that

$$1 + \chi_2(-4/k) = 1 + \chi_2(1/k) = \begin{cases} 0 \text{ if } k \text{ is not a square} \\ 2 \text{ if } k \text{ is a square mod } p, \end{cases}$$

hence $1 + \chi_2(-4/k) = \delta$. Therefore,

$$\#E(\mathbf{F}_p) = \#\{(u,v) \in \mathbf{F}_p \times \mathbf{F}_p \,|\, v^2 = (k/4)u^4 + 1\} + \delta$$
$$= p + 1 - \alpha - \overline{\alpha},$$

where

$$\alpha = -\chi_4(-4/k) J(\chi_2,\chi_4) \in \mathbf{Z}[i].$$

If we write $\alpha = a + bi$, then $\alpha + \overline{\alpha} = 2a$. Proposition 4.25 implies that $a^2 + b^2 = p$, so we have almost proved Theorem 4.21. It remains to evaluate $a \bmod 4$.

Let $x_1 + y_1 i, \; x_2 + y_2 i \in \mathbf{Z}[i]$. We say that

$$x_1 + y_1 i \equiv x_2 + y_2 i \pmod{2 + 2i}$$

if

$$(x_1 - x_2) + (y_1 - y_2)i = (x_3 + y_3 i)(2 + 2i)$$

for some $x_3 + y_3 i \in \mathbf{Z}[i]$. Clearly $-2i \equiv 2 \pmod{2 + 2i}$. Since $2i - 2 = i(2 + 2i)$ and $-2 = 2 + (-1 + i)(2 + 2i)$, we have

$$2i \equiv 2 \equiv -2 \equiv -2i \pmod{2 + 2i}.$$

It follows easily that

$$2\chi_4(a) \equiv 2 \pmod{2 + 2i} \tag{4.3}$$

for all a. Since $p - 1$ is a multiple of $4 = (1 - i)(2 + 2i)$, we have $p \equiv 1 \pmod{2 + 2i}$.

LEMMA 4.27
 Let $p \equiv 1 \pmod 4$ be prime. Then

$$J(\chi_2, \chi_4) \equiv -1 \pmod{2 + 2i}.$$

PROOF   Let $S = \{x \in \mathbf{F}_p^\times \mid x \neq 1\}$. Let

$$\tau : S \to S, \quad x \mapsto \frac{x}{x - 1}.$$

It is easy to check that $\tau(\tau(x)) = x$ for all $x \in S$ and that $x = 2$ is the only value of x such that $\tau(x) = x$. Put the elements of S, other than 2, into pairs $(x, \tau(x))$. Note that if x is paired with $y = \tau(x)$, then y is paired with $\tau(y) = \tau(\tau(x)) = x$. This divides S into $(p - 3)/2$ pairs plus the element 2, which is not in a pair. We have

$$J(\chi_2, \chi_4) = \sum_{a \neq 0,1} \chi_2(a)\chi_4(1 - a) =$$

$$\chi_2(2)\chi_4(1 - 2) + \sum_{(a, \tau(a))} \left(\chi_2(a)\chi_4(1 - a) + \chi_2\left(\frac{a}{a-1}\right)\chi_4\left(1 - \frac{a}{a-1}\right) \right),$$

where the sum is over pairs $(a, \tau(a))$. Note that since $\chi_2 \chi_4 = \chi_4^{-1}$, we have

$$\chi_2\left(\frac{a}{a-1}\right)\chi_4\left(1 - \frac{a}{a-1}\right) = \frac{\chi_2(a)}{\chi_2(a-1)}\frac{\chi_4(-1)}{\chi_4(a-1)}$$

$$= \chi_2(a)\chi_4(-1)\chi_4(a-1) = \chi_2(a)\chi_4(1 - a).$$

Therefore, since $\chi_2(2) = \chi_4(2)^2 = \chi_4(4)$,

$$J(\chi_2, \chi_4) = \chi_4(-4) + 2 \sum_{(a, \tau(a))} \chi_2(a)\chi_4(1 - a)$$

$$\equiv \chi_4(-4) + \sum_{(a, \tau(a))} 2 \quad \text{(by (4.3))}$$

$$\equiv \chi_4(-4) + (p - 3) \equiv \chi_4(-4) - 2 \pmod{2 + 2i}.$$

Suppose $p \equiv 1 \pmod 8$. Since $g^{(p-1)/2} \equiv -1 \pmod p$, we have that -1 is a fourth power. It is well known that 2 is a square mod p if and only if $p \equiv \pm 1 \pmod 8$ (this is one of the supplementary laws for quadratic reciprocity and is covered in most elementary number theory texts). Therefore 4 is a fourth power when $p \equiv 1 \pmod 8$. It follows that $\chi_4(-4) = 1$.

Now suppose $p \equiv 5 \pmod 8$. Then 2 is not a square mod p, so $2 \equiv g^j \pmod p$ with j odd. Therefore

$$-4 \equiv g^{2j+(p-1)/2} \pmod p.$$

Since $2j \equiv 2 \pmod 4$ and $(p-1)/2 \equiv 2 \pmod 4$, it follows that -4 is a fourth power mod p. Therefore, $\chi_4(-4) = 1$.

In both cases, we obtain $J(\chi_2, \chi_4) \equiv \chi_4(-4) - 2 \equiv -1 \pmod{2 + 2i}$. ∎

Since we just proved that $\chi_4(-4) = 1$, the lemma implies that

$$\alpha = -\chi_4(-4/k)J(\chi_2, \chi_4) = -\chi_4(1/k)J(\chi_2, \chi_4) \equiv \chi_4(k)^3 \pmod{2 + 2i}.$$

LEMMA 4.28
Let $\alpha = x + yi \in \mathbf{Z}[i]$.

1. If $\alpha \equiv 1 \pmod{2 + 2i}$, then x is odd and $x + y \equiv 1 \pmod 4$.

2. If $\alpha \equiv -1 \pmod{2 + 2i}$, then x is odd and $x + y \equiv 3 \pmod 4$.

3. If $\alpha \equiv \pm i \pmod{2 + 2i}$, then x is even.

PROOF   Suppose $\alpha \equiv 1 \pmod{2 + 2i}$, so $\alpha - 1 = (u + iv)(2 + 2i)$ for some u, v. Since $(1 - i)(2 + 2i) = 4$, we have

$$(x + y - 1) + (y + 1 - x)i = (1 - i)(\alpha - 1) = 4u + 4vi.$$

Therefore, $x + y \equiv 1 \pmod 4$ and $x - y \equiv 1 \pmod 4$. It follows that y is even. This proves (1). The proofs of (2) and (3) are similar. ∎

If k is a fourth power mod p, then $\chi_4(k) = 1$, so $\alpha \equiv 1 \pmod{2 + 2i}$. The lemma yields $\alpha = a + bi$ with b even and $a + b \equiv 1 \pmod 4$. This proves part of part (2) of Theorem 4.21. The other parts are proved similarly. This completes the proof of Theorem 4.21.

4.5 Schoof's Algorithm

In 1985, Schoof [80] published an algorithm for computing the number of points on elliptic curves over finite fields $\mathbf{F}_q$ that runs much faster than existing algorithms, at least for very large q. In particular, it requires at most a constant times $\log^8 q$ bit operations, in contrast to the $q^{1/4}$ used in Baby Step, Giant Step, for example. Subsequently, Atkin and Elkies refined and improved Schoof's method. It has now been used successfully when q has several hundred decimal digits. In the following, we'll give Schoof's method. For details of the method of Atkin and Elkies, see [7]. For a different method for counting points, see [78].

Suppose E is an elliptic curve given by $y^2 = x^3 + Ax + B$ over $\mathbf{F}_q$. We know, by Hasse's theorem, that

$$\#E(\mathbf{F}_q) = q + 1 - a, \quad \text{with } |a| \le 2\sqrt{q}.$$

Let $S = \{2, 3, 5, 7, \ldots, L\}$ be a set of primes such that

$$\prod_{\ell \in S} \ell > 4\sqrt{q}.$$

If we can determine $a \bmod \ell$ for each prime $\ell \in S$, then we know $a \bmod \prod \ell$, and therefore a is uniquely determined.

Let ℓ be prime. For simplicity, we assume $\ell \ne p$, where p is the characteristic of $\mathbf{F}_q$. We also assume that q is odd. We want to compute $a \pmod{\ell}$.

If $\ell = 2$, this is easy. If $x^3 + Ax + B$ has a root $e \in \mathbf{F}_q$, then $(e, 0) \in E[2]$ and $(e, 0) \in E(\mathbf{F}_q)$, so $E(\mathbf{F}_q)$ has even order. In this case, $q + 1 - a \equiv 0 \pmod{2}$, so a is even. If $x^3 + Ax + B$ has no roots in $\mathbf{F}_q$, then $E(\mathbf{F}_q)$ has no points of order 2, and a is odd. To determine whether $x^3 + Ax + B$ has a root in $\mathbf{F}_q$, we could try all the elements in $\mathbf{F}_q$, but there is a faster way. Recall (see Appendix C) that the roots of $x^q - x$ are exactly the elements of $\mathbf{F}_q$. Therefore, $x^3 + Ax + B$ has a root in $\mathbf{F}_q$ if and only if it has a root in common with $x^q - x$. The Euclidean algorithm, applied to polynomials, yields the gcd of the two polynomials.

If q is very large, the polynomial x^q has very large degree. Therefore, it is more efficient to compute $x_q \equiv x^q \pmod{x^3 + Ax + B}$ by successive squaring (cf. Section 2.2), and then use the result to compute

$$\gcd(x_q - x, \, x^3 + Ax + B) = \gcd(x^q - x, \, x^3 + Ax + B).$$

If the gcd is 1, then there is no common root and a is odd. If the gcd is not 1, then a is even. This finishes the case $\ell = 2$.

In the following, various expressions such as x^q and x^{q^2} will be used. They will always be computed mod a polynomial in a manner similar to that just done in the case $\ell = 2$

In Section 3.2, we defined the division polynomials ψ_n. When n is odd, ψ_n is a polynomial in x and, for $(x, y) \in E(\overline{\mathbf{F}}_q)$, we have

$$(x, y) \in E[n] \iff \psi_n(x) = 0.$$

These polynomials play a crucial role in Schoof's algorithm.

Let ϕ_q be the Frobenius endomorphism (not to be confused with the polynomials ϕ_n from Section 3.2, which are not used in this section), so

$$\phi_q(x, y) = (x^q, y^q).$$

By Theorem 4.10,

$$\phi_q^2 - a\phi_q + q = 0.$$

Let (x, y) be a point of order ℓ. Then

$$\left(x^{q^2}, y^{q^2}\right) + q(x, y) = a\,(x^q, y^q).$$

Let

$$q_\ell = q \pmod{\ell}, \quad |q_\ell| < \ell/2.$$

Then $q(x, y) = q_\ell(x, y)$, so

$$\left(x^{q^2}, y^{q^2}\right) + q_\ell(x, y) = a\,(x^q, y^q).$$

Since (x^q, y^q) is also a point of order ℓ, this relation determines $a \bmod \ell$. The idea is to compute all the terms except a in this relation, then determine a value of a that makes the relation hold. Note that if the relation holds for one point $(x, y) \in E[\ell]$, then we have determined $a \pmod{\ell}$; hence, it holds for all $(x, y) \in E[\ell]$.

Assume first that $\left(x^{q^2}, y^{q^2}\right) \neq \pm q_\ell(x, y)$ for some $(x, y) \in E[\ell]$. Then

$$(x', y') \stackrel{\text{def}}{=} \left(x^{q^2}, y^{q^2}\right) + q_\ell(x, y) \neq \infty,$$

so $a \not\equiv 0 \pmod{\ell}$. In this case, the x-coordinates of $\left(x^{q^2}, y^{q^2}\right)$ and $q_\ell(x, y)$ are distinct, so the sum of the two points is found by the formula using the line through the two points, rather than a tangent line or a vertical line. Write

$$j(x, y) = (x_j, y_j)$$

for integers j. We may compute x_j and y_j using division polynomials, as in Section 3.2. Moreover, $x_j = r_{1,j}(x)$ and $y_j = r_{2,j}(x)y$, as on page 47. We have

$$x' = \left(\frac{y^{q^2} - y_{q_\ell}}{x^{q^2} - x_{q_\ell}}\right)^2 - x^{q^2} - x_{q_\ell}.$$

Writing

$$\left(y^{q^2} - y_{q_\ell}\right)^2 = y^2 \left(y^{q^2-1} - r_{2,q_\ell}(x)\right)^2$$

$$= (x^3 + Ax + B)\left((x^3 + Ax + B)^{(q^2-1)/2} - r_{2,q_\ell}(x)\right)^2,$$

and noting that $x_{q\ell}$ is a function of x, we change x' into a rational function of x. We want to find j such that

$$(x', y') = (x_j^q, y_j^q).$$

First, we look at the x-coordinates. Starting with $(x, y) \in E[\ell]$, we have $(x', y') = \pm(x_j^q, y_j^q)$ if and only if $x' = x_j^q$. As pointed out above, if this happens for one point in $E[\ell]$, it happens for all (finite) points in $E[\ell]$. Since the roots of ψ_ℓ are the x-coordinates of the points in $E[\ell]$, this implies that

$$x' - x_j^q \equiv 0 \pmod{\psi_\ell} \tag{4.4}$$

(this means that the numerator of $x' - x_j^q$ is a multiple of ψ_ℓ). We are using here the fact that the roots of ψ_ℓ are simple (otherwise, we would obtain only that ψ_ℓ divides some power of $x' - x_j^q$). This is proved by noting that there are $\ell^2 - 1$ distinct points of order ℓ, since ℓ is assumed not to be the characteristic of $\mathbf{F}_q$. There are $(\ell^2 - 1)/2$ distinct x-coordinates of these points, and all of them are roots of ψ_ℓ, which has degree $(\ell^2 - 1)/2$. Therefore, the roots of ψ_ℓ must be simple.

Assume now that we have found j such that (4.4) holds. Then

$$(x', y') = \pm(x_j^q, y_j^q) = (x_j^q, \pm y_j^q).$$

To determine the sign, we need to look at the y-coordinates. Both y'/y and y_j^q/y can be written as functions of x. If

$$(y' - y_j^q)/y \equiv 0 \pmod{\psi_\ell},$$

then $a \equiv j \pmod{\ell}$. Otherwise, $a \equiv -j \pmod{\ell}$. Therefore, we have found $a \pmod{\ell}$.

It remains to consider the case where $\left(x^{q^2}, y^{q^2}\right) = \pm q(x, y)$ for all $(x, y) \in E[\ell]$. If

$$\phi_q^2(x, y) = \left(x^{q^2}, y^{q^2}\right) = q(x, y),$$

then

$$a\phi_q(x, y) = \phi_q^2(x, y) + q(x, y) = 2q(x, y),$$

hence

$$a^2 q(x, y) = a^2 \phi_q^2(x, y) = (2q)^2(x, y).$$

Therefore, $a^2 q \equiv 4q^2 \pmod{\ell}$, so q is a square mod ℓ. If q is not a square mod ℓ, then we cannot be in this case. If q is a square mod ℓ, let $w^2 \equiv q \pmod{\ell}$. We have

$$(\phi_q + w)(\phi_q - w)(x, y) = (\phi_q^2 - q)(x, y) = \infty$$

for all $(x, y) \in E[\ell]$. Let P be any point in $E[\ell]$. Then either $(\phi_q - w)P = \infty$, so $\phi_q P = wP$, or $P' = (\phi_q - w)P$ is a finite point with $(\phi_q + w)P' = \infty$. Therefore, in either case, there exists a point $P \in E[\ell]$ with $\phi_q P = \pm wP$.

Suppose there exists a point $P \in E[\ell]$ such that $\phi_q P = wP$. Then

$$\infty = (\phi_q^2 - a\phi_q + q)P = (q - aw + q)P,$$

so $aw \equiv 2q \equiv 2w^2 \pmod{\ell}$. Therefore, $a \equiv 2w \pmod{\ell}$. Similarly, if there exists P such that $\phi_q P = -wP$, then $a \equiv -2w \pmod{\ell}$. We can check whether we are in this case as follows. We need to know whether or not

$$(x^q, y^q) = \pm w(x, y) = \pm(x_w, y_w) = (x_w, \pm y_w)$$

for some $(x, y) \in E[\ell]$. Therefore, we compute $x^q - x_w$, which is a rational function of x. If

$$\gcd(\text{numerator}(x^q - x_w), \psi_\ell) \neq 1,$$

then there is some $(x, y) \in E[\ell]$ such that $\phi_q(x, y) = \pm w(x, y)$. If this happens, then use the y-coordinates to determine the sign.

Why do we use the gcd rather than simply checking whether we have 0 mod ψ_ℓ? The gcd checks for the existence of one point. Looking for 0 $\pmod{\psi_\ell}$ checks if the relation holds for all points simultaneously. The problem is that we are not guaranteed that $\phi_q P = \pm wP$ for all $P \in E[\ell]$. For example, the matrix representing ϕ_q on $E[\ell]$ might not be diagonalizable. It might be $\begin{pmatrix} w & 1 \\ 0 & w \end{pmatrix}$. In this case, the eigenvectors for ϕ_q form a one-dimensional subspace.

If we have $\gcd(\text{numerator}(x^q - x_w), \psi_\ell) = 1$, then we cannot be in the case $\left(x^{q^2}, y^{q^2}\right) = q(x, y)$, so the only remaining case is $\left(x^{q^2}, y^{q^2}\right) = -q(x, y)$. In this case, $aP = (\phi_q^2 + q)P = \infty$ for all $P \in E[\ell]$. Therefore, $a \equiv 0 \pmod{\ell}$.

We summarize Schoof's algorithm as follows. We start with an elliptic curve E over $\mathbf{F}_q$ given by $y^2 = x^3 + Ax + B$. We want to compute $\#E(\mathbf{F}_q) = q + 1 - a$.

1. Choose a set of primes $S = \{2, 3, 5, \ldots, L\}$ (with $p \notin S$) such that $\prod_{\ell \in S} \ell > 4\sqrt{q}$.

2. If $\ell = 2$, we have $a \equiv 0 \pmod{2}$ if and only if $\gcd(x^3 + Ax + B, x^q - x) \neq 1$.

3. For each odd prime $\ell \in S$, do the following.

   (a) Let $q_\ell \equiv q \pmod{\ell}$ with $|q_\ell| < \ell/2$.

   (b) Compute the x-coordinate x' of

   $$(x', y') = \left(x^{q^2}, y^{q^2}\right) + q_\ell(x, y) \bmod \psi_\ell.$$

   (c) For $j = 1, 2, \ldots, (\ell - 1)/2$, do the following.

      i. Compute the x-coordinate x_j of $(x_j, y_j) = j(x, y)$.

    ii. If $x' - x_j^q \equiv 0 \pmod{\psi_\ell}$, go to step (iii). If not, try the next value of j (in step (c)). If all values $1 \leq j \leq (\ell - 1)/2$ have been tried, go to step (d).

    iii. Compute y' and y_j. If $(y' - y_j)/y \equiv 0 \pmod{\psi_\ell}$, then $a \equiv j \pmod{\ell}$. If not, then $a \equiv -j \pmod{\ell}$.

(d) If all values $1 \leq j \leq (\ell - 1)/2$ have been tried without success, let $w^2 \equiv q \pmod{\ell}$. If w does not exist, then $a \equiv 0 \pmod{\ell}$.

(e) If $\gcd(\text{numerator}(x^q - x_w), \psi_\ell) = 1$, then $a \equiv 0 \pmod{\ell}$. Otherwise, compute

$$\gcd(\text{numerator}((y^q - y_w)/y), \psi_\ell).$$

If this gcd is not 1, then $a \equiv 2w \pmod{\ell}$. Otherwise, $a \equiv -2w \pmod{\ell}$.

4. Use the knowledge of $a \pmod{\ell}$ for each $\ell \in S$ to compute $a \pmod{\prod \ell}$. Choose the value of a that satisfies this congruence and such that $|a| \leq 2\sqrt{q}$. The number of points in $E(\mathbf{F}_q)$ is $q + 1 - a$.

Example 4.12

Let E be the elliptic curve $y^2 = x^3 + 2x + 1 \bmod 19$. Then

$$\#E(\mathbf{F}_{19}) = 19 + 1 - a.$$

We want to determine a. We'll show that

$$a \equiv \begin{cases} 1 & \pmod{2} \\ 2 & \pmod{3} \\ 3 & \pmod{5}. \end{cases}$$

Putting these together yields

$$a \equiv 23 \pmod{30}.$$

Since $|a| < 2\sqrt{19} < 9$, we must have $a = -7$.

    We start with $\ell = 2$. We compute

$$x^{19} \equiv x^2 + 13x + 14 \pmod{x^3 + 2x + 1}$$

by successive squaring (cf. Section 2.2) and then use the result to compute

$$\gcd(x^{19} - x, \, x^3 + 2x + 1) = \gcd(x^2 + 13x + 14, \, x^3 + 2x + 1) = 1.$$

It follows that $x^3 + 2x + 1$ has no roots in $\mathbf{F}_{19}$. Therefore, there is no 2-torsion in $E(\mathbf{F}_{19})$, so $a \equiv 1 \pmod{2}$.

For $\ell = 3$, we proceed as in Schoof's algorithm and eventually get to $j = 1$. We have $q^2 = 361$ and we have $q \equiv 1 \pmod 3$. Therefore, $q_\ell = 1$ and we need to check whether

$$(x^{361}, y^{361}) + (x, y) = \pm(x^{19}, y^{19})$$

for $(x, y) \in E[3]$. The third division polynomial is

$$\psi_3 = 3x^4 + 12x^2 + 12x - 4.$$

We compute the x-coordinate of $(x^{361}, y^{361}) + (x, y)$:

$$\left(\frac{y^{361} - y}{x^{361} - x}\right)^2 - x^{361} - x = (x^3 + 2x + 1)\left(\frac{(x^3 + 2x + 1)^{180} - 1}{x^{361} - x}\right)^2 - x^{361} - x,$$

where we have used the relation $y^2 = x^3 + 2x + 1$. We need to reduce this mod ψ_3. The natural way to start is to use the extended Euclidean algorithm to find the inverse of $x^{361} - x \pmod{\psi_3}$. However,

$$\gcd(x^{361} - x, \psi_3) = x - 8 \neq 1,$$

so the multiplicative inverse does not exist. We could remove $x - 8$ from the numerator and denominator of

$$\frac{(x^3 + 2x + 1)^{180} - 1}{x^{361} - x},$$

but this is unnecessary. Instead, we realize that since $x = 8$ is a root of ψ_3, the point $(8, 4) \in E(\mathbf{F}_{19})$ has order 3. Therefore,

$$\#E(\mathbf{F}_{19}) = 19 + 1 - a \equiv 0 \pmod 3,$$

so $a \equiv 2 \pmod 3$.

For $\ell = 5$, we follow Schoof's algorithm, eventually arriving at $j = 2$. Note that

$$19 \equiv 4 \equiv -1 \pmod 5,$$

so $q_\ell = -1$ and

$$19(x, y) = -(x, y) = (x, -y) \quad \text{for all } (x, y) \in E[5].$$

We need to check whether

$$(x_1, y_1) \overset{\text{def}}{=} (x^{361}, y^{361}) + (x, -y) \overset{?}{=} \pm 2(x^{19}, y^{19}) \overset{\text{def}}{=} \pm(x_2, y_2)$$

for all $(x, y) \in E[5]$. The recurrence of Section 3.2 shows that the fifth division polynomial is

$$\psi_5 = 32(x^3 + 2x + 1)^2(x^6 + 10x^4 + 20x^3 - 20x^2 - 8x - 8 - 8) - \psi_3^3$$
$$= 5x^{12} + 10x^{10} + 17x^8 + 5x^7 + x^6 + 9x^5 + 12x^4 + 2x^3 + 5x^2 + 8x + 8.$$

The equation for the x-coordinates yields

$$x_1 = \left(\frac{y^{361} + y}{x^{361} - x}\right)^2 - x^{361} - x \overset{?}{\equiv} \left(\frac{3x^{38} + 2}{2y^{19}}\right)^2 - 2x^{19} = x_2 \pmod{\psi_5}.$$

When y^2 is changed to $x^3 + 2x + 1$, this reduces to a polynomial relation in x, which is then verified. Therefore,

$$a \equiv \pm 2 \pmod 5.$$

To determine the sign, we look at the y-coordinates. The y-coordinate of $(x_1, y_1) = (x^{361}, y^{361}) + (x, -y)$ is computed to be

$$y(9x^{11} + 13x^{10} + 15x^9 + 15x^7 + 18x^6 + 17x^5 + 8x^4 + 12x^3 + 8x + 6) \pmod{\psi_5}.$$

The y-coordinate of $(x_2, y_2) = 2(x, y)$ is

$$y(13x^{10} + 15x^9 + 16x^8 + 13x^7 + 8x^6 + 6x^5 + 17x^4 + 18x^3 + 8x + 18) \pmod{\psi_5}.$$

A computation yields

$$(y_1 + y_2^{19})/y \equiv 0 \pmod{\psi_5}.$$

This means that

$$(x_1, y_1) \equiv (x_2^{19}, -y_2^{19}) = -2(x^q, y^q) \pmod{\psi_5}.$$

It follows that $a \equiv -2 \pmod 5$.

As we showed above, the information from $\ell = 2, 3, 5$ is sufficient to yield $a = -7$. Therefore, $\#E(\mathbf{F}_{19}) = 27$.   ◻

4.6 Supersingular Curves

An elliptic curve E in characteristic p is called **supersingular** if $E[p] = \{\infty\}$. In other words, there are no points of order p, even with coordinates in an algebraically closed field. Supersingular curves have many interesting properties, some of which we'll discuss in the present section.

Note: Supersingular curves are not singular curves in the sense of Section 2.4. The term "singular" was used classically to describe the j-invariants of elliptic curves with endomorphism rings larger than **Z**. These rings usually are subrings of quadratic extensions of the rationals. The term "supersingular" refers to j-invariants of curves with even larger rings of endomorphisms, namely, subrings of quaternion algebras. These ideas will be discussed in Chapter 10.

The following result is useful because it gives a simple way of determining whether or not an elliptic curve over a finite field is supersingular.

PROPOSITION 4.29

Let E be an elliptic curve over $\mathbf{F}_q$, where q is a power of the prime number p. Let $a = q + 1 - \#E(\mathbf{F}_q)$. Then E is supersingular if and only if $a \equiv 0 \pmod{p}$, which is if and only if $\#E(\mathbf{F}_q) \equiv 1 \pmod{p}$.

PROOF   Write $X^2 - aX + q = (X - \alpha)(X - \beta)$. Theorem 4.12 implies that

$$\#E(\mathbf{F}_{q^n}) = q^n + 1 - (\alpha^n + \beta^n).$$

Lemma 4.13 says that $s_n = \alpha^n + \beta^n$ satisfies the recurrence relation

$$s_0 = 2, \quad s_1 = a, \quad s_{n+1} = as_n - qa_{n-1}.$$

Suppose $a \equiv 0 \pmod{p}$. Then $s_1 = a \equiv 0 \pmod{p}$, and $s_{n+1} \equiv 0 \pmod{p}$ for all $n \geq 1$ by the recurrence. Therefore,

$$\#E(\mathbf{F}_{q^n}) = q^n + 1 - s_n \equiv 1 \pmod{p},$$

so there are no points of order p in $E(\mathbf{F}_{q^n})$ for any $n \geq 1$. Since $\overline{\mathbf{F}}_q = \cup_{n \geq 1} \mathbf{F}_{q^n}$, there are no points of order p in $E(\overline{\mathbf{F}}_q)$. Therefore, E is supersingular.

Now suppose $a \not\equiv 0 \pmod{p}$. The recurrence implies that $s_{n+1} \equiv as_n \pmod{p}$ for $n \geq 1$. Since $s_1 = a$, we have $s_n \equiv a^n \pmod{p}$ for all $n \geq 1$. Therefore

$$\#E(\mathbf{F}_{q^n}) = q^n + 1 - s_n \equiv 1 - a^n \pmod{p}.$$

By Fermat's little theorem, $a^{p-1} \equiv 1 \pmod{p}$. Therefore, $E(\mathbf{F}_{q^{p-1}})$ has order divisible by p, hence contains a point of order p. This means that E is not supersingular.

For the last part of the proposition, note that

$$\#E(\mathbf{F}_q) \equiv q + 1 - a \equiv 1 - a \pmod{p},$$

so $\#E(\mathbf{F}_q) \equiv 1 \pmod{p}$ if and only if $a \equiv 0 \pmod{p}$.   ∎

COROLLARY 4.30

Suppose $p \geq 5$ is a prime. Then E is supersingular if and only if $a = 0$, which is the case if and only if $\#E(\mathbf{F}_p) = p + 1$.

PROOF   If $a = 0$, then E is supersingular, by the proposition. Conversely, suppose E is supersingular but $a \neq 0$. Then $a \equiv 0 \pmod{p}$ implies that $|a| \geq p$. By Hasse's theorem, $|a| \leq 2\sqrt{p}$, so we have $p \leq 2\sqrt{p}$. This means that $p \leq 4$.   ∎

When $p = 2$ or $p = 3$, there are examples of supersingular curves with $a \neq 0$. See Exercise 4.7.

For general finite fields $\mathbf{F}_q$, it can be shown that if E defined over $\mathbf{F}_q$ is supersingular, then a^2 is one of 0, q, $2q$, $3q$, $4q$. See [81], [66], or Theorem 4.3.

In Section 3.1, we saw that the elliptic curve $y^2 + a_3 y = x^3 + a_4 x + a_6$ in characteristic 2 is supersingular. Also, in characteristic 3, the curve $y^2 = x^3 + a_2 x^2 + a_4 x + a_6$ is supersingular if and only $a_2 = 0$. Here is a way to construct supersingular curves in many other characteristics.

PROPOSITION 4.31

Suppose q is odd and $q \equiv 2$ (mod 3). Let $B \in \mathbf{F}_q^\times$. Then the elliptic curve E given by $y^2 = x^3 + B$ is supersingular.

PROOF   Let $\psi : \mathbf{F}_q^\times \to \mathbf{F}_q^\times$ be the homomorphism defined by $\psi(x) = x^3$. Since $q - 1$ is not a multiple of 3, there are no elements of order 3 in $\mathbf{F}_q^\times$, so the kernel of ψ is trivial. Therefore, ψ is injective, hence must be surjective since it is a map from a finite group to itself. In particular, every element of $\mathbf{F}_q$ has a unique cube root in $\mathbf{F}_q$.

For each $y \in \mathbf{F}_q$, there is exactly one $x \in \mathbf{F}_q$ such that (x, y) lies on the curve, namely, x is the unique cube root of $y^2 - B$. Since there are q values of y, we obtain q points. Including the point ∞ yields

$$\#E(\mathbf{F}_q) = q + 1.$$

Therefore, E is supersingular.   ∎

Later (Theorem 4.32), we'll see how to obtain all supersingular elliptic curves over an algebraically closed field.

An attractive feature of supersingular curves is that computations involving an integer times a point can sometimes be done faster than might be expected. Suppose E is a supersingular elliptic curve defined over $\mathbf{F}_q$ and let $P = (x, y)$ be a point in $E(\mathbf{F}_{q^n})$ for some $n \geq 1$. Usually n is large. Let k be a positive integer. We want to compute kP. This can be done quickly by successive doubling, but it is possible to do even better. Let's assume that $a = 0$. Then

$$\phi_q^2 + q = 0$$

by Theorem 4.10. Therefore

$$q(x, y) = -\phi_q^2(x, y) = \left(x^{q^2}, -y^{q^2} \right).$$

The calculations of x^{q^2} and y^{q^2} involve finite field arithmetic, which is generally faster than elliptic curve calculations. Moreover, if x and y are expressed in terms of a normal basis of $\mathbf{F}_{q^n}$ over $\mathbf{F}_q$, then x^{q^2} and y^{q^2} are computed by shift operations (see Appendix C). The procedure is now as follows:

1. Expand k in base q:

$$k = k_0 + k_1 q + k_2 q^2 + \cdots + k_r q^r,$$

with $0 \le k_i < q$.

2. Compute $k_i P = (x_i, y_i)$ for each i.

3. Compute $q^i k_i P = (x_i^{q^{2i}}, (-1)^i y_i^{q^{2i}})$.

4. Sum the points $q^i k_i P$ for $0 \le i \le r$.

The main savings is in step (3), where elliptic curve calculations are replaced by finite field computations.

We now show how to obtain all supersingular curves over $\overline{\mathbf{F}}_q$. Note that supersingularity means that there are no points of order p with coordinates in the algebraic closure; hence, it is really a property of an elliptic curve over an algebraically closed field. If we have two elliptic curves E_1 and E_2 defined over a field such that E_1 can be transformed into E_2 by a change of variables defined over some extension field, then E_1 is supersingular if and only if E_2 is supersingular.

For example, in Proposition 4.31, the curve $y_1^2 = x_1^3 + B$ can be changed into $y_2^2 = x_2^3 + 1$ via $x_2 = x_1/B^{1/3}$, $y_2 = y_1/B^{1/2}$. Therefore, it would have sufficed to prove the proposition for the curve $y^2 = x^3 + 1$.

Recall (Section 2.5.1) that an elliptic curve E over an algebraically closed field of characteristic not 2 can be put into the Legendre form $y^2 = x(x - 1)(x - \lambda)$ with $\lambda \ne 0, 1$.

THEOREM 4.32
Let p be an odd prime. Define the polynomial

$$H_p(T) = \sum_{i=0}^{(p-1)/2} \binom{(p-1)/2}{i}^2 T^i.$$

The elliptic curve E given by $y^2 = x(x-1)(x-\lambda)$ with $\lambda \in \overline{\mathbf{F}}_p$ is supersingular if and only if $H_p(\lambda) = 0$.

PROOF   Since $\overline{\mathbf{F}}_p = \cup_{n \ge 1} \mathbf{F}_{p^n}$, we have $\lambda \in \mathbf{F}_q = \mathbf{F}_{p^n}$ for some n. So E is defined over $\mathbf{F}_q$. To determine supersingularity, it suffices to count points in $E(\mathbf{F}_q)$, by Proposition 4.29. We know (Exercise 4.4) that

$$\left(\frac{x}{\mathbf{F}_q}\right) = x^{(q-1)/2}$$

in $\mathbf{F}_q$. Therefore, by Theorem 4.14,

$$\#E(\mathbf{F}_q) = q + 1 + \sum_{x \in \mathbf{F}_q} (x(x-1)(x-\lambda))^{(q-1)/2},$$

where this is now an equality in $\mathbf{F}_q$. The integers in this formula are regarded as elements of $\mathbf{F}_p \subseteq \mathbf{F}_q$. The following lemma allows us to simplify the sum.

LEMMA 4.33
Let $i > 0$ be an integer. Then

$$\sum_{x \in \mathbf{F}_q} x^i = \begin{cases} 0 & \text{if } q-1 \nmid i \\ -1 & \text{if } q-1 \mid i. \end{cases}$$

PROOF   If $q - 1 \mid i$ then $x^i = 1$ for all nonzero x, so the sum equals $q - 1$, which equals -1 in $\mathbf{F}_q$. The group $\mathbf{F}_q^\times$ is cyclic of order $q - 1$. Let g be a generator. Then every nonzero element of $\mathbf{F}_q$ can be written in the form g^j with $0 \le j \le q - 2$. Therefore, if $q - 1 \nmid i$,

$$\sum_{x \in \mathbf{F}_q} x^i = 0 + \sum_{x \in \mathbf{F}_q^\times} x^i = \sum_{j=0}^{q-2} (g^j)^i = \sum_{j=0}^{q-2} (g^i)^j = \frac{(g^i)^{q-1} - 1}{g^i - 1} = 0,$$

since $g^{q-1} = 1$. $\blacksquare$

Expand $(x(x - 1)(x - \lambda))^{(q-1)/2}$ into a polynomial of degree $3(q - 1)/2$. There is no constant term, so the only term x^i with $q - 1 \mid i$ is x^{q-1}. Let A_q be the coefficient of x^{q-1}. By the lemma,

$$\sum_{x \in \mathbf{F}_q} (x(x - 1)(x - \lambda))^{(q-1)/2} = -A_q,$$

since all the powers of x except for x^{q-1} sum to 0. Therefore,

$$\#E(\mathbf{F}_q) = 1 - A_q \quad \text{in } \mathbf{F}_q.$$

By Corollary 4.30, E is supersingular if and only if $A_q = 0$. The following lemma allows us to relate A_q to A_p.

LEMMA 4.34
Let $f(x) = x^3 + c_2 x^2 + c_1 x + c_0$ be a cubic polynomial with coefficients in a field of characteristic p. For each $r \ge 1$, let A_{p^r} be the coefficient of x^{p^r-1} in $f(x)^{(p^r-1)/2}$. Then

$$A_{p^r} = A_p^{1+p+p^2+\cdots+p^{r-1}}.$$

PROOF   We have

$$(f(x)^{(p-1)/2})^{p^r} = (x^{3(p-1)/2} + \cdots + A_p x^{p-1} + \cdots)^{p^r}$$
$$= x^{3(p-1)p^r/2} + \cdots + A_p^{p^r} x^{p^r(p-1)} + \cdots.$$

Therefore,

$$f(x)^{(p^{r+1}-1)/2} = f(x)^{(p^r-1)/2} \left(f(x)^{(p-1)/2} \right)^{p^r}$$
$$= (x^{3(p^r-1)/2} + \cdots + A_{p^r} x^{p^r-1} + \cdots)$$
$$\cdot (x^{3(p-1)p^r/2} + \cdots + A_p^{p^r} x^{p^r(p-1)} + \cdots).$$

To obtain the coefficient of $x^{p^{r+1}-1}$, choose indices i and j with $i + j = p^{r+1} - 1$, multiply the corresponding coefficients from the first and second factors in the above product, and sum over all such pairs i, j. A term with $0 \le i \le 3(p^r - 1)/2$ from the first factor requires a term with

$$p^{r+1} - 1 \ge j \ge (p^{r+1} - 1) - \frac{3}{2}(p^r - 1) > (p - 2)p^r$$

from the second factor. Since all of the exponents in the second factor are multiples of p^r, the only index j in this range that has a nonzero exponent is $j = (p - 1)p^r$. The corresponding index i is $p^r - 1$. The product of the coefficients yields

$$A_{p^{r+1}} = A_{p^r} A_p^{p^r}.$$

The formula of the lemma is trivially true for $r = 1$. It now follows by an easy induction for all r. ∎

From the lemma, we now see that E is supersingular if and only if $A_p = 0$. This is significant progress, since A_p depends on p but not on which power of p is used to get q.

It remains to express A_p as a polynomial in λ. The coefficient A_p of x^{p-1} in $(x(x - 1)(x - \lambda))^{(p-1)/2}$ is the coefficient of $x^{(p-1)/2}$ in

$$((x - 1)(x - \lambda))^{(p-1)/2}.$$

By the binomial theorem,

$$(x - 1)^{(p-1)/2} = \sum_i \binom{(p-1)/2}{i} x^i (-1)^{(p-1)/2-i}$$

$$(x - \lambda)^{(p-1)/2} = \sum_j \binom{(p-1)/2}{j} x^{(p-1)/2-j} (-\lambda)^j.$$

The coefficient A_p of $x^{(p-1)/2}$ in $(x - 1)^{(p-1)/2}(x - \lambda)^{(p-1)/2}$ is

$$(-1)^{(p-1)/2} \sum_{k=0}^{(p-1)/2} \binom{(p-1)/2}{k}^2 \lambda^k = (-1)^{(p-1)/2} H_p(\lambda).$$

Therefore, E is supersingular if and only if $H_p(\lambda) = 0$. This completes the proof of Theorem 4.32. ∎

It is possible to use the method of the preceding proof to determine when certain curves are supersingular.

PROPOSITION 4.35

Let $p \geq 5$ be prime. Then the elliptic curve $y^2 = x^3 + 1$ over $\mathbf{F}_p$ is supersingular if and only if $p \equiv 2 \pmod 3$, and the elliptic curve $y^2 = x^3 + x$ over $\mathbf{F}_p$ is supersingular if and only if $p \equiv 3 \pmod 4$.

PROOF   The coefficient of x^{p-1} in $(x^3 + 1)^{(p-1)/2}$ is 0 if $p \equiv 2 \pmod 3$ (since we only get exponents that are multiples of 3), and is $\binom{(p-1)/2}{(p-1)/3} \not\equiv 0 \pmod p$ when $p \equiv 1 \pmod 3$ (since the binomial coefficient contains no factors of p). Since the coefficient of x^{p-1} is zero mod p if and only if the curve is supersingular, this proves the first part.

The coefficient of x^{p-1} in $(x^3 + x)^{(p-1)/2}$ is the coefficient of $x^{(p-1)/2}$ in $(x^2 + 1)^{(p-1)/2}$. All exponents appearing in this last expression are even, so $x^{(p-1)/2}$ doesn't appear when $p \equiv 3 \pmod 4$. When $p \equiv 1 \pmod 4$, the coefficient is $\binom{(p-1)/2}{(p-1)/4} \not\equiv 0 \pmod p$. This proves the second part of the proposition.   ∎

If E is an elliptic curve defined over $\mathbf{Z}$ with complex multiplication (see Chapter 10) by a subring of $\mathbf{Q}(\sqrt{-d})$, and p is an odd prime number not dividing d for which $E \pmod p$ is an elliptic curve, then $E \pmod p$ is supersingular if and only if $-d$ is not a square mod p. Therefore, for such an E, the curve $E \pmod p$ is supersingular for approximately half of the primes. In the proposition, the curve $y^2 = x^3 + 1$ has complex multiplication by $\mathbf{Z}[(1 + \sqrt{-3})/2]$, and -3 is a square mod p if and only if $p \equiv 1 \pmod 3$. The curve $y^2 = x^3 + x$ has complex multiplication by $\mathbf{Z}[\sqrt{-1}]$, and -1 is a square mod p if and only if $p \equiv 1 \pmod 4$.

If E does not have complex multiplication, the set of primes for which $E \pmod p$ is supersingular is much more sparse. Elkies [27] proved in 1986 that, for each E, the set of such primes is infinite. Wan [104], improving on an argument of Serre, showed that, for each $\epsilon > 0$, the number of such $p < x$ for which $E \pmod p$ is supersingular is less than $C_\epsilon x / \ln^{2-\epsilon}(x)$ for some constant C_ϵ depending on ϵ. Since the number of primes less than x is approximately $x / \ln x$, this shows that substantially less than half of the primes are supersingular for E. It has been conjectured by Lang and Trotter that the number of supersingular p is asymptotic to $C'\sqrt{x} / \ln x$ (as $x \to \infty$) for some constant C' depending on E.

We now change our viewpoint and fix p and count supersingular E over $\overline{\mathbf{F}}_p$. This essentially amounts to counting distinct zeros of $H_p(T)$. The values $\lambda = 0, 1$ are not allowed in the Legendre form of an elliptic curve. Moreover, they also don't appear as zeros of $H_p(T)$. It is easy to see that $H_p(0) = 1$.

For $H_p(1)$, observe that the coefficient of $x^{(p-1)/2}$ in

$$(x+1)^{p-1} = (x+1)^{(p-1)/2}(x+1)^{(p-1)/2}$$

is

$$\binom{p-1}{(p-1)/2} = \sum_k \binom{(p-1)/2}{k}\binom{(p-1)/2}{(p-1)/2-k} = H_p(1),$$

(use the identity $\binom{n}{k} = \binom{n}{n-k}$). Since $\binom{p-1}{(p-1)/2}$ contains no factors p, it is nonzero mod p. Therefore, $H_p(1) \neq 0$.

PROPOSITION 4.36
$H_p(T)$ has $(p-1)/2$ distinct roots in $\overline{\mathbf{F}}_p$.

PROOF We claim that

$$4T(1-T)H_p''(T) + 4(1-2T)H_p'(T) - H_p(T) \equiv 0 \quad (\text{mod } p). \quad (4.5)$$

Write $H_p(T) = \sum_k b_k T^k$. The coefficient of T^k on the left side of (4.5) is

$$4(k+1)kb_{k+1} - 4k(k-1)b_k + 4(k+1)b_{k+1} - 8kb_k - b_k$$
$$= 4(k+1)^2 b_{k+1} - (2k+1)^2 b_k.$$

Using the fact that

$$b_{k+1} = \binom{(p-1)/2}{k+1}^2$$
$$= \left(\frac{((p-1)/2)!}{(k+1)!(((p-1)/2)-k-1)!}\right)^2$$
$$= \left(\frac{((p-1)/2)-k}{k+1}\right)^2 b_k,$$

we find that the coefficient of T^k is

$$\left(4(((p-1)/2)-k)^2 - (2k+1)^2\right)b_k = p(p-2-4k)b_k \equiv 0 \quad (\text{mod } p).$$

This proves the claim.

Suppose now that $H_p(\lambda) = 0$ with $\lambda \in \overline{\mathbf{F}}_p$. Since $H_p(0) \neq 0$ and $H_p(1) \neq 0$, we have $\lambda \neq 0, 1$. Write $H_p(T) = (T-\lambda)^r G(T)$ for some polynomial $G(T)$ with $G(\lambda) \neq 0$. Suppose $r \geq 2$. In (4.5), we have $(T-\lambda)^{r-1}$ dividing the last term and the middle term, but only $(T-\lambda)^{r-2}$ divides the term $4T(1-T)H_p''(T)$. Since the sum of the three terms is 0, this is impossible, so we must have $r = 1$. Therefore, λ is a simple root. (*Technical point*: Since the degree of $H_p(T)$ is less than p, we have $r < p$, so the first term of the derivative

$$H_p''(T) = r(r-1)(T-\lambda)^{r-2}G(T) + 2r(T-\lambda)^{r-1}G'(T) + (T-\lambda)^r G''(T)$$

does not disappear in characteristic p. Hence $(T - \lambda)^{r-1}$ does not divide the first term of (4.5).)   ∎

REMARK 4.37   The differential equation 4.5 is called a **Picard-Fuchs differential equation**. For a discussion of this equation in the study of families of elliptic curves in characteristic 0, see [18]. Once we know that $H_p(T)$ satisfies this differential equation, the simplicity of the roots follows from a characteristic p version of the uniqueness theorem for second order differential equations. If λ is a multiple root of $H_p(T)$, then $H_p(\lambda) = H'_p(\lambda) = 0$. Such a uniqueness theorem would say that $H_p(T)$ must be identically 0, which is a contradiction. Note that we must avoid $\lambda = 0, 1$ because of the coefficient $T(1 - T)$ for $H''_p(T)$.   ∎

COROLLARY 4.38
 Let $p \geq 5$ be prime. The number of $j \in \overline{\mathbf{F}}_p$ that occur as j-invariants of supersingular elliptic curves is

$$\left[\frac{p}{12}\right] + \epsilon_p,$$

where $\epsilon_p = 0, 1, 1, 2$ if $p \equiv 1, 5, 7, 11 \pmod{12}$, respectively.

PROOF   The j-invariant of $y^2 = x(x - 1)(x - \lambda)$ is

$$2^8 \frac{(\lambda^2 - \lambda + 1)^3}{\lambda^2(\lambda - 1)^2}$$

(see Exercise 2.9), so the values of λ yielding a given j are roots of the polynomial

$$P_j(\lambda) = 2^8(\lambda^2 - \lambda + 1)^3 - j\lambda^2(\lambda - 1)^2.$$

The discriminant of this polynomial is $2^{30}(j-1728)^3 j^4$, which is nonzero unless $j = 0$ or 1728. Therefore, there are 6 distinct values of $\lambda \in \overline{\mathbf{F}}_p$ corresponding to each value of $j \neq 0, 1728$. If one of these λ's is a root of $H_p(T)$, then all six must be roots, since the corresponding elliptic curves are all the same (up to changes of variables), and therefore all or none are supersingular.

Since the degree of $H_p(T)$ is $(p-1)/2$, we expect approximately $(p-1)/12$ supersingular j-invariants, with corrections needed for the cases when at least one of $j = 0$ or $j = 1728$ is supersingular.

When $j = 0$, the polynomial $P_j(\lambda)$ becomes $2^8(\lambda^2 - \lambda + 1)^3$, so there are two values of λ that give $j = 0$. When $j = 1728$, the polynomial becomes $2^8(\lambda - 2)^2(\lambda - \frac{1}{2})^2(\lambda + 1)^2$, so there are three values of λ yielding $j = 1728$.

A curve with j-invariant 0 can be put into the form $y^2 = x^3 + 1$ over an algebraically closed field. Theorem 4.32 therefore tells us that when $p \equiv 2 \pmod 3$, the two λ's yielding $j = 0$ are roots of $H_p(T)$. Similarly, when $p \equiv 3 \pmod 4$, the three λ yielding $j = 1728$ are roots of $H_p(T)$.

Putting everything together, the total count of roots of $H_p(T)$ is

$$6 \cdot \#\{\text{supersingular } j \neq 0, 1728\} + 2\delta_{2(3)} + 3\delta_{3(4)}$$
$$= \deg H_p(T) = (p-1)/2,$$

where $\delta_{i(j)} = 1$ if $p \equiv i \pmod{j}$ and $= 0$ otherwise.

Suppose that $p \equiv 5 \pmod{12}$. Then $\delta_{2(3)} = 1$ and $\delta_{3(4)} = 0$, so the number of supersingular $j \neq 0, 1728$ is

$$\frac{p-1}{12} - \frac{1}{3} = \left[\frac{p}{12}\right].$$

Adding 1 for the case $j = 0$ yields the number given in the proposition. The other cases of $p \pmod{12}$ are similar. ∎

Example 4.13
When $p = 23$, we have

$$H_{23}(T) = (T-3)(T-8)(T-21)(T-11)(T-13)(T-16)$$
$$\cdot (T-2)(T-12)(T+1)(T^2-T+1)$$

(this is a factorization over $\mathbf{F}_{23}$). The first 6 factors correspond to

$$\{\lambda, \frac{1}{\lambda}, 1-\lambda, \frac{1}{1-\lambda}, \frac{\lambda}{\lambda-1}, \frac{\lambda-1}{\lambda}\},$$

with $\lambda = 3$, hence to the curve $y^2 = x(x-1)(x-3)$. The next three factors correspond to $j = 1728$, hence to the curve $y^2 = x^3 + x$. The last factor corresponds to $j = 0$, hence to $y^2 = x^3 + 1$. Therefore, we have found the three supersingular curves over $\overline{\mathbf{F}}_{23}$. Of course, over $\mathbf{F}_{23}$, there are different forms of these curves. For example, $y^2 = x^3 + 1$ and $y^2 = x^3 + 2$ are different curves over $\mathbf{F}_{23}$, but are the same over $\overline{\mathbf{F}}_{23}$. ⧄

Example 4.14
When $p = 13$,

$$H_{13}(T) \equiv (T^2 + 4T + 9)(T^2 + 12T + 3)(T^2 + 7T + 1).$$

The six roots correspond to one value of j. Since $\lambda = -2 + \sqrt{8}$ is a root of the first factor, the corresponding elliptic curve is

$$y^2 = x(x-1)(x+2-\sqrt{8}).$$

⧄

The appearance of a square root such as $\sqrt{8}$ is fairly common. It is possible to show that a supersingular curve over a perfect field of characteristic p must have its j-invariant in $\mathbf{F}_{p^2}$ (see [90, Theorem V.3.1]). Therefore, a supersingular elliptic curve over $\overline{\mathbf{F}}_q$ can always be transformed via a change of variables (over $\overline{\mathbf{F}}_q$) into a curve defined over $\mathbf{F}_{p^2}$.

Exercises

4.1 Let E be the elliptic curve $y^2 = x^3 + x + 1 \pmod 5$.

  (a) Show that $3(0, 1) = (2, 1)$ on E.

  (b) Show that $(0, 1)$ generates $E(\mathbf{F}_5)$. (Use the fact that $E(\mathbf{F}_5)$ has order 9 (see Example 4.1), plus the fact that the order of any element of a group divides the order of the group.)

4.2 Let E be the elliptic curve $y^2 + y = x^3$ over $\mathbf{F}_2$. Show that

$$\#E(\mathbf{F}_{2^n}) = \begin{cases} 2^n + 1 & \text{if } n \text{ is odd} \\ 2^n + 1 - 2^{1 + \frac{1}{2}n} & \text{if } n \text{ is even.} \end{cases}$$

4.3 Let $\mathbf{F}_q$ be a finite field with q odd. Since $\mathbf{F}_q^\times$ is cyclic of even order $q-1$, half of the elements of $\mathbf{F}_q^\times$ are squares and half are nonsquares.

  (a) Let $u \in \mathbf{F}_q$. Show that

$$\sum_{x \in \mathbf{F}_q} \left(\frac{x + u}{\mathbf{F}_q} \right) = 0.$$

  (b) Let $f(x) = (x - r)^2 (x - s)$, where $r, s \in \mathbf{F}_q$ with q odd. Show that

$$\sum_{x \in \mathbf{F}_q} \left(\frac{f(x)}{\mathbf{F}_q} \right) = - \left(\frac{r - s}{\mathbf{F}_q} \right).$$

  (*Hint:* If $x \neq r$, then $(x - r)^2 (x - s)$ is a square exactly when $x - s$ is a square.)

4.4 Let $x \in \mathbf{F}_q$ with q odd. Show that

$$\left(\frac{x}{\mathbf{F}_q} \right) = x^{(q-1)/2}$$

as elements of $\mathbf{F}_q$. (*Remark:* Since the exponentiation on the right can be done quickly, for example, by successive squaring (this is the multiplicative version of the successive doubling in Section 2.2), this shows that the generalized Legendre symbol can be calculated quickly. Of course, the classical Legendre symbol can also be calculated quickly using quadratic reciprocity.)

4.5 Let $p \equiv 1 \pmod 4$ be prime and let E be given by $y^2 = x^3 - kx$, where $k \not\equiv 0 \pmod p$.

(a) Use Theorem 4.21 to show that $\#E(\mathbf{F}_p)$ is a multiple of 4 when k is a square mod p.

(b) Show that when k is a square mod p, then $E(\mathbf{F}_p)$ contains 4 points P satisfying $2P = \infty$. Conclude again that $\#E(\mathbf{F}_p)$ is a multiple of 4.

(c) Show that when k is not a square mod p, then $E(\mathbf{F}_p)$ contains no points of order 4.

(d) Let k be a square but not a fourth power mod p. Show that exactly one of the curves $y^2 = x^3 - x$ and $y^2 = x^3 - kx$ has a point of order 4 defined over $\mathbf{F}_p$.

4.6 Let E be an elliptic curve over $\mathbf{F}_q$ and suppose

$$E(\mathbf{F}_q) \simeq \mathbf{Z}_n \oplus \mathbf{Z}_{mn}.$$

(a) Use the techniques of the proof of Proposition 4.16 to show that $q = mn^2 + kn + 1$ for some integer k.

(b) Show that if n is sufficiently large, then $|k| \le 2\sqrt{m}$. Therefore, if m is fixed, q occurs as the value of one of finitely many quadratic polynomials.

(c) The prime number theorem implies that the number of prime powers less than x is approximately $x/\ln x$. Use this to show that most prime powers do not occur as values of the finite list of polynomials in (2).

(d) Use Hasse's theorem to show that $mn \ge \sqrt{m}(\sqrt{q} - 1)$.

(e) Show that if $m \ge 17$ and q is sufficiently large ($q \ge 1122$ suffices), then $E(\mathbf{F}_q)$ has a point of order greater than $4\sqrt{q}$.

(f) Show that for most values of q, an elliptic curve over $\mathbf{F}_q$ has a point of order greater than $4\sqrt{q}$.

4.7 (a) Let E be defined by $y^2 + y = x^3 + x$ over $\mathbf{F}_2$. Show that $\#E(\mathbf{F}_2) = 5$.

(b) Let E be defined by $y^2 = x^3 - x + 2$ over $\mathbf{F}_3$. Show that $\#E(\mathbf{F}_3) = 1$.

(c) Show that the curves in (a) and (b) are supersingular, but that, in each case, $a = p + 1 - \#E(\mathbf{F}_p) \ne 0$. This shows that the restriction to $p \ge 5$ is needed in Corollary 4.30.

4.8 Let $p \ge 5$ be prime. Use Theorem 4.21 to prove Hasse's theorem for the elliptic curve given by $y^2 = x^3 - kx$ over $\mathbf{F}_p$.

4.9 Let E be an elliptic curve over $\mathbf{F}_q$ with $q = p^{2m}$. Suppose that $\#E(\mathbf{F}_q) = q + 1 - 2\sqrt{q}$.

(a) Let ϕ_q be the Frobenius endomorphism. Show that $(\phi_q - p^m)^2 = 0$.

(b) Show that $\phi_q - p^m = 0$ (*Hint:* Theorem 2.21).

(c) Show that ϕ_q acts as the identity on $E[p^m - 1]$, and therefore that $E[p^m - 1] \subseteq E(\mathbf{F}_q)$.

(d) Show that $E(\mathbf{F}_q) \simeq \mathbf{Z}_{p^m-1} \oplus \mathbf{Z}_{p^m-1}$.

4.10 Let E be an elliptic curve over $\mathbf{F}_q$ with q odd. Write $\#E(\mathbf{F}_q) = q+1-a$. Let $d \in \mathbf{F}_q^\times$ and let $E^{(d)}$ be the twist of E, as in Exercise 2.18. Show that

$$\#E^{(d)}(\mathbf{F}_q) = q + 1 - \left(\frac{d}{\mathbf{F}_q}\right) a.$$

(*Hint:* Use Exercise 2.18(c) and Theorem 4.14.)

4.11 Let $\mathbf{F}_q$ be a finite field of odd characteristic and let $a, b \in \mathbf{F}_q$ with $a \neq \pm 2b$ and $b \neq 0$. Define the elliptic curve E by

$$y^2 = x^3 + ax^2 + b^2 x.$$

(a) Show that the points $(b, b\sqrt{a + 2b})$ and $(-b, -b\sqrt{a - 2b})$ have order 4.

(b) Show that at least one of $a + 2b$, $a - 2b$, $a^2 - 4b^2$ is a square in $\mathbf{F}_q$.

(c) Show that if $a^2 - 4b^2$ is a square in $\mathbf{F}_q$, then $E[2] \subseteq E(\mathbf{F}_q)$.

(d) (Suyama) Show that $\#E(\mathbf{F}_q)$ is a multiple of 4.

(e) Let E' be defined by $y'^2 = x'^3 - 2ax'^2 + (a^2 - 4b^2)x'$. Show that $E'[2] \subseteq E'(\mathbf{F}_q)$. Conclude that $\#E'(\mathbf{F}_q)$ is a multiple of 4.

The curve E' is isogenous to E via

$$(x', y') = (y^2/x^2,\ y(b^2 - x^2)/x^2)$$

(see the end of Section 8.6 and also [90, p. 301]). It can be shown that this implies that $\#E(\mathbf{F}_q) = \#E'(\mathbf{F}_q)$. This gives another proof of the result of part (d). The curve E has been used in certain elliptic curve factorization implementations (see [14]).

4.12 Let p be a prime and let E be a supersingular elliptic curve over the finite field $\mathbf{F}_p$. Let ϕ_p be the Frobenius endomorphism. Show that some power of ϕ_p is an integer. (*Note:* This is easy when $p \geq 5$. The cases $p = 2, 3$ can be done by a case-by-case calculation.)

Chapter 5

The Discrete Logarithm Problem

Let p be a prime and let a, b be integers that are nonzero mod p. Suppose we know that there exists an integer k such that

$$a^k \equiv b \pmod{p}.$$

The classical **discrete logarithm problem** is to find k. Since $k + (p-1)$ is also a solution, the answer k should be regarded as being defined mod $p - 1$, or mod a divisor d of $p - 1$ if $a^d \equiv 1 \pmod{p}$.

More generally, let G be any group, written multiplicatively for the moment, and let $a, b \in G$. Suppose we know that $a^k = b$ for some integer k. In this context, the discrete logarithm problem is again to find k. For example, G could be the multiplicative group $\mathbf{F}_q^\times$ of a finite field. Also, G could be $E(\mathbf{F}_q)$ for some elliptic curve, in which case a and b are points on E and we are trying to find an integer k with $ka = b$.

In Chapter 6, we'll meet several cryptographic applications of the discrete logarithm problem. The security of the cryptosystems will depend on the difficulty of solving the discrete log problem.

One way of attacking a discrete log problem is simple brute force: try all possible values of k until one works. This is impractical when the answer k can be an integer of several hundred digits, which is a typical size used in cryptography. Therefore, better techniques are needed.

In this chapter, we start by discussing an attack, called the index calculus, that can be used in $\mathbf{F}_p^\times$, and more generally in the multiplicative group of a finite field. However, it does not apply to general groups. Then we discuss the method of Pohlig-Hellman, the baby step, giant step method, and Pollard's ρ and λ methods. These work for general finite groups, in particular for elliptic curves. Finally, we show that for special classes of elliptic curves, namely supersingular and anomalous curves, it is possible to reduce the discrete log problem to easier discrete log problems (in the multiplicative group of a finite field and in the additive group of integers mod a prime, respectively).

5.1   The Index Calculus

Let p be a prime and let g be primitive root (see Appendix A) mod p, which means that g is a generator for the cyclic group $\mathbf{F}_p^\times$. In other words, every $h \not\equiv 0 \pmod{p}$ can be written in the form $h \equiv g^k$ for some integer k that is uniquely determined mod $p - 1$. Let $k = L(h)$ denote the **discrete logarithm** of h with respect to g and p, so

$$g^{L(h)} \equiv h \pmod{p}.$$

Suppose we have h_1 and h_2. Then

$$g^{L(h_1 h_2)} \equiv h_1 h_2 \equiv g^{L(h_1) + L(h_2)} \pmod{p},$$

which implies that

$$L(h_1 h_2) \equiv L(h_1) + L(h_2) \pmod{p - 1}.$$

Therefore, L changes multiplication into addition, just like the classical logarithm function.

The **index calculus** is a method for computing values of the discrete log function L. The idea is to compute $L(\ell)$ for several small primes ℓ, then use this information to compute $L(h)$ for arbitrary h. It is easiest to describe the method with an example.

Example 5.1

Let $p = 1217$ and $g = 3$. We want to solve $3^k \equiv 37 \pmod{1217}$. Most of our work will be precomputation that will be independent of the number 37. Let's choose a set of small primes, called the **factor base**, to be $B = \{2, 3, 5, 7, 11, 13\}$. First, we find relations of the form

$$3^x \equiv \pm \text{product of some primes in } B \pmod{1217}.$$

Eventually, we find the following:

$$3^1 \equiv 3 \pmod{1217}$$
$$3^{24} \equiv -2^2 \cdot 7 \cdot 13$$
$$3^{25} \equiv 5^3$$
$$3^{30} \equiv -2 \cdot 5^2$$
$$3^{54} \equiv -5 \cdot 11$$
$$3^{87} \equiv 13$$

These can be changed into equations for discrete logs, where now the congruences are all mod $p - 1 = 1216$. Note that we already know that $3^{(p-1)/2} \equiv -1$

(mod p), so $L(-1) = 608$.

$$1 \equiv L(3) \quad (\text{mod } 1216)$$
$$24 \equiv 608 + 2L(2) + L(7) + L(13)$$
$$25 \equiv 3L(5)$$
$$30 \equiv 608 + L(2) + 2L(5)$$
$$54 \equiv 608 + L(5) + L(11)$$
$$87 \equiv L(13)$$

The first equation yields $L(3) \equiv 1$. The third yields $L(5) \equiv 819$ (mod 1216). The sixth yields $L(13) \equiv 87$. The fourth gives

$$L(2) \equiv 30 - 608 - 2 \cdot 819 \equiv 216 \quad (\text{mod } 1216).$$

The fifth yields $L(11) \equiv 54 - 608 - L(5) \equiv 1059$. Finally, the second gives

$$L(7) \equiv 24 - 608 - 2L(2) - L(13) \equiv 113 \quad (\text{mod } 1216).$$

We now know the discrete logs of all the elements of the factor base.

Recall that we want to solve $3^k \equiv 37$ (mod 1216). We compute $3^j \cdot 37$ (mod p) for several random values of j until we obtain an integer that can be factored into a product of primes in B. In our case, we find that

$$3^{16} \cdot 37 \equiv 2^3 \cdot 7 \cdot 11 \quad (\text{mod } 1217).$$

Therefore,

$$L(37) \equiv 3L(2) + L(7) + L(11) - 16 \equiv 588 \quad (\text{mod } 1217),$$

and $3^{588} \equiv 37$ (mod 1217). $\square$

The choice of the size of the factor base B is important. If B is too small, then it will be very hard to find powers of g that factor with primes in B. If B is too large, it will be easy to find relations, but the linear algebra needed to solve for the logs of the elements of B will be unwieldy. An example that was completed in 2001 by A. Joux and R. Lercier used the first 1 million primes to compute discrete logs mod a 120-digit prime.

There are various methods that produce relations of the form $g^x \equiv$ product of primes in B. A popular one uses the number field sieve. See [45].

The expected running time of the index calculus is approximately a constant times $\exp(\sqrt{2 \ln p \ln \ln p})$ (see [67, p. 129]), which means that it is a subexponential algorithm. The algorithms in Section 5.2, which are exponential algorithms, run in time approximately $\sqrt{p} = exp(\frac{1}{2} \ln p)$. Since $\sqrt{2 \ln p \ln \ln p}$ is much smaller than $\frac{1}{2} \ln p$ for large p, the index calculus is generally much faster when it can be used.

Note that the index calculus depends heavily on the fact that integers can be written as products of primes. An analogue of this is not available for arbitrary groups.

There is a generalization of the index calculus that works for finite fields, but it requires some algebraic number theory, so we do not discuss it here.

5.2   General Attacks on Discrete Logs

In this section, we discuss attacks that work for arbitrary groups. Since our main focus is elliptic curves, we write our group G additively. Therefore, we are given $P, Q \in G$ and we are trying to solve $kP = Q$ (we always assume that k exists). Let N be the order of G. Usually, we assume N is known.

5.2.1   Baby Step, Giant Step

This method, developed by D. Shanks [88], requires approximately $\sqrt{N}$ steps and around $\sqrt{N}$ storage. Therefore it only works well for moderate sized N. The procedure is as follows.

1. Fix an integer $m \geq \sqrt{N}$ and compute mP.

2. Make and store a list of iP for $0 \leq i < m$.

3. Compute the points $Q - jmP$ for $j = 0, 1, \cdots m - 1$ until one matches an element from the stored list.

4. If $iP = Q - jmP$, we have $Q = kP$ with $k \equiv i + jm \pmod{N}$.

Why does this work? Since $m^2 > N$, we may assume the answer k satisfies $0 \leq k < m^2$. Write $k = k_0 + mk_1$ with $k_0 = k \pmod{m}$ and $0 \leq k_0 < m$ and let $k_1 = (k - k_0)/m$. Then $0 \leq k_1 < m$. When $i = k_0$ and $j = k_1$, we have

$$Q - k_1 mP = kP - k_1 mP = k_0 P,$$

so there is a match.

The point iP is calculated by adding P (a "**baby step**") to $(i-1)P$. The point $Q - jmP$ is computed by adding $-mP$ (a "**giant step**") to $Q - (j-1)mP$. The method was developed by Shanks for computations in algebraic number theory.

Note that we did not need to know the exact order N of G. We only required an upper bound for N. Therefore, for elliptic curves over $\mathbf{F}_q$, we could use this method with $m^2 \geq q + 1 + 2\sqrt{q}$, by Hasse's theorem.

A slight improvement of the method can be made for elliptic curves by computing and storing only the points iP for $0 \le i \le m/2$ and checking whether $Q - jmP = \pm iP$ (see Exercise 5.1).

Example 5.2

Let $G = E(\mathbf{F}_{41})$, where E is given by $y^2 = x^3 + 2x + 1$. Let $P = (0, 1)$ and $Q = (30, 40)$. By Hasse's theorem, we know that the order of G is at most 54, so we let $m = 8$. The points iP for $0 \le i \le 7$ are

$$(0, 1), (1, 39), (8, 23), (38, 38), (23, 23), (20, 28), (26, 9).$$

We calculate $Q - jmP$ for $j = 0, 1, 2$ and obtain

$$(30, 40), \ (9, 25), \ (26, 9),$$

at which point we stop since this third point matches $7P$. Since $j = 2$ yielded the match, we have

$$(30, 40) = (7 + 2 \cdot 8)P = 23P.$$

Therefore $k = 23$.   □

5.2.2   Pollard's ρ and λ Methods

A disadvantage of the Baby Step, Giant Step method is that it requires a lot of storage. Pollard's ρ and λ methods [71] run in approximately the same time as Baby Step, Giant Step, but require very little storage. First, we'll discuss the ρ method, then its generalization to the λ method.

Let G be a finite group of order N. Choose a function $f : G \to G$ that behaves rather randomly. Then start with a random element P_0 and compute the iterations $P_{i+1} = f(P_i)$. Since G is a finite set, there will be some indices $i_0 < j_0$ such that $P_{i_0} = P_{j_0}$. Then

$$P_{i_0+1} = f(P_{i_0}) = f(P_{j_0}) = P_{j_0+1},$$

and, similarly, $P_{i_0+\ell} = P_{j_0+\ell}$ for all $\ell \ge 0$. Therefore, the sequence P_i is periodic with period $j_0 - i_0$ (or possibly a divisor of $j_0 - i_0$). The picture describing this process (see Figure 5.1) looks like the Greek letter ρ, which is why it is called **Pollard's ρ method**. If f is a randomly chosen random function (we'll not make this precise), then we expect to find a match with j_0 at most a constant times $\sqrt{N}$. For an analysis of the running time for various choices of function f, see [100].

A naive implementation of the method stores all the points P_i until a match is found. This takes around $\sqrt{N}$ storage, which is similar to Baby Step, Giant

Step. However, as R. W. Floyd has pointed out, it is possible to do much better at the cost of a little more computation. The key idea is that once there is a match for two indices differing by d, all subsequent indices differing by d will yield matches. This is just the periodicity mentioned above. Therefore, we can compute pairs (P_i, P_{2i}) for $i = 1, 2, \ldots$, but only keep the current pair; we don't store the previous pairs. These can be calculated by the rules

$$P_{i+1} = f(P_i), \quad P_{2(i+1)} = f(f(P_{2i})).$$

Suppose $i \geq i_0$ and i is a multiple of d. Then the indices $2i$ and i differ by a multiple of d and hence yield a match: $P_i = P_{2i}$. Since $d \leq j_0$ and $i_0 < j_0$, it follows easily that there is a match for $i \leq j_0$. Therefore, the number of steps to find a match is expected to be at most a constant multiple of $\sqrt{N}$.

Another method of finding a match is to store only those points P_i that satisfy a certain property (call them "distinguished points"). For example, we could require the last k bits of the binary representation of the x-coordinate to be 0. We then store, on the average, one out of every 2^k points P_i. Suppose there is a match $P_i = P_j$ but P_i is not one of these distinguished points. We expect $P_{i+\ell}$ to be a distinguished point for some ℓ with $1 \leq \ell \leq 2^k$, approximately. Then $P_{j+\ell} = P_{i+\ell}$, so we find a match between distinguished points with only a little more computation.

The problem remains of how to choose a suitable function f. Besides having f act randomly, we need to be able to extract useful information from a match. Here is one way of doing this. Divide G into s disjoint subsets $S_1, S_2, \ldots, S_s$ of approximately the same size. A good choice for s seems to be around 20. Choose $2s$ random integers $a_i, b_i \bmod N$. Let

$$M_i = a_i P + b_i Q.$$

Finally, define

$$f(g) = g + M_i \quad \text{if } g \in S_i.$$

The best way to think of f is as giving a random walk in G, with the possible steps being the elements M_i.

Finally, choose random integers a_0, b_0 and let $P_0 = a_0 P + b_0 Q$ be the starting point for the random walk. While computing the points P_j, we also record how these points are expressed in terms of P and Q. If $P_j = u_j P + v_j Q$ and $P_{j+1} = P_j + M_i$, then $P_{j+1} = (u_j + a_i)P + (v_j + b_j)Q$, so $(u_{j+1}, v_{j+1}) = (u_j, v_j) + (a_i, b_i)$. When we find a match $P_{j_0} = P_{i_0}$, then we have

$$u_{j_0} P + v_{j_0} Q = u_{i_0} P + v_{i_0} Q, \quad \text{hence } (u_{i_0} - u_{j_0})P = (v_{j_0} - v_{i_0})Q.$$

If $\gcd(v_{j_0} - v_{i_0}, N) = d$, we have

$$k \equiv (v_{j_0} - v_{i_0})^{-1}(u_{i_0} - u_{j_0}) \pmod{N/d}.$$

This gives us d choices for k. Usually, d will be small, so we can try all possibilities until we have $Q = kP$.

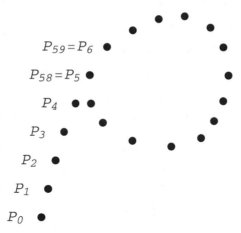

$P_{59} = P_6$

$P_{58} = P_5$

P_4

P_3

P_2

P_1

P_0

Figure 5.1
Pollard's Rho Method

In cryptographic applications, N is often prime, in which case, $d = 1$ or N. If $d = N$, we have a trivial relation (the coefficients of both P and Q are multiples of N), so we start over. If $d = 1$, we obtain k.

Example 5.3
Let $G = E(\mathbf{F}_{1093})$, where E is the elliptic curve given by $y^2 = x^3 + x + 1$. We'll use $s = 3$. Let $P = (0, 1)$ and $Q = (413, 959)$. It can be shown that the order of P is 1067. We want to find k such that $kP = Q$. Let

$$P_0 = 3P + 5Q, \quad M_0 = 4P + 3Q, \quad M_1 = 9P + 17Q, \quad M_2 = 19P + 6Q.$$

Let $f : E(\mathbf{F}_{1093}) \to E(\mathbf{F}_{1093})$ be defined by

$$f(x, y) = (x, y) + M_i \quad \text{if } x \equiv i \pmod 3.$$

Here the number x is regarded as an integer $0 \le x < 1093$ and is then reduced mod 3. For example,

$$f(P_0) = P_0 + M_2 = (727, 589),$$

since $P_0 = (326, 69)$ and $326 \equiv 2 \pmod 3$.

We can define $f(\infty) = \infty$ if we want. However, if we encounter $f(\infty)$, we have found a relation of the form $aP + bQ = \infty$ and can find k easily (if the relation isn't something trivial like $1067P + 2134Q = \infty$). Therefore, we don't worry about ∞.

If we compute P_0, $P_1 = f(P_0)$, $P_2 = f(P_1), \ldots$, we obtain

$$P_0 = (326, 69), \ P_1 = (727, 589), \ P_2 = (560, 365), \ P_3 = (1070, 260),$$
$$P_4 = (473, 903), \ P_5 = (1006, 951), \ P_6 = (523, 938), \ldots,$$
$$P_{57} = (895, 337), \ P_{58} = (1006, 951), \ P_{59} = (523, 938), \ldots.$$

Therefore, the sequence starts repeating at $P_5 = P_{58}$.

If we keep track of the coefficients of P and Q in the calculations, we find that

$$P_5 = 88P + 46Q \quad \text{and} \quad P_{58} = 685P + 620Q.$$

Therefore,

$$\infty = P_{58} - P_5 = 597P + 574Q.$$

Since P has order 1067, we calculate

$$-574^{-1}597 \equiv 499 \pmod{1067}.$$

Therefore, $Q = 499P$, so $k = 499$.

We stored all of the points $P_0, P_1, \ldots, P_{58}$ until we found a match. Instead, let's repeat the computation, but compute the pairs (P_i, P_{2i}) and store nothing except the current pair. We then find that for $i = 53$ there is the match $P_{53} = P_{106}$. This yields

$$620P + 557Q = P_{53} = P_{106} = 1217P + 1131Q.$$

Therefore, $597P + 574Q = \infty$, which yields $k = 499$, as before.   ☐

Pollard's λ method uses a function f as in the ρ method, but several random starting points $P_0^{(1)}, \ldots, P_0^{(r)}$ are used. We then get sequences defined by

$$P_{i+1}^{(\ell)} = f(P_i^{(\ell)}), \quad 1 \le \ell \le r, \quad i = 0, 1, 2, \ldots.$$

These can be computed by several computers in parallel. Points satisfying certain conditions are called distinguished and are reported to a central computer. When a match is found among the inputs from the various computers, we have a relation that should allow us to solve the discrete log problem, as in the ρ method. When there is a match between two sequences, these two sequences will always match from that point on. We only need to look at distinguished points because distinguished points should occur soon after a match occurs.

When there are only two random starting points, we have two random walks. Eventually they will have a point in common, and therefore they will coincide thereafter. The picture of this process resembles the Greek letter λ, hence the name.

Sometimes the λ method is described in terms of kangaroos jumping around a field (this is the random walk). A variant of the λ method with two random walks records every 10th point, for example, in the first sequence and then checks whether the second sequence matches any of these points. In this case, the first sequence is called a tame kangaroo, and the second is called a wild kangaroo. The idea is to use the tame kangaroo to catch the wild kangaroo.

The λ method is expected to find a match in at most a constant times $\sqrt{N}$ steps. If it is run in parallel with many starting points, the running time can be improved significantly.

Finally, we should point out a difference between the baby step, giant step method and the ρ and λ methods. The baby step, giant step method is **deterministic**, which means that it is guaranteed to finish within the predicted time of a constant times $\sqrt{N}$. On the other hand, the ρ and λ methods are **probabilistic**, which means that there is a very high probability that they will finish within the predicted time, but this is not guaranteed.

5.2.3   The Pohlig-Hellman Method

As before, P, Q are elements in a group G and we want to find an integer k with $Q = kP$. We also know the order N of P and we know the prime factorization

$$N = \prod_i q_i^{e_i}$$

of N. The idea of Pohlig-Hellman is to find k (mod $q_i^{e_i}$) for each i, then use the Chinese Remainder theorem to combine these and obtain k (mod N).

Let q be a prime, and let q^e be the exact power of q dividing N. Write k in its base q expansion as

$$k = k_0 + k_1 q + k_2 q^2 + \cdots$$

with $0 \leq k_i < q$. We'll evaluate k (mod q^e) by successively determining $k_0, k_1, \ldots, k_{e-1}$. The procedure is as follows.

1. Compute $T = \left\{ j \left(\frac{N}{q} P \right) \mid 0 \leq j \leq q - 1 \right\}$.

2. Compute $\frac{N}{q} Q$. This will be an element $k_0 \left(\frac{N}{q} P \right)$ of T.

3. If $e = 1$, stop. Otherwise, continue.

4. Let $Q_1 = Q - k_0 P$.

5. Compute $\frac{N}{q^2} Q_1$. This will be an element $k_1 \left(\frac{N}{q} P \right)$ of T.

6. If $e = 2$, stop. Otherwise, continue.

7. Suppose we have computed $k_0, k_1, \ldots, k_{r-1}$, and $Q_1, \ldots, Q_{r-1}$.

8. Let $Q_r = Q_{r-1} - k_{r-1}q^{r-1}P$.

9. Determine k_r such that $\frac{N}{q^{r+1}}Q_r = k_r\left(\frac{N}{q}P\right)$.

10. If $r = e - 1$, stop. Otherwise, return to step (7).

Then

$$k \equiv k_0 + k_1 q + \cdots + k_{e-1}q^{e-1} \pmod{q^e}.$$

Why does this work? We have

$$\frac{N}{q}Q = \frac{N}{q}(k_0 + k_1 q + \cdots)P$$

$$= k_0\frac{N}{q}P + (k_1 + k_2 q + \cdots)NP = k_0\frac{N}{q}P,$$

since $NP = \infty$. Therefore, step (2) finds k_0. Then

$$Q_1 = Q - k_0 P = (k_1 q + k_2 q^2 + \cdots)P,$$

so

$$\frac{N}{q^2}Q_1 = (k_1 + k_2 q + \cdots)\frac{N}{q}P$$

$$= k_1\frac{N}{q}P + (k_2 + k_3 q + \cdots)NP = k_1\frac{N}{q}P.$$

Therefore, we find k_1. Similarly, the method produces $k_2, k_3, \ldots$. We have to stop after $r = e - 1$ since N/q^{e+1} is no longer an integer, and we cannot multiply Q_e by the non-integer N/q^{e+1}. Besides, we do not need to continue because we now know $k \bmod q^e$.

Example 5.4

Let $G = E(\mathbf{F}_{599})$, where E is the elliptic curve given by $y^2 = x^3 + 1$. Let $P = (60, 19)$ and $Q = (277, 239)$. The methods of Section 4.3.3 can be used to show that P has order $N = 600$. We want to solve $Q = kP$ for k. The prime factorization of N is

$$600 = 2^3 \cdot 3 \cdot 5^2.$$

We'll compute $k \bmod 8$, $\bmod 3$, and $\bmod 25$, then recombine to obtain $k \bmod 600$ (the Chinese Remainder Theorem allows us to do this).

$k \bmod 8$. We compute $T = \{\infty, (598, 0)\}$. Since

$$(N/2)Q = \infty = 0 \cdot \left(\frac{N}{2}P\right),$$

we have $k_0 = 0$. Therefore,

$$Q_1 = Q - 0P = Q.$$

Since $(N/4)Q_1 = 150Q_1 = (598, 0) = 1 \cdot \frac{N}{2}P$, we have $k_1 = 1$. Therefore,

$$Q_2 = Q_1 - 1 \cdot 2 \cdot P = (35, 243).$$

Since $(N/8)Q_2 = 75Q_2 = \infty = 0 \cdot \frac{N}{2}P$, we have $k_2 = 0$. Therefore,

$$k = 0 + 1 \cdot 2 + 0 \cdot 4 + \cdots \equiv 2 \pmod 8.$$

k mod 3. We have $T = \{\infty, (0, 1), (0, 598)\}$. Since

$$(N/3)Q = (0, 598) = 2 \cdot \frac{N}{3}P,$$

we have $k_0 = 2$. Therefore,

$$k \equiv 2 \pmod 3.$$

k mod 25. We have

$$T = \{\infty, (84, 179), (491, 134), (491, 465), (84, 420)\}.$$

Since $(N/5)Q = (84, 179)$, we have $k_0 = 1$. Then

$$Q_1 = Q - 1 \cdot P = (130, 129).$$

Since $(N/25)Q_1 = (491, 465)$, we have $k_1 = 3$. Therefore,

$$k = 1 + 3 \cdot 5 + \cdots \equiv 16 \pmod{25}.$$

We now have the simultaneous congruences

$$\begin{cases} x \equiv 2 \pmod 8 \\ x \equiv 2 \pmod 3 \\ x \equiv 16 \pmod{25} \end{cases}.$$

These combine to yield $k \equiv 266 \pmod{600}$, so $k = 266$. $\square$

The Pohlig-Hellman method works well if all of the prime numbers dividing N are small. However, if q is a large prime dividing N, then it is difficult to list the elements of T, which contains q elements. We could try to find the k_i without listing the elements; however, finding k_i is a discrete log problem in the group generated by $(N/q)P$, which has order q. If q is of the same order of magnitude as N (for example, $q = N$ or $q = N/2$), then the Pohlig-Hellman method is of little use. For this reason, if a cryptographic system is based on discrete logs, the order of the group should be chosen so it contains a large prime factor.

If N contains some small prime factors, then the Pohlig-Hellman method can be used to obtain partial information on the value of k, namely a congruence modulo a product of these small prime factors. In certain cryptographic situations, this could be undesirable. Therefore, the group G is often chosen to be of large prime order. This can be accomplished by starting with a group that has a large prime q in its order. Pick a random point P_1 and compute its order. With high probability (at least $1 - 1/q$; cf. Remark 5.2), the order of P_1 is divisible by q, so in a few tries, we can find such a point P_1. Write the order of P_1 as qm. Then $P = mP_1$ will have order q. As long as q is sufficiently large, discrete log problems in the cyclic group generated by P will resist the Pohlig-Hellman attack.

5.3   The MOV Attack

The MOV attack, named after Menezes, Okamoto, and Vanstone [66], uses the Weil pairing to convert a discrete log problem in $E(\mathbf{F}_q)$ to one in $\mathbf{F}_{q^m}^{\times}$. Since discrete log problems in finite fields can be attacked by index calculus methods, they can be solved faster than elliptic curve discrete log problems, as long as the field $\mathbf{F}_{q^m}$ is not much larger than $\mathbf{F}_q$. For supersingular curves, we can usually take $m = 2$, so discrete logarithms can be computed more easily for these curves than for arbitrary elliptic curves. This is unfortunate from a cryptographic standpoint since an attractive feature of supersingular curves is that calculations can often be done quickly on them (see Section 4.6).

Recall that for an elliptic curve E defined over $\mathbf{F}_q$, we let $E[N]$ denote the set of points of order dividing N with coordinates in the algebraic closure. If $\gcd(q, N) = 1$ and $S, T \in E[N]$, then the Weil pairing $e_N(S, T)$ is an Nth root of unity and can be computed fairly quickly. The pairing is bilinear, and if $\{S, T\}$ is a basis for $E[N]$, then $e_N(S, T)$ is a primitive Nth root of unity. For any S, $e_N(S, S) = 1$. For more properties of the Weil pairing, see Sections 3.3 and 11.2.

Let E be an elliptic curve over $\mathbf{F}_q$. Let $P, Q \in E(\mathbf{F}_q)$. Let N be the order of P. Assume that

$$\gcd(N, q) = 1.$$

We want to find k such that $Q = kP$. First, it's worthwhile to check that k exists.

LEMMA 5.1
There exists k such that $Q = kP$ if and only if $NQ = \infty$ and the Weil paring $e_N(P, Q) = 1$.

PROOF   If $Q = kP$, then $NQ = kNP = \infty$. Also,

$$e_N(P, Q) = e_N(P, P)^k = 1^k = 1.$$

Conversely, if $NQ = \infty$, then $Q \in E[N]$. Since $\gcd(N, q) = 1$, we have $E[N] \simeq \mathbf{Z}_N \oplus \mathbf{Z}_N$, by Theorem 3.2. Choose a point R such that $\{P, R\}$ is a basis of $E[N]$. Then

$$Q = aP + bR$$

for some integers a, b. By Corollary 3.10, $e_N(P, R) = \zeta$ is a primitive Nth root of unity. Therefore, if $e_N(P, Q) = 1$, we have

$$1 = e_N(P, Q) = e_N(P, P)^a e_N(P, R)^b = \zeta^b.$$

This implies that $b \equiv 0 \pmod{N}$, so $bR = \infty$. Therefore, $Q = aP$, as desired.
∎

The idea used to prove the lemma yields the MOV attack on discrete logs for elliptic curves. Choose m so that

$$E[N] \subseteq E(\mathbf{F}_{q^m}).$$

Since all the points of $E[N]$ have coordinates in $\overline{\mathbf{F}}_q = \cup_{j \geq 1} \mathbf{F}_{q^j}$, such an m exists. By Corollary 3.11, the group μ_N of Nth roots of unity is contained in $\mathbf{F}_{q^m}$. All of our calculations will be done in $\mathbf{F}_{q^m}$. The algorithm is as follows.

1. Choose a random point $T \in E(\mathbf{F}_{q^m})$.

2. Compute the order M of T.

3. Let $d = \gcd(M, N)$, and let $T_1 = (M/d)T$. Then T_1 has order d, which divides N, so $T_1 \in E[N]$.

4. Compute $\zeta_1 = e_N(P, T_1)$ and $\zeta_2 = e_N(Q, T_1)$. Then both ζ_1 and ζ_2 are in $\mu_d \subseteq \mathbf{F}_{q^m}^{\times}$.

5. Solve the discrete log problem $\zeta_2 = \zeta_1^k$ in $\mathbf{F}_{q^m}^{\times}$. This will give $k \pmod{d}$.

6. Repeat with random points T until the least common multiple of the various d's obtained is N. This determines $k \pmod{N}$.

REMARK 5.2   At first, it might seem that $d = 1$ will occur very often. However, the opposite is true because of the structure of $E(\mathbf{F}_{q^m})$. Recall that

$$E(\mathbf{F}_{q^m}) \simeq \mathbf{Z}_{n_1} \oplus \mathbf{Z}_{n_2}$$

for some integers n_1, n_2 with $n_1 | n_2$ (possibly, $n_1 = 1$, in which case the group is cyclic). Then $N | n_2$, since n_2 is the largest possible order of an element of the group. Let B_1, B_2 be points of orders n_1, n_2, respectively, such that B_1, B_2 generate $E(\mathbf{F}_{q^m})$. Then $T = a_1 B_1 + a_2 B_2$. Let ℓ^e be a prime power dividing N. Then $\ell^f | n_2$ with $f \geq e$. If $\ell \nmid a_2$, then ℓ^f divides M, the order of T. Therefore, $\ell^e | d = \gcd(M, N)$. Since the probability that $\ell \nmid a_2$ is $1 - 1/\ell$, the probability is at least this high that the full power ℓ^e is in d. After a few choices of T, this should be the case. (Note that our probability estimates are low, since we never included the possible contribution of the $a_1 B_1$ term.) Therefore, a few iterations of the algorithm should yield k. ∎

Potentially, the integer m could be large, in which case the discrete log problem in the group $\mathbf{F}_{q^m}^\times$, which has order $q^m - 1$, is just as hard as the original discrete log problem in the smaller group $E(\mathbf{F}_q)$, which has order approximately q, by Hasse's theorem. However, for supersingular curves, we can usually take $m = 2$, as the next result shows.

Let E be an elliptic curve over $\mathbf{F}_q$, where q is a power of the prime number p. Then

$$\#E(\mathbf{F}_q) = q + 1 - a$$

for some integer a. The curve E is called **supersingular** if $a \equiv 0 \pmod{p}$. Corollary 4.30 says that this is equivalent to $a = 0$ when $q = p \geq 5$.

PROPOSITION 5.3

Let E be an elliptic curve over $\mathbf{F}_q$ and suppose $a = q + 1 - \#E(\mathbf{F}_q) = 0$. Let N be a positive integer. If there exists a point $P \in E(\mathbf{F}_q)$ of order N, then $E[N] \subseteq E(\mathbf{F}_{q^2})$.

PROOF   The Frobenius endomorphism ϕ_q satisfies $\phi_q^2 - a\phi_q + q = 0$. Since $a = 0$, this reduces to

$$\phi_q^2 = -q.$$

Let $S \in E[N]$. Since $\#E(\mathbf{F}_q) = q + 1$, and since there exists a point of order N, we have $N | q + 1$, or $-q \equiv 1 \pmod{N}$. Therefore

$$\phi_q^2(S) = -qS = 1 \cdot S.$$

By Lemma 4.5, $S \in E(\mathbf{F}_{q^2})$, as claimed.   ∎

Therefore, discrete log problems over $\mathbf{F}_q$ for supersingular curves with $a = 0$ can be reduced to discrete log calculations in $\mathbf{F}_{q^2}^\times$. These are much easier.

When E is supersingular but $a \neq 0$, the above ideas work, but possibly $m = 3, 4$, or 6 (see [66] and Exercise 5.11). This is still small enough to speed up discrete log computations.

5.4   Anomalous Curves

The reason the MOV attack works is that it is possible to use the Weil pairing. In order to avoid this, it was suggested that elliptic curves E over $\mathbf{F}_q$ with

$$\#E(\mathbf{F}_q) = q$$

be used. Such curves are called **anomalous**. Unfortunately, the discrete log problem for the group $E(\mathbf{F}_q)$ can be solved quickly. However, as we'll see below, anomalous curves are potentially useful when considered over extensions of $\mathbf{F}_q$, since they permit a speed-up in certain calculations in $E(\overline{\mathbf{F}}_q)$.

The Weil pairing is not defined on $E[p]$ (or, if we defined it, it would be trivial since $E[p]$ is cyclic and also since there are no nontrivial pth roots of unity in characteristic p). Therefore, it was hoped that this would be a good way to avoid the MOV attack. However, it turns out that there is a different attack for anomalous curves that works even faster for these curves than the MOV attack works for supersingular curves.

In the following, we show how to compute discrete logs in the case $q = p$. Procedures for doing this have been developed in [79], [84], and [96]. Similar ideas work for subgroups of p-power order in $E(\mathbf{F}_q)$ when q is a power of p (but in Proposition 5.4 we would need to lift E to a curve defined over a larger ring than $\mathbf{Z}$).

Warning: The property of being anomalous depends on the base field. If E is anomalous over $\mathbf{F}_q$, it is not necessarily anomalous over any $\mathbf{F}_{q^n}$ for $n \geq 2$. See Exercises 5.5 and 5.6. This is in contrast to supersingularity, which is independent of the base field and is really a property of the curve over the algebraic closure (since supersingular means that there are no points of order p with coordinates in the algebraic closure of the base field).

The first thing we need to do is lift the curve E and the points P, Q to an elliptic curve over $\mathbf{Z}$.

PROPOSITION 5.4

Let E be an elliptic curve over $\mathbf{F}_p$ and let $P, Q \in E(\mathbf{F}_p)$. We assume E is in Weierstrass form $y^2 = x^3 + Ax + B$. Then there exist integers $\tilde{A}, \tilde{B}, x_1, x_2, y_1, y_2$ and an elliptic curve $\tilde{E}$ given by

$$y^2 = x^3 + \tilde{A}x + \tilde{B}$$

such that $\tilde{P} = (x_1, y_1),\ \tilde{Q} = (x_2, y_2) \in \tilde{E}(\mathbf{Q})$ and such that

$$A \equiv \tilde{A}, \quad B \equiv \tilde{B}, \quad P \equiv \tilde{P}, \quad Q \equiv \tilde{Q} \pmod{p}.$$

PROOF   Choose integers x_1 and x_2 such that $x_1, x_2 \pmod{p}$ give the x-coordinates of P, Q. First, assume that $x_1 \not\equiv x_2 \pmod{p}$. Choose an integer

y_1 such that $\tilde{P} = (x_1, y_1)$ reduces to P mod p. Now choose y_2 such that

$$y_2^2 \equiv y_1^2 \pmod{x_2 - x_1} \text{ and } (x_2, y_2) \equiv Q \pmod{p}.$$

This is possible by the Chinese Remainder Theorem, since $\gcd(p, x_2 - x_1) = 1$ by assumption.

Consider the simultaneous equations

$$y_1^2 = x_1^3 + \tilde{A}x_1 + \tilde{B}$$
$$y_2^2 = x_2^3 + \tilde{A}x_2 + \tilde{B}.$$

We can solve these for $\tilde{A}, \tilde{B}$:

$$\tilde{A} = \frac{y_2^2 - y_1^2}{x_2 - x_1} - \frac{x_2^3 - x_1^3}{x_2 - x_1}, \quad \tilde{B} = y_1^2 - x_1^3 - \tilde{A}x_1.$$

Since $y_2^2 - y_1^2$ is divisible by $x_2 - x_1$, and since x_1, x_2, y_1, y_2 are integers, it follows that $\tilde{A}$, and therefore $\tilde{B}$, are integers. The points $\tilde{P}$ and $\tilde{Q}$ lie on the curve $\tilde{E}$ we obtain.

If $x_1 \equiv x_2 \pmod{p}$, then $P = \pm Q$. In this case, take $x_1 = x_2$. Then choose y_1 that reduces mod p to the y-coordinate of P. Choose an integer $\tilde{A} \equiv A \pmod{p}$ and let $\tilde{B} = y_1^2 - x_1^3 - \tilde{A}x_1$. Then $\tilde{P} = (x_1, y_1)$ lies on $\tilde{E}$. Let $\tilde{Q} = \pm\tilde{P}$. Then $\tilde{Q}$ reduces to $\pm P = Q$ mod p.

Finally, $4\tilde{A}^3 + 27\tilde{B}^2 \equiv 4A^3 + 27B^2 \not\equiv 0 \pmod{p}$, since E is an elliptic curve. It follows that $4\tilde{A}^3 + 27\tilde{B}^2 \neq 0$. Therefore $\tilde{E}$ is an elliptic curve. ∎

REMARK 5.5   If we start with $Q = kP$ for some integer k, it is very unlikely that this relation still holds on $\tilde{E}$. In fact, usually $\tilde{P}$ and $\tilde{Q}$ are independent points. However, if they are dependent, so $a\tilde{P} = b\tilde{Q}$ for some nonzero integers a, b, then $aP = bQ$, which allows us to find k (unless $bP = \infty$). The amazing thing about the case of anomalous curves is that even when $\tilde{P}$ and $\tilde{Q}$ are independent, we can extract enough information to find k. ∎

Let $a/b \neq 0$ be a rational number, where a, b are relatively prime integers. Write $a/b = p^r a_1/b_1$ with $p \nmid a_1 b_1$. Define the *p-adic valuation* to be

$$v_p(a/b) = r.$$

For example,

$$v_2(7/40) = -3, \quad v_5(50/3) = 2, \quad v_7(1/2) = 0.$$

Define $v_p(0) = +\infty$ (so $v_p(0) > n$ for every integer n).

Let $\tilde{E}$ be an elliptic curve over $\mathbf{Z}$ given by $y^2 = x^3 + \tilde{A}x + \tilde{B}$. Let $r \geq 1$ be an integer. Define

$$\tilde{E}_r = \{(x, y) \in \tilde{E}(\mathbf{Q}) \mid v_p(x) \leq -2r, \ v_p(y) \leq -3r\} \cup \{\infty\}.$$

These are the points such that x has at least p^{2r} in its denominator and y has at least p^{3r} in its denominator. These should be thought of as the points that are close to ∞ mod powers of p (that is, p-adically close to ∞).

THEOREM 5.6
Let $\tilde{E}$ be given by $y^2 = x^3 + \tilde{A}x + \tilde{B}$, with $\tilde{A}, \tilde{B} \in \mathbf{Z}$. Let p be prime and let r be a positive integer. Then

1. *$\tilde{E}_1$ is a subgroup of $\tilde{E}(\mathbf{Q})$.*

2. *If $(x, y) \in \tilde{E}(\mathbf{Q})$, then $v_p(x) < 0$ if and only if $v_p(y) < 0$. In this case, there exists an integer $r \geq 1$ such that $v_p(x) = -2r$, $v_p(y) = -3r$.*

3. *The map*

$$\lambda_r : \tilde{E}_r / \tilde{E}_{5r} \to \mathbf{Z}_{p^{4r}}$$
$$(x, y) \mapsto p^{-r}x/y \pmod{p^{4r}}$$
$$\infty \mapsto 0$$

   is an injective homomorphism (where $\mathbf{Z}_{p^{4r}}$ is a group under addition).

4. *If $(x, y) \in \tilde{E}_r$ but $(x, y) \notin \tilde{E}_{r+1}$, then $\lambda_r(x, y) \not\equiv 0 \pmod{p}$.*

This will be proved in Section 8.1. The map λ_r should be regarded as a logarithm for the group $\tilde{E}_r / \tilde{E}_{r+1}$ since it changes the law of composition in the group to addition in $\mathbf{Z}_{p^{4r}}$, just as the classical logarithm changes the composition law in the multiplicative group of positive real numbers to addition in $\mathbf{R}$.

We need one more fact, which is contained in Corollary 2.32: the reduction mod p map

$$\mathrm{red}_p : \tilde{E}(\mathbf{Q}) \longrightarrow \tilde{E} \pmod{p}$$
$$(x, y) \mapsto (x, y) \pmod{p} \quad \text{when } (x, y) \notin \tilde{E}_1$$
$$\tilde{E}_1 \to \{\infty\}$$

is a homomorphism. The kernel of red_p is $\tilde{E}_1$.

We are now ready for a theoretical version of the algorithm. We start with an elliptic curve E over $\mathbf{F}_p$ in Weierstrass form, and we have points P and Q on E. We want to find an integer k such that $Q = kP$ (assume $k \neq 0$). The crucial assumption is that E is anomalous, so $\#E(\mathbf{F}_p) = p$. Perform the following steps.

1. Lift E, P, Q to $\mathbf{Z}$ to obtain $\tilde{E}, \tilde{P}, \tilde{Q}$, as in Proposition 5.4.

2. Let $\tilde{P}_1 = p\tilde{P}, \tilde{Q}_1 = p\tilde{Q}$. Note that $\tilde{P}_1, \tilde{Q}_1 \in \tilde{E}_1$ since $\mathrm{red}_p(p\tilde{P}) = p \cdot \mathrm{red}_p(\tilde{P}) = \infty$ (this is where we use the fact that E is anomalous).

3. If $\tilde{P}_1 \in \tilde{E}_2$, choose new $\tilde{E}, \tilde{P}, \tilde{Q}$ and try again. Otherwise, let $\ell_1 = \lambda_1(\tilde{P}_1)$ and $\ell_2 = \lambda_1(\tilde{Q}_1)$. We have $k \equiv \ell_2/\ell_1 \pmod{p}$.

Why does this work? Let $\tilde{K} = k\tilde{P} - \tilde{Q}$. We have

$$\infty = kP - Q = \mathrm{red}_p(k\tilde{P} - \tilde{Q}) = \mathrm{red}_p(\tilde{K}).$$

Therefore $\tilde{K} \in \tilde{E}_1$, so $\lambda_1(\tilde{K})$ is defined and

$$\lambda_1(p\tilde{K}) = p\lambda_1(\tilde{K}) \equiv 0 \pmod{p}.$$

Therefore,

$$k\ell_1 - \ell_2 = \lambda_1(k\tilde{P}_1 - \tilde{Q}_1) = \lambda_1(kp\tilde{P} - p\tilde{Q}) = \lambda_1(p\tilde{K}) \equiv 0 \pmod{p}.$$

This means that $k \equiv \ell_2/\ell_1 \pmod{p}$, as claimed.

Note that the assumption that E is anomalous is crucial. If $E(\mathbf{F}_p)$ has order N, we need to multiply by N to put $\tilde{P}, \tilde{Q}$ into $\tilde{E}_1$, where λ_1 is defined. The difference $\tilde{K} = k\tilde{P} - \tilde{Q}$ gets multiplied by N, also. When N is a multiple of p, we have $\lambda_1(N\tilde{K}) \equiv 0 \pmod{p}$, so the contribution from $\tilde{K}$ disappears from our calculations.

If we try to implement the above algorithm, we soon encounter difficulties. If p is a large prime, the point $\tilde{P}_1$ has coordinates whose numerators and denominators are too large to work with. For example, the numerator and denominator of the x-coordinate usually have approximately p^2 digits (see Section 8.3). However, we are only looking for $x/y \pmod{p}$. As we shall see, it suffices to work with numbers mod p^2.

Let's try calculating on $\tilde{E} \pmod{p^2}$. When we compute $(x, y) = \tilde{P}_1 = p\tilde{P}$, we run into problems. Usually $\tilde{P}_1 \notin \tilde{E}_2$, so p, but not p^2, is in the denominator of x. Therefore, $\tilde{P}_1$ is "halfway" to infinity on $\tilde{E} \pmod{p^2}$ and we cannot write $\tilde{P}_1$ as a finite point $(x, y) \pmod{p^2}$ or as ∞. We could work in projective coordinates, as in Section 2.10. Instead, we calculate $(p-1)\tilde{P} \pmod{p^2}$, then add it to $\tilde{P}$, keeping track of p in denominators.

The procedure is the following.

1. Lift E, P, Q to $\mathbf{Z}$ to obtain $\tilde{E}, \tilde{P} = (x_1, y_1), \tilde{Q} = (x_2, y_2)$, as in Proposition 5.4.

2. Calculate
$$\tilde{P}_2 = (p-1)\tilde{P} \equiv (x', y') \pmod{p^2}.$$

The rational numbers in the calculation of $\tilde{P}_2$ should not have p in their denominators, so the denominators can be inverted mod p^2 to obtain integers x', y'.

3. Calculate $\tilde{Q}_2 = (p-1)\tilde{Q} \equiv (x'', y'') \pmod{p^2}$.

4. Compute
$$m_1 = p\frac{y' - y_1}{x' - x_1}, \quad m_2 = p\frac{y'' - y_2}{x'' - x_2}.$$

5. If $v_p(m_2) < 0$ or $v_p(m_1) < 0$, then try another $\tilde{E}$. Otherwise, $Q = kP$, where $k \equiv m_1/m_2 \pmod{p}$.

Example 5.5
Let E be the elliptic curve given by $y^2 = x^3 + 108x + 4$ over $\mathbf{F}_{853}$. Let $P = (0, 2)$ and $Q = (563, 755)$. It can be shown that $853P = \infty$. Since 853 is prime, the order of P is 853, so $853 | \#E(\mathbf{F}_{853})$. Hasse's theorem implies that $\#E(\mathbf{F}_{853}) = 853$, as in Section 4.3.3. Therefore, E is anomalous. Proposition 5.4 yields

$$\tilde{E} : y^2 = x^3 + 714069x + 4, \quad \tilde{P} = (0, 2), \quad \tilde{Q} = (563, 755).$$

We have

$$\tilde{P}_2 = 852\tilde{P} \equiv (525448, 365082) \pmod{853^2}$$
$$\tilde{Q}_2 = 852\tilde{Q} \equiv (543924, 505074) \pmod{853^2}.$$

Note that even with a prime as small as 853, writing $\tilde{P}_2$ without reducing mod 853^3 would require more than 100 thousand digits. We now calculate

$$m_1 = 853\frac{365082 - 2}{525448 - 0} = \frac{45635}{77} \quad \text{and} \quad m_2 = 853\frac{505074 - 755}{543924 - 563} = \frac{504319}{637}.$$

Therefore, $k \equiv m_1/m_2 \equiv 234 \pmod{853}$.   □

Let's prove this algorithm works (the proof consists mostly of keeping track of powers of p, and can be skipped without much loss). The following notation is useful. We write $O(p^k)$ to represent a rational number of the form $p^k z$ with $v_p(z) \geq 0$. Therefore, if $a, b \in \mathbf{Z}$ and $k > 0$, then $a = b + O(p^k)$ simply means that $a \equiv b \pmod{p^k}$. But we are allowing rational numbers and we are allowing negative k. For example,

$$\frac{1}{49} = \frac{23}{98} + O(7^{-1})$$

since

$$\frac{23}{98} = \frac{1}{49} + \frac{1}{7}\frac{3}{2}.$$

The following rule is useful:

$$\frac{a}{b + O(p^k)} = \frac{a}{b} + O(p^k) \quad \text{when } v_p(b) = 0, \ v_p(a) \geq 0, \text{ and } k > 0.$$

To prove it, simply rewrite the difference $\frac{a}{b + p^k z} - \frac{a}{b}$.

Write $\tilde{P}_2 = (p - 1)\tilde{P} = (u, v)$, with $u, v \in \mathbf{Q}$ (this is not yet mod p^2). Then

$$u = x' + O(p^2), \quad v = y' + O(p^2).$$

Let

$$(x, y) = \tilde{P}_1 = p\tilde{P} = \tilde{P} + \tilde{P}_2 = (x_1, y_1) + (u, v).$$

Then

$$x = \left(\frac{v - y_1}{v - x_1}\right)^2 - u - x_1 = \left(\frac{y' - y_1 + O(p^2)}{x' - x_1 + O(p^2)}\right)^2 - u - x_1.$$

We have $\tilde{P}_1 \in \tilde{E}_1$ and usually we have $\tilde{P}_1 \notin \tilde{E}_2$. This means that $x' - x_1$ is a multiple of p, but not of p^2 (note: $y' \not\equiv y_1 \pmod{p}$ since otherwise $(p-1)P = P$, which is not the case). We'll assume this is the case. Then

$$\frac{y' - y_1 + O(p^2)}{x' - x_1 + O(p^2)} = \frac{1}{p}\left(\frac{y' - y_1 + O(p^2)}{\frac{x' - x_1}{p} + O(p)}\right)$$

$$= \frac{1}{p}\left(\frac{y' - y_1}{\frac{x' - x_1}{p}} + O(p)\right)$$

$$= \frac{1}{p}m_1 + O(p^0).$$

Note that $v_p(m_1) = 0$. Since $v_p(u) \geq 0$ and $v_p(x_1) \geq 0$, we obtain

$$x = \left(\frac{1}{p}m_1 + O(p^0)\right)^2 - u - x_1 = \frac{m_1^2}{p^2} + O(p^{-1}).$$

Similarly, the y-coordinate of $\tilde{P}_1$ satisfies

$$y = -\frac{m_1^3}{p^3} + O(p^{-2}).$$

Therefore,

$$\ell_1 = \lambda_1(\tilde{P}_1) = \lambda_1(x, y) = p^{-1}\frac{x}{y} = -\frac{1}{m_1} + O(p) \equiv -\frac{1}{m_1} \pmod{p}.$$

Similarly,

$$\ell_2 = \lambda_1(\tilde{Q}_1) \equiv -\frac{1}{m_2} \pmod{p}.$$

If $v_p(m_2) < 0$, then $\tilde{Q}_1 \in \tilde{E}_2$ by Theorem 5.6, hence either $\tilde{P}_1 \in \tilde{E}_2$ or $k = 0$. We are assuming these cases do not happen, and therefore the congruence just obtained makes sense. Therefore,

$$k \equiv \frac{\ell_2}{\ell_1} \equiv \frac{m_1}{m_2} \pmod{p},$$

as claimed. This shows that the algorithm works.

  Anomalous curves are attractive from a computational viewpoint since calculating an integer multiple of a point in $E(\overline{\mathbf{F}}_q)$ can be done efficiently. In designing a cryptosystem, one therefore starts with an anomalous curve E over a small finite field $\mathbf{F}_q$ and works in $E(\mathbf{F}_{q^k})$ for a large k. Usually it is best to work with the subgroup generated by a point whose order ℓ is a large prime number. In particular, ℓ will be much larger than p, hence not equal to p. Therefore, the above attack on anomalous curves does not apply to the present situation.

Let E be an elliptic curve over $\mathbf{F}_q$ such that $\#E(\mathbf{F}_q) = q$. Then the trace of the Frobenius ϕ_q is $a = 1$, so

$$\phi_q^2 - \phi_q + q = 0.$$

This means that $q = \phi_q - \phi_q^2$. Therefore

$$q(x, y) = (x^q, y^q) + (x^{q^2}, -y^{q^2}) \text{ for all } (x, y) \in E(\overline{\mathbf{F}}_q).$$

The calculation of x^q, for example, can be done quickly in a finite field. Therefore, the expense of multiplying by q is little more than the expense of one addition of points. The standard method of computing $q(x, y)$ (see Section 2.2) involves more point additions (except when $q = 2$; but see Exercise 5.7). To calculate $k(x, y)$ for some integer k, expand $k = k_0 + k_1 q + k_2 q^2 + \cdots$ in base q. Compute $k_i P$ for each i, then compute $q^i k_i P$. Finally, add these together to obtain kP.

5.5   The Tate-Lichtenbaum Pairing

THEOREM 5.7

Let E be an elliptic curve over $\mathbf{F}_q$. Let n be an integer such that $n|q-1$. Let $E(\mathbf{F}_q)[n]$ denote the elements of $E(\mathbf{F}_q)$ of order dividing n, and let $\mu_n = \{x \in \mathbf{F}_q \mid x^n = 1\}$. Assume $E(\mathbf{F}_q)$ contains an element of order n. Then there are nondegenerate bilinear pairings

$$\langle \cdot, \cdot \rangle_n : E(\mathbf{F}_q)[n] \times E(\mathbf{F}_q)/nE(\mathbf{F}_q) \to \mathbf{F}_q^{\times}/(\mathbf{F}_q^{\times})^n$$

and

$$\tau_n : E(\mathbf{F}_q)[n] \times E(\mathbf{F}_q)/nE(\mathbf{F}_q) \to \mu_n.$$

The first pairing of the theorem is called the **Tate-Lichtenbaum pairing**. We'll call τ_n the **modified Tate-Lichtenbaum pairing**. The pairing τ_n is better suited for computations since it gives a definite answer, rather than a coset in $\mathbf{F}_q^{\times}$ mod nth powers. The pairings will be constructed in Section 11.3. They can be computed quickly (using at most a constant times $\log q$ point additions on E). See Section 11.4.

Technically, we should write $\tau_n(P, Q)$ as $\tau_n(P, Q+nE(\mathbf{F}_q))$, since an element of $E(\mathbf{F}_q)/nE(\mathbf{F}_q)$ has the form $Q + nE(\mathbf{F}_q)$. However, we'll simply write $\tau_n(P, Q)$ and similarly for $\langle P, Q \rangle_n$. The fact that τ_n is nondegenerate means that if $\tau_n(P, Q) = 1$ for all Q then $P = \infty$, and if $\tau_n(P, Q) = 1$ for all P then $Q \in nE(\mathbf{F}_q)$. Bilinearity means that

$$\tau_n(P_1 + P_1, Q) = \tau_n(P_1, Q)\tau_n(P_2, Q)$$

and

$$\tau_n(P, Q_1 + Q_2) = \tau_n(P, Q_1)\tau_n(P, Q_2).$$

The Tate-Lichtenbaum pairing is closely related to the Weil pairing. This will be discussed in Section 11.4. However, the Tate-Lichtenbaum pairing can be used in some situations where the Weil pairing does not apply. The Weil pairing needs $E[n] \subseteq E(\mathbf{F}_q)$, which implies that $\mu_n \subseteq \mathbf{F}_q^\times$, by Corollary 3.11. The Tate-Lichtenbaum pairing requires that $\mu_n \subseteq \mathbf{F}_q^\times$, but only needs a point of order n, rather than all of $E[n]$, to be in $E(\mathbf{F}_q)$.

Frey and Rück showed that in some situations, the Tate-Lichtenbaum pairing can be used to solve discrete logarithm problems (see [30] and also [29]). First, we need the following.

LEMMA 5.8

Let ℓ be a prime with $\ell | q - 1$, $\ell | \#E(\mathbf{F}_q)$, and $\ell^2 \nmid \#E(\mathbf{F}_q)$. Let P be a generator of $E(\mathbf{F}_q)[\ell]$. Then $\tau_\ell(P, P)$ is a primitive ℓth root of unity.

PROOF   If $\tau_\ell(P, P) = 1$, then $\tau_\ell(uP, P) = 1^u = 1$ for all $u \in \mathbf{Z}$. Since τ_ℓ is nondegenerate, $P \in \ell E(\mathbf{F}_q)$. Write $P = \ell P_1$. Then $\ell^2 P_1 = \ell P = \infty$. Since $\ell^2 \nmid \#E(\mathbf{F}_q)$, there are no points of order ℓ^2. Therefore P_1 must have order 1 or ℓ. In particular, $P = \ell P_1 = \infty$, which is a contradiction. Therefore $\tau_\ell(P, P) \neq 1$, so it must be a primitive ℓth root of unity.   ∎

Let $E(\mathbf{F}_q)$ and P be as in the lemma, and suppose $Q = kP$. Compute

$$\tau_\ell(P, Q) = \tau_\ell(P, P)^k.$$

Since $\tau_\ell(P, P)$ is a primitive ℓth root of unity, this determines $k \pmod{\ell}$. We have therefore reduced the discrete log problem to one in the multiplicative group of the finite field $\mathbf{F}_q$. Such discrete log problems are usually easier to solve.

Therefore, to choose a situation where the discrete log problem is hard, we should choose a situation where there is a point of order ℓ, where ℓ is a large prime, and such that $\ell \nmid q - 1$. In fact, as we show below, we should arrange that $q^k \not\equiv 1 \pmod{\ell}$ for small values of k.

Suppose $E(\mathbf{F}_q)$ has a point of order n, but $n \nmid q - 1$. We can extend our field to $\mathbf{F}_{q^m}$ so that $n | q^m - 1$. Then the Tate-Lichtenbaum pairing can be

used. However, the following proposition from [6] shows, at least in the case
n is prime, that the Weil pairing also can be used.

PROPOSITION 5.9

*Let E be an elliptic curve over $\mathbf{F}_q$. Let ℓ be a prime such that $\ell | \#E(\mathbf{F}_q)$,
$E[\ell] \not\subseteq E(\mathbf{F}_q)$, and $\ell \nmid q(q-1)$. Then*

$$E[\ell] \subseteq E(\mathbf{F}_{q^m}) \text{ if and only if } q^m \equiv 1 \pmod{\ell}.$$

PROOF If $E[\ell] \subseteq E(\mathbf{F}_{q^m})$, then $\mu_\ell \subseteq \mathbf{F}_{q^m}$ by Corollary 3.11, hence $q^m \equiv 1$
(mod ℓ).

Conversely, suppose $q^m \equiv 1 \pmod{\ell}$. Let $P \in E(\mathbf{F}_q)$ have order ℓ and let
$Q \in E[\ell]$ with $Q \notin E(\mathbf{F}_q)$. We claim that P and Q are independent points of
order ℓ. If not, then $uP = vQ$ for some integers $u, v \not\equiv 0 \pmod{p}$. Multiplying
by $v^{-1} \pmod{p}$, we find that $Q = v^{-1}uP \in E(\mathbf{F}_q)$, which is a contradiction.
Therefore $\{P, Q\}$ is a basis for $E[\ell]$.

Let ϕ_q be the Frobenius map. The action of ϕ_q on the basis $\{P, Q\}$ of
$E[\ell]$ gives us a matrix $(\phi_q)_\ell$, as in Section 3.1. Since $P \in E(\mathbf{F}_q)$, we have
$\phi_q(P) = P$. Let $\phi_q(Q) = bP + dQ$. Then

$$(\phi_q)_\ell = \begin{pmatrix} 1 & b \\ 0 & d \end{pmatrix}.$$

From Theorem 4.10, we know that

$$\mathrm{Trace}((\phi_q)_\ell) \equiv a = q + 1 - \#E(\mathbf{F}_q) \pmod{\ell}.$$

Since $\#E(\mathbf{F}_q) \equiv 0 \pmod{\ell}$ by assumption, we have

$$1 + d \equiv q + 1 \pmod{\ell},$$

so $d \equiv q \pmod{\ell}$. An easy induction shows that

$$\begin{pmatrix} 1 & b \\ 0 & q \end{pmatrix}^m = \begin{pmatrix} 1 & b\frac{q^m-1}{q-1} \\ 0 & q^m \end{pmatrix}.$$

Since $q \not\equiv 1 \pmod{\ell}$, by assumption, we have

$$\phi_q^m = 1 \text{ on } E[\ell] \iff \phi^m \equiv I \pmod{\ell} \iff q^m \equiv 1 \pmod{\ell}.$$

Since $E[\ell] \subseteq E(\mathbf{F}_{q^m})$ if and only if $\phi_q^m = 1$ on $E[\ell]$, by Lemma 4.5, this proves
the proposition. ∎

If we have $E[n] \subseteq E(\mathbf{F}_{q^m})$, then we can use the MOV attack or we can
use the Tate-Lichtenbaum pairing to reduce discrete log problems in $E(\mathbf{F}_{q^m})$
to discrete log problems in $\mathbf{F}_{q^m}^\times$. The Tate-Lichtenbaum pairing is generally
faster (see [32]). In both cases, we pick arbitrary points R and compute their
pairings with P and kP. With high probability (as in Section 5.3), we obtain
k after using only a few values of R.

5.6   Other Attacks

For arbitrary elliptic curves, Baby Step/Giant Step and the Pollard ρ and λ methods seem to be the best algorithms. There are a few cases where index calculus techniques can be used in the jacobians of higher genus curves to solve discrete logarithm problems on certain elliptic curves, but it is not clear how generally their methods apply. See [33], [34], [65]. See also [94] for a discussion of some other index calculus ideas and elliptic curves.

An interesting approach due to Silverman [93] is called the **xedni calculus**. Suppose we want to find k such that $Q = kP$ on a curve E over $\mathbf{F}_p$. Proposition 5.4 shows that we can lift E, P, and Q to an elliptic curve $\tilde{E}$ over $\mathbf{Z}$ with points $\tilde{P}$ and $\tilde{Q}$. If we can find k' with $\tilde{Q} = k'\tilde{P}$, then $Q = k'P$. However, it is usually the case that $\tilde{P}$ and $\tilde{Q}$ are independent, so no k' exists. Silverman's idea was to start with several (up to 9) points of the form $a_i P + b_i Q$ and lift them to a curve over $\mathbf{Q}$. This is possible: Choose a lift to $\mathbf{Z}$ for each of the points. Write down an arbitrary cubic curve containing lifts of the points. The fact that a point lies on the curve gives a linear equation in the coefficients of the cubic equation. Use linear algebra to solve for these coefficients. This curve can then be converted to Weierstrass form (see Section 2.5.2). Since most curves over $\mathbf{Q}$ tend to have at most 2 independent points, the hope was that there would be relations among the lifted points. These could then be reduced mod p to obtain relations between P and Q, thus solving the discrete log problem. Unfortunately, the curves obtained tend to have many independent points and no relations. Certain modifications that should induce the curve to have fewer independent points do not seem to work. For an analysis of the algorithm and why it probably is not successful, see [42].

Exercises

5.1   Suppose G is a subgroup of order N of the points on an elliptic curve over a field. Show that the following algorithm finds k such that $kP = Q$:

(a) Fix an integer $m \geq \sqrt{N}$.

(b) Compute and store a list of the x-coordinates of iP for $0 \leq i \leq m/2$.

(c) Compute the points $Q - jP$ for $j = 0, 1, 2, \cdots, m$ until the x-coordinate of one of them matches an element from the stored list.

(d) Decide whether $Q - jP = iP$ or $= -iP$.

(e) If $\pm iP = Q - jmP$, we have $Q = kP$ with $k \equiv \pm i + jm \pmod{N}$.

This requires a little less computation and half as much storage as the baby step, giant step algorithm in the text. It is essentially the same as the method used in Section 4.3.4 to find the order of $E(\mathbf{F}_q)$.

5.2 Let G be the additive group $\mathbf{Z}_n$. Explain why the discrete logarithm problem for G means solving $ka \equiv b \pmod{n}$ for k and describe how this can be solved quickly. This shows that the difficulty of a discrete logarithm problem depends on the group.

5.3 Let G be a group and let p be a prime. Suppose we have a fast algorithm for solving the discrete log problem for elements of order p (that is, given $g \in G$ of order p and $h = g^k$, there is a fast way to find k). Show that there is a fast algorithm for solving the discrete log problem for elements of order a power of p. (This is essentially what the Pohlig-Hellman method does. Since Pohlig-Hellman works with small primes, the fast algorithm for elements of order p in this case is simply brute force search.)

5.4 Let $p \geq 7$ be prime. Show that if E is an elliptic curve over $\mathbf{F}_p$ such that $E(\mathbf{F}_p)$ contains a point of order p, then $\#E(\mathbf{F}_p) = p$.

5.5 Show that if E is anomalous over $\mathbf{F}_q$ then it is not anomalous over $\mathbf{F}_{q^2}$.

5.6 Show that if E is anomalous over $\mathbf{F}_2$ then it is anomalous over $\mathbf{F}_{16}$.

5.7 Suppose E is anomalous over $\mathbf{F}_2$, so $\phi_2^2 - \phi_2 + 2 = 0$. Show that

(a) $4 = -\phi_2^3 - \phi_2^2$

(b) $8 = -\phi_2^3 + \phi_2^5$

(c) $16 = \phi_2^4 - \phi_2^8$

These equations were discovered by Koblitz [48], who pointed out that multiplication by each of 2, 4, 8, 16 in $E(\overline{\mathbf{Q}})$ can be accomplished by applying suitable powers of ϕ_2 (this is finite field arithmetic and is fast) and then performing only one point addition. This is faster than successive doubling for 4, 8, and 16.

5.8 Let E be defined over $\mathbf{F}_q$.

(a) Show that a map from $E(\mathbf{F}_q)$ to itself is injective if and only if it is surjective.

(b) Show that if $E(\mathbf{F}_q)$ has no point of order n, then $E(\mathbf{F}_q)/nE(\mathbf{F}_q) = 0$ (in which case, the Tate-Lichtenbaum pairing is trivial).

5.9 (a) Let ψ be a homomorphism from a finite group G to itself. Show that the index of $\psi(G)$ in G equals the order of the kernel of ψ.

(b) Let E be defined over $\mathbf{F}_q$. Show that $E(\mathbf{F}_q)[n]$ and $E(\mathbf{F}_q)/nE(\mathbf{F}_q)$ have the same order. (This can be proved from the nondegeneracy of the Tate-Lichtenbaum pairing. The point of the present exercise is to prove it without using this fact.)

5.10 This exercise gives a way to attack discrete logarithms using the Tate-Lichtenbaum pairing, even when there is a point of order ℓ^2 in $E(\mathbf{F}_q)$ (cf. Lemma 5.8). Assume ℓ is a prime such that $\ell | \#E(\mathbf{F}_q)$ and $\ell | q - 1$, and suppose that the ℓ-power torsion in $E(\mathbf{F}_q)$ is cyclic of order ℓ^i, with $i \geq 1$. Let P_i have order ℓ^i and let P have order ℓ.

(a) Show that $\tau_\ell(P, P_i)$ is a primitive ℓth root of unity.

(b) Suppose $Q = kP$. Show how to use (a) to reduce the problem of finding k to a discrete logarithm problem in $\mathbf{F}_q^\times$.

(c) Let $N = \#E(\mathbf{F}_q)$. Let R be a random point in $E(\mathbf{F}_q)$. Explain why $(N/\ell^i)R$ is very likely to be a point of order ℓ^i. This shows that finding a suitable point P_i is not difficult.

5.11 Let E be defined by $y^2 + y = x^3 + x$ over $\mathbf{F}_2$. Exercise 4.7 showed that $\#E(\mathbf{F}_2) = 5$, so E is supersingular and $\phi_2^2 + 2\phi_2 + 2 = 0$.

(a) Show that $\phi_2^4 = -4$.

(b) Show that $E[5] \subseteq E(\mathbf{F}_{16})$.

(c) Show that $\#E(\mathbf{F}_4) = 5$ and $\#E(\mathbf{F}_{16}) = 25$.

This example shows that Proposition 5.3 can fail when $a \neq 0$.

Chapter 6

Elliptic Curve Cryptography

In this chapter, we'll discuss several cryptosystems based on elliptic curves, especially on the discrete logarithm problem for elliptic curves. We'll also treat various related ideas, such as digital signatures.

One might wonder why elliptic curves are used in cryptographic situations. The reason is that elliptic curves provide security equivalent to classical systems while using fewer bits. For example, it is estimated in [7] that a key size of 4096 bits for RSA gives the same level of security as 313 bits in an elliptic curve system. This means that implementations of elliptic curve cryptosystems require smaller chip size, less power consumption, etc. Daswani and Boneh [9] performed experiments using 3Com's PalmPilot, which is a small hand-held device that is larger than a smart card but smaller than a laptop computer. They found that generating a 512-bit RSA key took 3.4 minutes, while generating a 163-bit ECC-DSA key to 0.597 seconds. Though certain procedures, such as signature verifications, were slightly faster for RSA, the elliptic curve methods such as ECC-DSA clearly offer great increases in speed in many situations.

6.1 The Basic Setup

Alice wants to send a message, often called the **plaintext**, to Bob. In order to keep the eavesdropper Eve from reading the message, she encrypts it to obtain the **ciphertext**. When Bob receives the ciphertext, he decrypts it and reads the message. In order to encrypt the message, Alice uses an **encryption key**. Bob uses a **decryption key** to decrypt the ciphertext. Clearly, the decryption key must be kept secret from Eve.

There are two basic types of encryption. In **symmetric encryption**, the encryption key and decryption key are the same, or one can be easily deduced from the other. Popular symmetric encryption methods include the Data Encryption Standard (DES) and the Advanced Encryption Standard (AES, often referred to by its original name *Rijndael*). In this case, Alice and Bob need to have some way of establishing a key. For example, Bob could send a messenger to Alice several days in advance. Then, when it is time to send the message, they both will have the key. Clearly this is impractical in many situations.

The other type of encryption is **public key encryption**, or asymmetric encryption. In this case, Alice and Bob do not need to have prior contact. Bob publishes a public encryption key, which Alice uses. He also has a private decryption key that allows him to decrypt ciphertexts. Since everyone knows the encryption key, it should be infeasible to deduce the decryption key from the encryption key. The most famous public key system is known as RSA and is based on the difficulty of factoring integers into primes. Another well-known system is due to ElGamal and is based on the difficulty of the discrete logarithm problem.

Generally, public key systems are slower than good symmetric systems. Therefore, it is common to use a public key system to establish a key that is then used in a symmetric system. The improvement in speed is important when massive amounts of data are being transmitted.

6.2   Diffie-Hellman Key Exchange

Alice and Bob want to agree on a common key that they can use for exchanging data via a symmetric encryption scheme such as DES or AES. For example, Alice and Bob could be banks that want to transmit financial data. It is impractical and time-consuming to use a courier to deliver the key. Moreover, we assume that Alice and Bob have had no prior contact and therefore the only communication channels between them are public. One way to establish a secret key is the following method, due to Diffie and Hellman (actually, they used multiplicative groups of finite fields).

1. Alice and Bob agree on an elliptic curve E over a finite field $\mathbf{F}_q$ such that the discrete logarithm problem is hard in $E(\mathbf{F}_q)$. They also agree on a point $P \in E(\mathbf{F}_q)$ such that the subgroup generated by P has large order (usually, the curve and point are chosen so that the order is a large prime).

2. Alice chooses a secret integer a, computes $P_a = aP$, and sends P_a to Bob.

3. Bob chooses a secret integer b, computes $P_b = bP$, and sends P_b to Alice.

4. Alice computes $aP_b = abP$.

5. Bob computes $bP_a = baP$.

6. Alice and Bob use some publicly agreed on method to extract a key from abP. For example, they could use the last 256 bits of the x-coordinate of abP as the key. Or they could evaluate a hash function at the x-coordinate.

The only information that the eavesdropper Eve sees is the curve E, the finite field $\mathbf{F}_q$, and the points P, aP, and bP. She therefore needs to solve the following:

DIFFIE-HELLMAN PROBLEM
Given P, aP, and bP in $E(\mathbf{F}_q)$, compute abP.

If Eve can solve discrete logs in $E(\mathbf{F}_q)$, then she can use P and aP to find a. Then she can compute $a(bP)$ to get abP. However, it is not known whether there is some way to compute abP without first solving a discrete log problem.

A related question is the following:

DECISION DIFFIE-HELLMAN PROBLEM
Given P, aP, and bP in $E(\mathbf{F}_q)$, and given a point $Q \in E(\mathbf{F}_q)$ determine whether or not $Q = abP$.

In other words, if Eve receives an anonymous tip telling her abP, can she verify that the information is correct?

The Diffie-Hellman problem and the Decision Diffie-Hellman problem can be asked for arbitrary groups. Originally, they appeared in the context of multiplicative groups $\mathbf{F}_q^{\times}$ of finite fields.

For elliptic curves, the Weil pairing can be used to solve the Decision Diffie-Hellman problem in some cases. We give one such example.

Let E be the curve $y^2 = x^3 + 1$ over $\mathbf{F}_q$, where $q \equiv 2 \pmod{3}$. By Proposition 4.31, E is supersingular. Let $\omega \in \mathbf{F}_{q^2}$ be a primitive third root of unity. Note that $\omega \notin \mathbf{F}_q$ since the order of $\mathbf{F}_q^{\times}$ is $q - 1$, which is not a multiple of 3. Define a map

$$\beta : E(\overline{\mathbf{F}}_q) \to E(\overline{\mathbf{F}}_q), \quad (x, y) \mapsto (\omega x, y), \quad \beta(\infty) = \infty.$$

It is straightforward to show, using the formulas for the addition law, that β is an isomorphism (Exercise 6.1).

Suppose $P \in E(\overline{\mathbf{F}}_q)$ has order n. Then $\beta(P)$ also has order n. Define the modified Weil pairing

$$\tilde{e}_n(P_1, P_2) = e_n(P_1, \beta(P_2)),$$

where e_n is the usual Weil pairing and $P_1, P_2 \in E[n]$.

LEMMA 6.1
Assume $3 \nmid n$. If $P \in E(\mathbf{F}_q)$ has order exactly n, then $\tilde{e}_n(P, P)$ is a primitive nth root of unity.

PROOF   Suppose $uP = v\beta(P)$ for some integers u, v. Then

$$\beta(vP) = v\beta(P) = uP \in E(\mathbf{F}_q).$$

If $vP = \infty$, then $uP = \infty$, so $u \equiv 0 \pmod{n}$. If $vP \neq \infty$, write $vP = (x, y)$ with $x, y \in \mathbf{F}_q$. Then

$$(\omega x, y) = \beta(vP) \in E(\mathbf{F}_q).$$

Since $\omega \notin \mathbf{F}_q$, we must have $x = 0$. Therefore $vP = (0, \pm 1)$, which has order 3. This is impossible since we have assumed that $3 \nmid n$. It follows that the only relation of the form $uP = v\beta(P)$ has $u, v \equiv 0 \pmod{n}$, so P and $\beta(P)$ form a basis of $E[n]$. By Corollary 3.10, $\tilde{e}_n(P, P) = e_n(P, \beta(P))$ is a primitive nth root of unity.   ∎

   Suppose now that we know P, aP, bP, Q and we want to decide whether or not $Q = abP$. First, use the usual Weil pairing to decide whether or not Q is a multiple of P. By Lemma 5.1, Q is a multiple of P if and only if $e_n(P, Q) = 1$. Assume this is the case, so $Q = tP$ for some t. We have

$$\tilde{e}_n(aP, bP) = \tilde{e}_n(P, P)^{ab} = \tilde{e}_n(P, abP) \quad \text{and} \quad \tilde{e}_n(Q, P) = \tilde{e}_n(P, P)^t.$$

Assume $3 \nmid n$. Then $\tilde{e}_n(P, P)$ is a primitive nth root of unity, so

$$Q = abP \iff t \equiv ab \pmod{n} \iff \tilde{e}_n(aP, bP) = \tilde{e}_n(Q, P).$$

This solves the Decision Diffie-Hellman problem in this case. Note that we did not need to compute any discrete logs, even in finite fields. All that was needed was to compute the Weil pairing.

   The above method was pointed out by Joux and Nguyen. For more on the Decision Diffie-Hellman problem, see [8].

   Joux [43] has given another application of the modified Weil pairing to what is known as **tripartite Diffie-Hellman** key exchange. Suppose Alice, Bob, and Chris want to establish a common key. The standard Diffie-Hellman procedure requires two rounds of interaction. The modified Weil pairing allows this to be cut to one round. As above, let E be the curve $y^2 = x^3 + 1$ over $\mathbf{F}_q$, where $q \equiv 2 \pmod{3}$. Let P be a point of order n. Usually, n should be chosen to be a large prime. Alice, Bob, and Chris do the following.

1. Alice, Bob, and Chris choose secret integers a, b, c mod n, respectively.

2. Alice broadcasts aP, Bob broadcasts bP, and Chris broadcasts cP.

3. Alice computes $\tilde{e}_n(bP, cP)^a$, Bob computes $\tilde{e}_n(aP, cP)^b$, and Chris computes $\tilde{e}_n(aP, bP)^c$.

4. Since each of the three users has computed the same number, they use this number to produce a key, using some publicly prearranged method.

Recall that, since E is supersingular, the discrete log problem on E can be reduced to a discrete log problem for $\mathbf{F}_{q^2}^{\times}$ (see Section 5.3). Therefore, q should be chosen large enough that this discrete log problem is hard.

For more on cryptographic applications of pairings, see [44].

6.3   Massey-Omura Encryption

Alice wants to send a message to Bob over public channels. They have not yet established a private key. One way to do this is the following. Alice puts her message in a box and puts her lock on it. She sends the box to Bob. Bob puts his lock on it and sends it back to Alice. Alice then takes her lock off and sends the box back to Bob. Bob then removes his lock, opens the box, and reads the message.

This procedure can be implemented mathematically as follows.

1. Alice and Bob agree on an elliptic curve E over a finite field $\mathbf{F}_q$ such that the discrete log problem is hard in $E(\mathbf{F}_q)$. Let $N = \#E(\mathbf{F}_q)$.

2. Alice represents her message as a point $M \in E(\mathbf{F}_q)$. (We'll discuss how to do this below.)

3. Alice chooses a secret integer m_A with $\gcd(m_A, N) = 1$, computes $M_1 = m_A M$, and sends M_1 to Bob.

4. Bob chooses a secret integer m_B with $\gcd(m_B, N) = 1$, computes $M_2 = m_B M_1$, and sends M_2 to Alice.

5. Alice computes $m_A^{-1} \in \mathbf{Z}_N$. She computes $M_3 = m_A^{-1} M_2$ and sends M_3 to Bob.

6. Bob computes $m_B^{-1} \in \mathbf{Z}_N$. He computes $M_4 = m_B^{-1} M_3$. Then $M_4 = M$ is the message.

Let's show that M_4 is the original message M. Formally, we have

$$M_4 = m_B^{-1} m_A^{-1} m_B m_A M = M,$$

but we need to justify the fact that m_A^{-1}, which is an integer representing the inverse of $m_A \mod N$, and m_A cancel each other. We have $m_A^{-1} m_A \equiv 1$

(mod N), so $m_A^{-1} m_A = 1 + kN$ for some k. The group $E(\mathbf{F}_q)$ has order N, so Lagrange's theorem implies that $NR = \infty$ for any $R \in E(\mathbf{F}_q)$. Therefore,

$$m_A^{-1} m_A R = (1 + kN)R = R + k\infty = R.$$

Applying this to $R = m_B M$, we find that

$$M_3 = m_A^{-1} m_B m_A M = m_B M.$$

Similarly, m_B^{-1} and m_B cancel, so

$$M_4 = m_B^{-1} M_3 = m_B^{-1} m_B M = M.$$

The eavesdropper Eve knows $E(\mathbf{F}_q)$ and the points $m_A M$, $m_B m_A M$, and $m_B M$. Let $a = m_A^{-1}, b = m_B^{-1}, P = m_A m_B M$. Then we see that Eve knows P, bP, aP and wants to find abP. This is the Diffie-Hellman problem (see Section 6.2).

The above procedure works in any finite group. It seems that the method is rarely used in practice.

It remains to show how to represent a message as a point on an elliptic curve. We use a method proposed by Koblitz. Suppose E is an elliptic curve given by $y^2 = x^3 + Ax + B$ over $\mathbf{F}_p$. The case of an arbitrary finite field $\mathbf{F}_q$ is similar. Let m be a message, expressed as a number $0 \le m < p/100$. Let $x_j = 100m + j$ for $0 \le j < 100$. For $j = 0, 1, 2, \ldots, 99$, compute $s_j = x_j^3 + Ax_j + B$. If $s_j^{(p-1)/2} \equiv 1 \pmod{p}$, then s_j is a square mod p, in which case we do not need to try any more values of j. When $p \equiv 3 \pmod 4$, a square root of s_j is then given by $y_j \equiv s_j^{(p+1)/4} \pmod p$ (see Exercise 6.7). When $p \equiv 1 \pmod 4$, a square root of s_j can also be computed, but the procedure is more complicated (see [19]). We obtain a point (x_j, y_j) on E. To recover m from (x_j, y_j), simply compute $[x_j/100]$ (= the greatest integer less than or equal to $x_j/100$). Since s_j is essentially a random element of $\mathbf{F}_p^\times$, which is cyclic of even order, the probability is approximately $1/2$ that s_j is a square. So the probability of not being able to find a point for m after trying 100 values is around 2^{-100}.

6.4   ElGamal Public Key Encryption

Alice wants to send a message to Bob. First, Bob establishes his public key as follows. He chooses an elliptic curve E over a finite field $\mathbf{F}_q$ such that the discrete log problem is hard for $E(\mathbf{F}_q)$. He also chooses a point P on E (usually, it is arranged that the order of P is a large prime). He chooses a secret integer s and computes $B = sP$. The elliptic curve E, the finite field

$\mathbf{F}_q$, and the points P and B are Bob's public key. They are made public. Bob's private key is the integer s.

To send a message to Bob, Alice does the following:

1. Downloads Bob's public key.

2. Expresses her message as a point $M \in \mathbf{E}(\mathbf{F}_q)$.

3. Chooses a secret random integer k and computes $M_1 = kP$.

4. Computes $M_2 = M + kB$.

5. Sends M_1, M_2 to Bob.

Bob decrypts by calculating

$$M = M_2 - sM_1.$$

This decryption works because

$$M_2 - sM_1 = (M + kB) - s(kP) = M + k(sP) - skP = M.$$

The eavesdropper Eve knows Bob's public information and the points M_1 and M_2. If she can calculate discrete logs, she can use P and B to find s, which she can then use to decrypt the message as $M_2 - sM_1$. Also, she could use P and M_1 to find k. Then she can calculate $M = M_2 - kB$. If she cannot calculate discrete logs, there does not appear to be a way to find M.

It is important for Alice to use a different random k each time she sends a message to Bob. Suppose Alice uses the same k for both M and M'. Eve recognizes this because then $M_1 = M_1'$. She then computes $M_2' - M_2 = M' - M$. Suppose M is a sales announcement that is made public a day later. Then Eve finds out M, so she calculates $M' = M - M_2 + M_2'$. Therefore, knowledge of one plaintext M allows Eve to deduce another plaintext M' in this case.

6.5   ElGamal Digital Signatures

Alice wants to sign a document. The classical way is to write her signature on a piece of paper containing the document. Suppose, however, that the document is electronic, for example, a computer file. The naive solution would be to digitize Alice's signature and append it to the file containing the document. In this case, evil Eve can copy the signature and append it to another document. Therefore, steps must be taken to tie the signature to the document in such a way that it cannot be used again. However, it must be possible for someone to verify that the signature is valid, and it should be possible to show that Alice must have been the person who signed the document. One solution to the problem relies on the difficulty of discrete logs. Classically, the algorithm was developed for the multiplicative group of a finite field. In fact, it applies to any finite group. We'll present it for elliptic curves.

Alice first must establish a public key. She chooses an elliptic curve E over a finite field $\mathbf{F}_q$ such that the discrete log problem is hard for $E(\mathbf{F}_q)$. She also chooses a point $A \in E(\mathbf{F}_q)$. Usually the choices are made so that the order N of A is a large prime. Alice also chooses a secret integer a and computes $B = aA$. Finally, she chooses a function

$$f : E(\mathbf{F}_q) \to \mathbf{Z}.$$

For example, if $\mathbf{F}_q = \mathbf{F}_p$, then she could use $f(x, y) = x$, where x is regarded as an integer, $0 \leq x < p$. The function f needs no special properties, except that its image should be large and only a small number of inputs should produce any given output (for example, for $f(x, y) = x$, at most two points (x, y) yield a given output x).

Alice's public information is E, $\mathbf{F}_q$, f, A, and B. She keeps a private. The integer N does not need to be made public. Its secrecy does not affect our analysis of the security of the system. To sign a document, Alice does the following:

1. Represents the document as an integer m (if $m > N$, choose a larger curve, or use a hash function (see below)).

2. Chooses a random integer k with $\gcd(k, N) = 1$ and computes $R = kA$.

3. Computes $s \equiv k^{-1}(m - af(R)) \pmod{N}$.

The signed message is (m, R, s). Note that m, s are integers, while R is a point on E. Also, note that Alice is not trying to keep the document m secret. If she wants to do that, then she needs to use some form of encryption. Bob verifies the signature as follows:

1. Downloads Alice's public information.

2. Computes $V_1 = f(R)B + sR$ and $V_2 = mA$.

3. If $V_1 = V_2$, he declares the signature valid.

If the signature is valid, then $V_1 = V_2$ since

$$V_1 = f(R)B + sR = f(R)aA + skA = f(R)aA + (m - af(R))A = mA = V_2.$$

We have used the fact that $sk \equiv m - af(R)$, hence $sk = m - af(R) + zN$ for some integer z. Therefore,

$$skA = (m - af(R))A + zNA = (m - af(R))A + \infty = (m - af(R))A.$$

This is why the congruence defining s was taken mod N.

If Eve can calculate discrete logs, then she can use A and B to find a. In this case, she can put Alice's signature on any message. Alternatively, Eve can use A and R to find k. Since she knows $s, f(R), m$, she can then use $ks \equiv m - af(R) \pmod{N}$ to find a. If $d = \gcd(f(R), N) \neq 1$, then $af(R) \equiv m - ks \pmod{N}$ has d solutions for a. As long as d is small, Eve can try each possibility until she obtains $B = aA$. Then she can use a, as before, to forge Alice's signature on arbitrary messages.

As we just saw, Alice must keep a and k secret. Also, she must use a different random k for each signature. Suppose she signs m and m' using the same k to obtain signed messages (m, R, s) and (m', R, s'). Eve immediately recognizes that k has been used twice since R is the same for both signatures. The equations for s, s' yield the following:

$$ks \equiv m - af(R) \pmod{N}$$
$$ks' \equiv m' - af(R) \pmod{N}$$

Subtracting yields $k(s-s') \equiv m-m' \pmod{N}$. Let $d = \gcd(s-s', N)$. There are d possible values for k. Eve tries each one until $R = kA$ is satisfied. Once she knows k, she can find a, as above.

It is perhaps not necessary for Eve to solve discrete log problems in order to forge Alice's signature on another message m. All Eve needs to do is produce R, s such that the verification equation $V_1 = V_2$ is satisfied. This means that she needs to find $R = (x, y)$ and s such that

$$f(R)B + sR = mA.$$

If she chooses some point R (there is no need to choose an integer k), she needs to solve the discrete log problem $sR = mA - f(R)B$ for the integer s. If, instead, she chooses s, then she must solve an equation for $R = (x, y)$. This equation appears to be at least as complex as a discrete log problem, though it has not been analyzed as thoroughly. Moreover, no one has been able to rule out the possibility of using some procedure that finds R and s simultaneously. There are ways of using a valid signed message to produce another valid signed message (see Exercise 6.2). However, the messages produced are unlikely to be meaningful messages.

The general belief is that the security of the ElGamal system is very close to the security of discrete logs for the group $E(\mathbf{F}_q)$.

A disadvantage of the ElGamal system is that the signed message (m, R, s) is approximately three times as long as the original message (it is not necessary to store the full y-coordinate of R since there are only two choices for y for a given x). A more efficient method is to choose a public hash function H and sign $H(m)$. A **cryptographic hash function** is a function that takes inputs of arbitrary length, sometimes a message of billions of bits, and outputs values of fixed length, for example, 160 bits. A hash function H should have the following properties:

1. Given a message m, the value $H(m)$ can be calculated very quickly.

2. Given y, it is computationally infeasible to find m with $H(m) = y$. (This says that H is **preimage resistant**.)

3. It is computationally infeasible to find distinct messages m_1 and m_2 with $h(m_1) = h(m_2)$. (This says that H is **strongly collision-free**.)

The reason for (2) and (3) is to prevent Eve from producing messages with a desired hash value, or two messages with the same hash value. This helps prevent forgery. There are several good hash functions available, for example, MD5 (due to Rivest; it produces a 128-bit output) and the Secure Hash Algorithm (from NIST; it produces a 160-bit output) We won't discuss these here. For details, see [67].

If Alice uses a hash function, the signed message is then

$$(m, R_H, s_H),$$

where $(H(m), R_H, s_H)$ is a valid signature. To verify that the signature (m, R_H, s_H) is valid, Bob does the following:

1. Downloads Alice's public information.

2. Computes $V_1 = f(R)B + sR$ and $V_2 = H(m)A$.

3. If $V_1 = V_2$, he declares the signature valid.

The advantage is that a very long message m containing billions of bits has a signature that requires only a few thousand extra bits. As long as the discrete log problem is hard for $E(\mathbf{F}_q)$, Eve will be unable to put Alice's signature on another message. The use of a hash function also guards against certain other forgeries (see Exercise 6.2).

6.6   The Digital Signature Algorithm

The Digital Signature Standard [1],[70] is based on the Digital Signature Algorithm (DSA). The original version used multiplicative groups of finite fields. A more recent elliptic curve version (ECDSA) uses elliptic curves. The algorithm is a variant on the ElGamal signature scheme, with some modifications. We sketch the algorithm here.

Alice wants to sign a document m, which is an integer (actually, she usually signs the hash of the document, as in Section 6.5). Alice chooses an elliptic curve over a finite field $\mathbf{F}_q$ such that $\#E(\mathbf{F}_q) = fr$, where r is a large prime and f is a small integer, usually 1,2, or 4 (f should be small in order to keep the algorithm efficient). She chooses a base point G in $E(\mathbf{F}_q)$ of order r.

Finally, Alice chooses a secret integer a and computes $Q = aG$. Alice makes public the following information:

$$\mathbf{F}_q, \quad E, \quad r, \quad G, \quad Q.$$

(There is no need to keep f secret; it can be deduced from q and r using Hasse's theorem by the technique in Examples 4.6 and 4.7.) To sign the message m Alice does the following:

1. Chooses a random integer k with $1 \le k < r$ and computes $R = kG = (x, y)$.

2. Computes $s = k^{-1}(m + ax) \pmod{r}$.

The signed document is

$$(m, R, s).$$

To verify the signature, Bob does the following.

1. Computes $u_1 = s^{-1}m \pmod{r}$ and $u_2 = s^{-1}x \pmod{r}$.

2. Computes $V = u_1 G + u_2 Q$.

3. Declares the signature valid if $V = R$.

If the message is signed correctly, the verification equation holds:

$$V = u_1 G + u_2 Q = s^{-1}mG + s^{-1}xQ = s^{-1}(mG + xaG) = kG = R.$$

The main difference between the ECDSA and the ElGamal system is the verification procedure. In the ElGamal system, the verification equation $f(R)B + sR = mA$ requires three computations of an integer times a point. These are the most expensive parts of the algorithm. In the ECDSA, only two computations of an integer times a point are needed. If many verifications are going to be made, then the improved efficiency of the ECDSA is valuable.

6.7 A Public Key Scheme Based on Factoring

Most cryptosystems using elliptic curves are based on the discrete log problem, in contrast to the situation for classical systems, which are sometimes based on discrete logs and sometimes based on the difficulty of factorization. The most famous public key cryptosystem is called RSA (for Rivest-Shamir-Adleman) and proceeds as follows. Alice wants to send a message to Bob. Bob secretly chooses two large primes p, q and multiplies them to obtain $n = pq$. Bob also chooses integers e and d with $ed \equiv 1 \pmod{(p-1)(q-1)}$. He makes n and e public and keeps d secret. Alice's message is a number $m \pmod{n}$.

She computes $c \equiv m^e \pmod{n}$ and sends c to Bob. Bob computes $m \equiv c^d$ (mod n) to obtain the message. If Eve can find p and q, then she can solve $ed \equiv 1 \pmod{(p-1)(q-1)}$ to obtain d. It can be shown (by methods similar to those used in the elliptic curve scheme below; see [101]) that if Eve can find the decryption exponent d, then she probably can factor n. Therefore, the difficulty of factoring n is the key to the security of the RSA system.

A natural question is whether there is an elliptic curve analogue of RSA. In the following, we present one such system, due to Koyama-Maurer-Okamoto-Vanstone. It does not seem to be used much in practice.

Alice want to send a message to Bob. They do the following.

1. Bob chooses two distinct large primes p, q with $p \equiv q \equiv 2 \pmod 3$ and computes $n = pq$.

2. Bob chooses integers e, d with $ed \equiv 1 \pmod{\mathrm{lcm}(p+1, q+1)}$. (He could use $(p+1)(q+1)$ in place of $\mathrm{lcm}(p+1, q+1)$.)

3. Bob makes n and e public (they form his public key) and he keeps d, p, q private.

4. Alice represents her message as a pair of integers $(m_1, m_2) \pmod n$. She regards (m_1, m_2) as a point M on the elliptic curve E given by

$$y^2 = x^3 + b \mod n,$$

where $b = m_2^2 - m_1^3 \pmod n$ (she does not need to compute b).

5. Alice adds M to itself e times on E to obtain $C = (c_1, c_2) = eM$. She sends C to Bob.

6. Bob computes $M = dC$ on E to obtain M.

We'll discuss the security of the system shortly. But, first, there are several points that need to be discussed.

1. Note that the formulas for the addition law on E never use the value of b. Therefore, Alice and Bob never need to compute it. Eve can compute it, if she wants, as $b = c_2^2 - c_1^3$.

2. The computation of eM and dC on E are carried out with the formulas for the group law on an elliptic curve, with all of the computations being done mod n. Several times during the computation, expressions such as $(y_2 - y_1)/(x_2 - x_1)$ are encountered. These are changed to integers mod n by finding the multiplicative inverse of $(x_2 - x_1)$ mod n. This requires $\gcd(x_2 - x_1, n) = 1$. If the gcd is not 1, then it is p, q, or n. If we assume it is very hard to factor n, then we regard the possibility of the gcd being p or q as very unlikely. If the gcd is n, then the slope is infinite and the sum of the points in question is ∞. The usual rules

for working with ∞ are followed. For technical details of working with elliptic curves mod n, see Section 2.10.

By the Chinese Remainder Theorem, an integer mod n may be regarded as a pair of integers, one mod p and one mod q. Therefore, we can regard a point on E in $\mathbf{Z}_n$ as a pair of points, one on E mod p and the other on E mod q. In this way, we have

$$E(\mathbf{Z}_n) = E(\mathbf{F}_p) \oplus E(\mathbf{F}_q). \tag{6.1}$$

For example, the point $(11, 32)$ on $y^2 = x^3 + 8 \bmod 35$ can be regarded as the pair of points

$$(1, 2) \quad \bmod 5, \quad (4, 4) \quad \bmod 7.$$

Any such pair of points can be combined to obtain a point mod n. There is a technicality with points at infinity, which is discussed in Section 2.10.

3. Using Equation 6.1, we see that the order of $E(\mathbf{Z}_n)$ is $\#E(\mathbf{F}_p) \cdot \#E(\mathbf{F}_q)$. By Proposition 4.31, E is supersingular mod p and mod q, so we find (by Corollary 4.30) that

$$\#E(\mathbf{F}_p) = p + 1 \text{ and } \#E(\mathbf{F}_q) = q + 1.$$

Therefore, $(p+1)M = \infty \pmod{p}$ and $(q+1)M = \infty \pmod{q}$. This means that the decryption works: Write $de = 1 + k(p+1)$ for some integer k. Then

$$dC = deM = (1+k(p+1))M = M+k(p+1)M = M+\infty = M \quad (\bmod\ p),$$

and similarly mod q. Therefore, $dC = M$.

4. A key point of the procedure is that the group order is independent of b. If Bob chooses a random elliptic curve $y^2 = x^3 + Ax + B$ over Z_n, then he has to compute the group order, perhaps by computing it mod p and mod q. This is infeasible if p and q are chosen large enough to make factoring n infeasible. Also, if Bob fixes the elliptic curve, Alice will have difficulty finding points M on the curve. If she does the procedure of first choosing the x-coordinate as the message, then solving $y^2 \equiv m^3 + Am + B \pmod{n}$ for y, she is faced with the problem of computing square roots mod n. This is computationally equivalent to factoring n (see [101]). If Bob fixes only A (the formulas for the group operations depend only on A) and allows Alice to choose B so that her point lies on the curve, then his choice of e, d requires that the group order be independent of B. This is the situation in the above procedure.

If Eve factors n as pq, then she knows $(p+1)(q+1)$, so she can find d with $ed \equiv 1 \pmod{(p+1)(q+1)}$. Therefore, she can decrypt Alice's message.

Suppose that Eve does not yet know the factorization of n, but she finds out the decryption exponent d. We claim that she can, with high probability, factor n. She does the following:

1. Writes $ed - 1 = 2^k v$ with v odd and with $k \geq 1$ ($k \neq 0$ since $p+1$ divides $ed - 1$).

2. Picks a random pair of integers $R = (r_1, r_2) \mod n$, lets $b' = r_2^2 - r_1^3$, and regards R as a point on the elliptic curve E' given by $y^2 = x^3 + b'$.

3. Computes $R_0 = vR$. If $R_0 = \infty \mod n$, start over with a new R. If R_0 is $\infty \mod$ exactly one of p, q, then Eve has factored n (see below).

4. For $i = 0, 1, 2, \ldots, k$, computes $R_{i+1} = 2R_i$.

5. If for some i, the point R_{i+1} is $\infty \mod$ exactly one of p, q, then $R_i = (x_i, y_i)$ with $y_i \equiv 0 \mod$ one of p, q. Therefore, $\gcd(y_i, n) = p$ or q. In this case, Eve stops, since she has factored n.

6. If for some i, $R_{i+1} = \infty \mod n$, then Eve starts over with a new random point.

In a few iterations, this should factor n. Since $ed - 1$ is a multiple of $\#E(\mathbf{Z}_n)$,

$$R_k = (ed - 1)R = edR - R = \infty.$$

Therefore, each iteration of the procedure will eventually end with a point R_j that is $\infty \mod$ at least one of p, q. Let $2^{k'}$ be the highest power of 2 dividing $p+1$. If we take a random point P in $E(\mathbf{F}_p)$, then the probability is $1/2$ that the order of P is divisible by $2^{k'}$. This follows easily from the fact that $E(\mathbf{F}_p)$ is cyclic (see Exercise 6.6). In this case, $R_{k'-1} = 2^{k'-1}vP \neq \infty \pmod{p}$, while $R_{k'} = 2^{k'}vP = \infty \pmod{p}$. If the order is not divisible by $2^{k'}$, then $R_{k'-1} = \infty \pmod{p}$. Similarly, if $2^{k''}$ is the highest power of 2 dividing $q+1$, then $R_{k''-1} = \infty \pmod{q}$ half the time, and $\neq \infty \pmod{q}$ half the time. Since $\mod p$ and $\mod q$ are independent, it is easy to see that the sequence $R_0, R_1, R_2, \ldots$ reaches $\infty \mod p$ and $\mod q$ at different indices i at least half the time. This means that for at least half of the choices of random starting points R, we obtain a factorization of n.

If $R_0 = \infty \mod p$, but not $\mod q$, then somewhere in the calculation of R_0 there was a denominator of a slope that was infinite $\mod p$ but not $\mod q$. The gcd of this denominator with n yields p. A similar situation occurs if p and q are switched. Therefore, if R_0 is infinite $\mod$ exactly one of the primes, Eve obtains a factorization, as claimed in step (3).

We conclude that knowledge of the decryption exponent d is computationally equivalent to knowledge of the factorization of n.

6.8   A Cryptosystem Based on the Weil Pairing

In Chapter 5, we saw how the Weil pairing could be used to reduce the discrete log problem on certain elliptic curves to the discrete log problem for the multiplicative group of a finite field. In the present section, we'll present a method, due to Boneh and Franklin, that uses the Weil pairing on these curves to obtain a cryptosystem. The reader may wonder why we use these curves, since the discrete log problem is easier on these curves. The reason is that the properties of the Weil pairing are used in an essential way. The fact that the Weil pairing can be computed quickly is vital for the present algorithm. This fact was also important in reducing the discrete log problem to finite fields. However, note that the discrete log problem in the finite field is still not trivial as long as the finite field is large enough.

For simplicity, we'll consider a specific curve, namely the one discussed in Section 6.2. Let E be defined by $y^2 = x^3 + 1$ over $\mathbf{F}_p$, where $p \equiv 2 \pmod 3$. Let $\omega \in \mathbf{F}_{p^2}$ be a primitive third root of unity. Define a map

$$\beta : E(\mathbf{F}_{p^2}) \to E(\mathbf{F}_{p^2}), \quad (x, y) \mapsto (\omega x, y), \quad \beta(\infty) = \infty.$$

Suppose P has order n. Then $\beta(P)$ also has order n. Define the modified Weil pairing

$$\tilde{e}_n(P_1, P_2) = e_n(P_1, \beta(P_2)),$$

where e_n is the usual Weil pairing and $P_1, P_2 \in E[n]$. We showed in Lemma 6.1 that if $3 \nmid n$ and if $P \in E(\mathbf{F}_p)$ has order exactly n, then $\tilde{e}_n(P, P)$ is a primitive nth root of unity.

Since E is supersingular, by Proposition 4.31, $E(\mathbf{F}_p)$ has order $p+1$. We'll add the further assumption that $p = 6\ell - 1$ for some prime ℓ. Then $6P$ has order ℓ or 1 for each $P \in E(\mathbf{F}_p)$.

In the system we'll describe, each user has a public key based on her or his identity, such as an email address. A central trusted authority assigns a corresponding private key to each user. In most public key systems, when Alice wants to send a message to Bob, she looks up Bob's public key. However, she needs some way of being sure that this key actually belongs to Bob, rather than someone such as Eve who is masquerading as Bob. In the present system, the authentication happens in the initial communication between Bob and the trusted authority. After that, Bob is the only one who has the information necessary to decrypt messages that are encrypted using his public identity.

A natural question is why RSA cannot be used to produce such a system. For example, all users could share the same common modulus n, whose factorization is known only to the trusted authority (TA). Bob's identity, call it *bobid*, would be his encryption exponent. The TA would then compute Bob's secret decryption exponent and communicate it to him. When Alice sends Bob a message m, she encrypts it as $m^{idbob} \pmod n$. Bob then decrypts using the secret exponent provided by the TA. However, anyone such as Bob who knows an encryption and decryption exponent can find the factorization of n (using a variation of the method of Section 6.7), and thus read all messages in the system. Therefore, the system would not protect secrets. If, instead, a different n is used for each user, some type of authentication procedure is needed for a communication in order to make sure that the n is the correct one. This brings us back to the original problem.

The system described in the following gives the basic idea, but is not secure against certain attacks. For ways to strengthen the system, see [10].

To set up the system, the trusted authority does the following:

1. Chooses a large prime $p = 6\ell - 1$ as above.

2. Chooses a point P of order ℓ in $E(\mathbf{F}_p)$.

3. Chooses hash functions H_1 and H_2. The function H_1 takes a string of bits of arbitrary length and outputs a point of order ℓ on E (see Exercise 6.8). The function H_2 inputs an element of order ℓ in $\mathbf{F}_{p^2}^{\times}$ and outputs a binary string of length n, where n is the length of the messages that will be sent.

4. Chooses a secret random $s \in \mathbf{F}_{\ell}^{\times}$ and computes $P_{pub} = sP$.

5. Makes $p, H_1, H_2, n, P, P_{pub}$ public, while keeping s secret.

If a user with identity ID wants a private key, the trusted authority does the following:

1. Computes $Q_{ID} = H_1(ID)$. This is a point on E.

2. Lets $D_{ID} = sQ_{ID}$.

3. After verifying that ID is the identification for the user with whom he is communicating, sends D_{ID} to this user.

If Alice wants to send a message M to Bob, she does the following:

1. Looks up Bob's identity, for example, $ID = bob@computer.com$ (written as a binary string) and computes $Q_{ID} = H_1(ID)$.

2. Chooses a random $r \in \mathbf{F}_{\ell}^{\times}$.

3. Computes $g_{ID} = \tilde{e}(Q_{ID}, P_{pub})$.

4. Lets the ciphertext be the pair

$$c = (rP,\ M \oplus H_2(g_{ID}^r)),$$

where $\oplus$ denotes XOR (= bitwise addition mod 2).

Bob decrypts a ciphertext (u, v) as follows:

1. Uses his private key D_{ID} to compute $h_{ID} = \tilde{e}(D_{ID}, u)$.

2. Computes $m = v \oplus H_2(h_{ID})$.

The decryption works because

$$\tilde{e}(D_{ID}, u) = \tilde{e}(sQ_{ID}, rP) = \tilde{e}(Q_{ID}, P)^{sr} = \tilde{e}(Q_{ID}, P_{pub})^r = g_{ID}^r.$$

Therefore,

$$m = v \oplus H_2(\tilde{e}(D_{ID}, u)) = (M \oplus H_2(g_{ID}^r)) \oplus H_2(g_{ID}^r) = M.$$

Exercises

6.1 Show that the map β in Section 6.2 is an isomorphism (it is clearly bijective; the main point is that it is a homomorphism).

6.2 (a) Suppose that the ElGamal signature scheme is used to produce the valid signed message (m, R, s), as in Section 6.5. Let h be an integer with $\gcd(h, N) = 1$. Assume $\gcd(f(R), N) = 1$. Let

$$R' = hR, \quad s' \equiv sf(R')f(R)^{-1}h^{-1} \pmod{N},$$
$$m' \equiv mf(R')f(R)^{-1} \pmod{N}.$$

Show that (m', R', s') is a valid signed message (however, it is unlikely that m' is a meaningful message, so this procedure does not affect the security of the system).

(b) Suppose a hash function is used, so the signed messages are of the form (m, R_H, s_H). Explain why this prevents the method of (a) from working.

6.3 Use the notation of Section 6.5. Let u, v be two integers with $\gcd(v, N) = 1$ and let $R = uA + vB$. Let $s \equiv -v^{-1}f(R) \pmod{N}$ and $m \equiv su \pmod{N}$.

(a) Show that (m, R, s) is a valid signed message for the ElGamal signature scheme. (However, it is unlikely that m is a meaningful message).

(b) Suppose a hash function is used, so the signed messages are of the form (m, R_H, s_H). Explain why this prevents the method of (a) from working.

6.4 Let E be an elliptic curve over $\mathbf{F}_q$ and let $N = \#E(\mathbf{F}_q)$. Alice has a message that she wants to sign. She represents the message as a point $M \in E(\mathbf{F}_q)$. Alice has a secret integer a and makes public points A and B in $E(\mathbf{F}_q)$ with $B = aA$, as in the ElGamal signature scheme. There is a public function $f : E(\mathbf{F}_q) \to \mathbf{Z}/N\mathbf{Z}$. Alice performs the following steps.

(a) She chooses a random integer k with $\gcd(k, N) = 1$.

(b) She computes $R = M - kA$.

(c) She computes $s \equiv k^{-1}(1 - f(R)a) \pmod{N}$.

(d) The signed message is (M, R, s).

Bob verifies the signature as follows.

(a) He computes $V_1 = sR - f(R)B$ and $V_2 = sM - A$.

(b) He declares the signature valid if $V_1 = V_2$.

Show that if Alice performs the required steps correctly, then the verification equation $V_1 = V_2$ holds. (This signature scheme is a variant of one due to Nyberg and Rueppel (see [7]). An interesting feature is that the message appears as an element of the group $E(\mathbf{F}_q)$ rather than as an integer.)

6.5 Let p, q be prime numbers and suppose you know the numbers $m = (p+1)(q+1)$ and $n = pq$. Show that p, q are the roots of the quadratic equation
$$x^2 - (m - n - 1)x + n = 0$$
(so p, q can be found using the quadratic formula).

6.6 Let E be the elliptic curve $y^2 = x^3 + b \bmod p$, where $p \equiv 2 \pmod 3$.

(a) Suppose $E[n] \subseteq E(\mathbf{F}_p)$. Show that $n|p-1$ and $n^2|p+1$. Conclude that $n \leq 2$.

(b) Show that $E[2] \not\subseteq E(\mathbf{F}_p)$.

(c) Show that $E(\mathbf{F}_p)$ is cyclic (of order $p + 1$).

6.7 Let $p \equiv 3 \pmod 4$ be a prime number. Suppose $x \equiv y^2 \pmod p$.

(a) Show that $(y^{(p+1)/2})^2 \equiv y^2 \pmod p$.

(b) Show that $y^{(p+1)/2} \equiv \pm y \pmod p$.

(c) Show that $x^{(p+1)/4}$ is a square root of $x \pmod p$.

(d) Suppose z is not a square mod p. Using the fact that -1 is not a square mod p, show that $-z$ is a square mod p.

(e) Show that $z^{(p+1)/4}$ is a square root of $-z \pmod p$.

6.8 Let $p = 6\ell - 1$ and E be as in Section 6.8. The hash function H_1 in that section inputs a string of bits of arbitrary length and outputs a point of order ℓ on E. One way to do this is as follows.

(a) Choose a hash function H that outputs integers mod p. Input a binary string B. Let the output of H be the y coordinate of a point: $y = II(B)$. Show that there is a unique x mod p such that (x, y) lies on E.

(b) Let $H_1(B) = 6(x, y)$. Show that $H_1(B)$ is a point of order ℓ or 1 on E. Why is it very unlikely that $H_1(B)$ has order 1?

Chapter 7

Other Applications

In the 1980s, about the same time that elliptic curves were being introduced into cryptography, two related applications of elliptic curves were found, one to factoring and one to primality testing. These are generalizations of classical methods that worked with multiplicative groups $\mathbf{Z}_n^\times$. The main advantage of elliptic curves stems from the fact that there are many elliptic curves mod a number n, so if one elliptic curve doesn't work, another can be tried.

The problems of factorization and primality testing are related, but are very different in nature. The largest announced factorization up to the year 2002 was of an integer with 155 digits. However, it was at that time possible to prove primality of primes of more than 1000 digits.

It is possible to prove that a number is composite without finding a factor. One way is to show that $a^{n-1} \not\equiv 1 \pmod{n}$ for some a with $\gcd(a, n) = 1$. Fermat's little theorem says that if n is prime and $\gcd(a, n) = 1$, then $a^{n-1} \equiv 1 \pmod{n}$, so it follows that n must be composite, even though we have not produced a factor. Of course, if $a^{n-1} \equiv 1 \pmod{n}$ for several random choices of a, we might suspect that n is probably prime. But how can we actually prove n is prime? If n has only a few digits, we can divide n by each of the primes up to $\sqrt{n}$. However, if n has hundreds of digits, this method will take too long (much longer than the predicted life of the universe). In Section 7.2, we discuss efficient methods for proving primality. Similarly, suppose we have proved that a number is composite. How do we find the factors? This is a difficult computational problem. If the smallest prime factor of n has more than a few digits, then trying all prime factors up to $\sqrt{n}$ cannot work. In Section 7.1, we give a method that works well on numbers n of around 60 digits.

7.1 Factoring Using Elliptic Curves

In the mid 1980s, Hendrik Lenstra [60] gave new impetus to the study of elliptic curves by developing an efficient factoring algorithm that used elliptic curves. It turned out to be very effective for factoring numbers of around 60 decimal digits, and, for larger numbers, finding prime factors having around 20 to 30 decimal digits.

We start with an example.

Example 7.1

We want to factor 4453. Let E be the elliptic curve $y^2 = x^3 + 10x - 2$ mod 4453 and let $P = (1, 3)$. Let's try to compute $3P$. First, we compute $2P$. The slope of the tangent line at P is

$$\frac{3x^2 + 10}{2y} = \frac{13}{6} \equiv 3713 \quad (\text{mod } 4453).$$

We used the fact that $\gcd(6, 4453) = 1$ to find $6^{-1} \equiv 3711 \pmod{4453}$. Using this slope, we find that $2P = (x, y)$, with

$$x \equiv 3713^2 - 2 \equiv 4332, \quad y \equiv -3713(x - 1) - 3 \equiv 3230.$$

To compute $3P$, we add P and $2P$. The slope is

$$\frac{3230 - 3}{4332 - 1} = \frac{3227}{4331}.$$

But $\gcd(4331, 4453) = 61 \neq 1$. Therefore, we cannot find $4331^{-1} \pmod{4453}$, and we cannot evaluate the slope. However, we have found the factor 61 of 4453, and therefore $4453 = 61 \cdot 73$.

Recall (Section 2.10) that

$$E(\mathbf{Z}_{4453}) = E(\mathbf{F}_{61}) \oplus E(\mathbf{F}_{73}).$$

If we look at the multiples of P mod 61 we have

$$P \equiv (1, 3), \ 2P \equiv (1, 58), \ 3P \equiv \infty, \ 4P \equiv (1, 3), \ \ldots \quad (\text{mod } 61).$$

However, the multiples of P mod 73 are

$$P \equiv (1, 3), \ 2P \equiv (25, 18), \ 3P \equiv (28, 44), \ \ldots, \ 64P \equiv \infty \quad (\text{mod } 73).$$

Therefore, when we computed $3P$ mod 4453, we obtained ∞ mod 61 and a finite point mod 73. This is why the slope had a 61 in the denominator and was therefore infinite mod 61. If the order of P mod 73 had been 3 instead of 64, the slope would have had 0 mod 4553 in its denominator and the gcd would have been 4553, which would have meant that we did not obtain the factorization of 4453. But the probability is low that the order of a point mod 61 is exactly the same as the order of a point mod 73, so this situation will usually not cause us much trouble. If we replace 4453 with a much larger composite number n and work with an elliptic curve mod n and a point P

on E, then the main problem we'll face is finding some integer k such that $kP = \infty$ mod one of the factors of n. In fact, we'll often not obtain such an integer k. But if we work with enough curves E, it is likely that at least one of them will allow us to find such a k. This is the key property of the elliptic curve factorization method. ☐

Before we say more about elliptic curves, let's look at the classical $p - 1$ **factorization method**. We start with a composite integer n that we want to factor. Choose a random integer a and a large integer B. Compute

$$a_1 \equiv a^{B!} \pmod{n}, \text{ and } \gcd(a_1 - 1, n).$$

Note that we do not compute $a^{B!}$ and then reduce mod n, since that would overflow the computer. Instead, we can compute $a^{B!}$ mod n recursively by $a^{b!} \equiv \left(a^{(b-1)!}\right)^{b} \pmod{n}$, for $b = 2, 3, 4, \ldots, B$. Or we can write $B!$ in binary and do modular exponentiation by successive squaring.

We say that an integer m is **B-smooth** if all of the prime factors of m are less than or equal to B. For simplicity, assume $n = pq$ is the product of two large primes. Suppose that $p - 1$ is B-smooth. Since $B!$ contains all of the primes up to B, it is likely that $B!$ is a multiple of $p - 1$ (the main exception is when $p - 1$ is divisible by the square of a prime that is between $B/2$ and B). Therefore,

$$a_1 \equiv a^{B!} \equiv 1 \pmod{p}$$

by Fermat's little theorem (we ignore the very unlikely case that $p|a$).

Now suppose $q - 1$ is divisible by a prime $\ell > B$. Among all the elements in the cyclic group $\mathbf{Z}_q^{\times}$, there are at most $(q-1)/\ell$ that have order not divisible by ℓ and at least $(\ell - 1)(q - 1)/\ell$ that have order divisible by ℓ. (These numbers are exact if $\ell^2 \nmid q - 1$.) Therefore, it is very likely that the order of a is divisible by ℓ, and therefore

$$a_1 \equiv a^{B!} \not\equiv 1 \pmod{q}.$$

Therefore, $a_1 - 1$ is a multiple of p but is not a multiple of q, so

$$\gcd(a_1 - 1, pq) = p.$$

If all the prime factors of $q - 1$ are less than B, we usually obtain $\gcd(a_1 - 1, n) = n$. In this case, we can try a smaller B, or use various other procedures (similar to the one in Section 6.7). The main problem is choosing B so that $p - 1$ (or $q - 1$) is B-smooth. If we choose B small, the probability of this is low. If we choose B very large, then the computation of a_1 becomes too lengthy. So we need to choose B of medium size, maybe around 10^8. But what if both $p - 1$ and $q - 1$ have prime factors of around 20 decimal digits? We could keep trying various random choices of a, hoping to get lucky. But the above calculation shows that if there is a prime ℓ' with $\ell'|p-1$ but $\ell' > B$, then the chance that $a_1 \equiv 1 \pmod{p}$ is at most $1/\ell'$. This is very small if $\ell' \approx 10^{20}$. There seems to be no way to get the method to work. The elliptic curve method has a much better chance of success in this case because it allows us to change groups.

In the elliptic curve factorization method, we will need to choose random elliptic curves mod n and random points on these curves. A good way to do this is as follows. Choose a random integer A mod n and a random pair of integers $P = (u, v)$ mod n. Then choose C (the letter B is currently being used for the bound) such that

$$C = v^2 - u^3 - Au \pmod{n}.$$

This yields an elliptic curve $y^2 = x^3 + Ax + C$ with a point (u, v). This is much more efficient than the naive method of choosing A, C, u, then trying to find v. In fact, since being able to find square roots mod n is computationally equivalent to factoring n, this naive method will almost surely fail.

Here is the **elliptic curve factorization method**. We start with a composite integer n (assume n is odd) that we want to factor and do the following.

1. Choose several (usually around 10 to 20) random elliptic curves E_i : $y^2 = x^3 + A_i x + B_i$ and points P_i mod n.

2. Choose an integer B (perhaps around 10^8) and compute $(B!)P_i$ on E_i for each i.

3. If step 2 fails because some slope does not exist mod n, then we have found a factor of n.

4. If step 2 succeeds, increase B or choose new random curves E_i and points P_i and start over.

Steps 2, 3, 4 can often be done in parallel using all of the curves E_i simultaneously.

The elliptic curve method is very successful in finding a prime factor p of n when $p < 10^{40}$. Suppose we have a random integer n of around 100 decimal digits, and we know it is composite (perhaps, for example, $2^{n-1} \not\equiv 1 \pmod{n}$, so Fermat's little theorem implies that n is not prime). If we cannot find a small prime factor (by testing all of the primes up to 10^7, for example), then the elliptic curve method is worth trying since there is a good chance that n will have a prime factor less than 10^{40}.

Values of n that are used in cryptographic applications are now usually chosen as $n = pq$ with both p and q large (at least 75 decimal digits). For such numbers, the quadratic sieve and the number field sieve factorization methods outperform the elliptic curve method. However, the elliptic curve method is sometimes used inside these methods to look for medium sized prime factors of numbers that appear in intermediate steps.

Why does the elliptic curve method work? For simplicity, assume $n = pq$. A random elliptic curve E mod n can be regarded as an elliptic curve mod p and an elliptic curve mod q. We know, by Hasse's theorem, that

$$p + 1 - 2\sqrt{p} < \#E(\mathbf{F}_p) < p + 1 + 2\sqrt{p}.$$

In fact, each integer in the interval $(p + 1 - 2\sqrt{p},\ p + 1 + 2\sqrt{p})$ occurs for some elliptic curve. If B is of reasonable size, then the density of B-smooth integers in this interval is high enough, and the distribution of orders of random elliptic curves is sufficiently uniform. Therefore, if we choose several random E, at least one will probably have B-smooth order. In particular, if P lies on this E, then it is likely that $(B!)P = \infty \pmod{p}$ (as in the $p - 1$ method, the main exception occurs when the order is divisible by the square of a prime near B). It is unlikely that the corresponding point P on E mod q will satisfy $(B!)P = \infty \pmod{q}$. (If it does, choose a smaller B or use the techniques of Section 6.7 to factor n.) Therefore, when computing $(B!)P \pmod{n}$, we expect to obtain a slope whose denominator is divisible by p but not by q. The gcd of this denominator with n yields the factor p.

In summary, the difference between the $p - 1$ method and the elliptic curve method is the following. In the $p - 1$ method, there is a reasonable chance that $p - 1$ is B-smooth, but if it is not, there is not much we can do. In the elliptic curve method, there is a reasonable chance that $\#E(\mathbf{F}_p)$ is B-smooth, but if it is not we can choose another elliptic curve E.

It is interesting to note that the elliptic curve method, when applied to singular curves (see Section 2.9), yields classical factorization methods.

First, let's consider the curve E given by $y^2 = x^2(x + 1)$ mod n. We showed in Theorem 2.30 that the map

$$(x, y) \mapsto \frac{x + y}{x - y}$$

is an isomorphism from $E_{ns} = E(\mathbf{Z}_n) \setminus (0, 0)$ to $\mathbf{Z}_n^\times$. (Actually, we only showed this for fields. But it is true mod p and mod q, so the Chinese Remainder Theorem allows us to get the result mod $n = pq$.) A random point P on E_{ns} corresponds to a random $a \in \mathbf{Z}_n^\times$. Calculating $(B!)P$ corresponds to computing $a_1 \equiv a^{B!} \pmod{n}$. We have $(B!)P = \infty \pmod{p}$ if and only if $a_1 \equiv 1 \pmod{p}$, since ∞ and 1 are the identity elements of their respective groups. Fortunately, we have ways to extract the prime factor p of n in both cases. The first is by computing the gcd in the calculation of a slope. The second is by computing $\gcd(a_1 - 1, n)$. Therefore, we see that the elliptic curve method for the singular curve $y^2 = x^2(x - 1)$ is really the $p - 1$ method in disguise.

If we consider $y^2 = x^2(x + a)$ when a is not a square mod p, then we get the classical $p + 1$ factoring method (see Exercise 7.1).

Now let's consider E given by $y^2 = x^3$. By Theorem 2.29, the map

$$(x, y) \mapsto \frac{x}{y}$$

is an isomorphism from $E_{ns} = E(\mathbf{Z}_n) \setminus (0,0)$ to $\mathbf{Z}_n$, regarded as an additive group. A random point P in E_{ns} corresponds to a random integer $a \bmod n$. Computing $(B!)P$ corresponds to computing $(B!)a \pmod{n}$. We have $(B!)P = \infty \pmod{p}$ if and only if $(B!)a \equiv 0 \pmod{p}$, which occurs if and only if $p \leq B$ (note that this is much less likely than having $p - 1$ be B-smooth). Essentially, this reduces to the easiest factorization method: divide n by each of the primes up to B. This method is impractical if the smallest prime factor of n is not small. But at least it is almost an efficient way to do it. If we replace $B!$ by the product Q of primes up to B, then computing $\gcd(Q, n)$ is often faster than trying each prime separately.

7.2   Primality Testing

Suppose n is an integer of several hundred decimal digits. It is usually easy to decide with reasonable certainty whether n is prime or composite. But suppose we actually want to prove that our answer is correct. If n is composite, then usually either we know a nontrivial factor (so the proof that n is composite consists of giving the factor) or n failed a pseudoprimality test (for example, perhaps $a^{n-1} \not\equiv 1 \pmod{n}$ for some a). Therefore, when n is composite, it is usually easy to prove it, and the proof can be stated in a form that can be checked easily. But if n is prime, the situation is more difficult. Saying that n passed several pseudoprimality tests indicates that n is probably prime, but does not prove that n is prime. Saying that a computer checked all primes up to $\sqrt{n}$ is not very satisfying (and is not believable when n has several hundred digits). Cohen and Lenstra developed methods involving Jacobi sums that work well for primes of a few hundred digits. However, for primes of a thousand digits or more, the most popular method currently in use involves elliptic curves. (*Note*: For primes restricted to special classes, such as Mersenne primes, there are special methods. However, we are considering randomly chosen primes.)

The elliptic curve primality test is an elliptic curve version of the classical **Pocklington-Lehmer primality test**. Let's look at it first.

PROPOSITION 7.1

Let $n > 1$ be an integer, and let $n - 1 = rs$ with $r \geq \sqrt{n}$. Suppose that, for each prime $\ell | r$, there exists an integer a_ℓ with

$$a_\ell^{n-1} \equiv 1 \pmod{n} \quad \text{and} \quad \gcd\left(a_\ell^{(n-1)/\ell} - 1, n\right) = 1.$$

Then n is prime.

PROOF   Let p be a prime factor of n and let ℓ^e be the highest power of ℓ dividing r. Let $b \equiv a_\ell^{(n-1)/\ell^e}$ (mod p). Then

$$b^{\ell^e} \equiv a_\ell^{n-1} \equiv 1 \pmod{p} \quad \text{and} \quad b^{\ell^{e-1}} \equiv a_\ell^{(n-1)/\ell} \not\equiv 1 \pmod{p},$$

since $\gcd\left(a_\ell^{(n-1)/\ell} - 1, n\right) = 1$. It follows that the order of b (mod p) is ℓ^e. Therefore, $\ell^e | p - 1$. Since this is true for every prime power factor ℓ^e of r, we have $r | p - 1$. In particular,

$$p > r \geq \sqrt{n}.$$

If n is composite, it must have a prime factor at most $\sqrt{n}$. We have shown this is not the case, so n is prime.   ∎

REMARK 7.2   A converse of Proposition 7.1 is true. See Exercise 7.2. ∎

Example 7.2
Let $n = 153533$. Then $n - 1 = 4 \cdot 131 \cdot 293$. Let $r = 4 \cdot 131$. The primes dividing r are $\ell = 2$ and $\ell = 131$. We have

$$2^{n-1} \equiv 1 \pmod{n} \quad \text{and} \quad \gcd\left(2^{(n-1)/2} - 1, n\right) = 1,$$

so we can take $a_2 = 2$. Also,

$$2^{n-1} \equiv 1 \pmod{n} \quad \text{and} \quad \gcd\left(2^{(n-1)/131} - 1, n\right) = 1,$$

so we can take $a_{131} = 2$, also. The hypotheses of Proposition 7.1 are satisfied, so we have proved that 153533 is prime. The fact that $a_2 = a_{131}$ can be regarded as coincidence. In fact, we could take $a_2 = a_{131} = a_{293} = 2$, which shows that 2 is a primitive root mod 153533 (see Appendix A). So, in a sense, the calculations for the Pocklington-Lehmer test can be regarded as progress towards showing that there is a primitive root mod n (see Exercise 7.2).

Of course, to make the proof complete, we should prove that 2 and 131 are primes. We leave the case of 2 as an exercise and look at 131. We'll use the Pocklington-Lehmer test again. Write $130 = 2 \cdot 5 \cdot 13$. Let $r = 13$, so we have only one prime ℓ, namely $\ell = 13$. We have

$$2^{130} \equiv 1 \pmod{131} \quad \text{and} \quad \gcd\left(2^{10} - 1, 131\right) = 1.$$

Therefore, we can take $a_{13} = 2$. The Pocklington-Lehmer test implies that 131 is prime. Of course, we need the fact that 13 is prime, but 13 is small enough to check by trying possible factors.   ⧠

We can compactly record the proof that an integer n is prime by stating the values of the prime factors ℓ of r and the corresponding integers a_ℓ. We should also include proofs of primality of each of these primes ℓ. And we should include proofs of primality of the auxiliary primes used in the proofs for each ℓ, etc. Anyone can use this information to verify our proof. We never need to say how we found the numbers a_ℓ, nor how we factored r.

What happens if we cannot find enough factors of $n-1$ to obtain $r \geq \sqrt{n}$ such that we know all the prime factors ℓ of r? This is clearly a possibility if we are working with n of a thousand digits. As in the case of the $p-1$ factoring method in Section 7.1, an elliptic curve analogue comes to the rescue. Note that the number $n-1$ that we need to factor is the order of the group $\mathbf{Z}_n^\times$. If we can use elliptic curves, we can replace $n-1$ with a group order near n, but there will be enough choices for the elliptic curve that we can probably find a number that can be partially factored. The following is due to Goldwasser and Kilian [35]. Recall that a finite point in $E(\mathbf{Z}_n)$ is a point (x, y) with $x, y \in \mathbf{Z}_n$. This is in contrast to the points in $E(\mathbf{Z}_n)$ that are infinite mod some of the factors of n and therefore cannot be expressed using coordinates in $\mathbf{Z}_n$. See Section 2.10.

THEOREM 7.3

Let $n > 1$ and let E be an elliptic curve mod n. Suppose there exist distinct prime numbers $\ell_1, \ldots, \ell_k$ and finite points $P_i \in E(\mathbf{Z}_n)$ such that

1. *$\ell_i P_i = \infty$ for $1 \leq i \leq k$*

2. *$\prod_{i=1}^k \ell_i > \left(n^{1/4} + 1\right)^2$.*

Then n is prime.

PROOF   Let p be a prime factor of n. Write $n = p^f n_1$ with $p \nmid n_1$. Then

$$E(\mathbf{Z}_n) = E(\mathbf{Z}_{p^f}) \oplus E(\mathbf{Z}_{n_1}).$$

Since P_i is a finite point in $E(\mathbf{Z}_n)$, it yields a finite point in $E(\mathbf{Z}_{p^f})$, namely $P_i \mod p^f$. We can further reduce and obtain a finite point $P_{i,p} = P_i \mod p$ in $E(\mathbf{F}_p)$. Since $\ell_i P_i = \infty \mod n$, we have $\ell_i P_i = \infty \mod$ every factor of n. In particular, $\ell_i P_{i,p} = \infty$ in $E(\mathbf{F}_p)$, which means that $P_{i,p}$ has order ℓ_i. It follows that

$$\ell_i \mid \#E(\mathbf{F}_p)$$

for all i, so $\#E(\mathbf{F}_p)$ is divisible by $\prod \ell_i$. Therefore,

$$\left(n^{1/4} + 1\right)^2 < \prod_{i=1}^k \ell_i \leq \#E(\mathbf{F}_p) < p + 1 + 2\sqrt{p} = \left(p^{1/2} + 1\right)^2,$$

so $p > \sqrt{n}$. Since all prime factors of n are greater than $\sqrt{n}$, it follows that n is prime.   ∎

Example 7.3

Let $n = 907$. Let E be the elliptic curve $y^2 = x^3 + 10x - 2 \bmod n$. Let $\ell = 71$. Then

$$\ell > \left(907^{1/4} + 1\right)^2 \approx 42.1.$$

Let $P = (819, 784)$. Then $71P = \infty$. Theorem 7.3 implies that 907 is prime. Of course, we needed the fact that 71 is prime, which could also be proved using Theorem 7.3, or by direct calculation.

How did we find E and P? First, we looked at a few elliptic curves mod 907 until we found one whose order was divisible by a prime ℓ that was slightly larger than 42.1. (If we had chosen $\ell \approx 907$ then we wouldn't have made much progress, since we would still have needed to prove the primality of ℓ). In fact, to find the order of the curve, we started with curves where we knew a point. In the present case, E has the point $(1, 3)$. Using Baby Step, Giant Step, we found the order of $(1, 3)$ to be $923 = 13 \cdot 71$. Then we took $P = 13(1, 3)$, which has order 71. $\quad \square$

For large n, the hardest part of the algorithm is finding an elliptic curve E with a suitable number of points. One possibility is to choose random elliptic curves mod n and compute their orders, for example, using Schoof's algorithm, until an order is found that has a suitable prime factor ℓ. A more efficient procedure, due to Atkin and Morain (see [4]), uses the theory of complex multiplication to find suitable curves.

As in the Pocklington-Lehmer test, once a proof of primality is found, it can be recorded rather compactly. The Goldwasser-Kilian test has been used to prove the primality of numbers of more than 1000 decimal digits.

Exercises

7.1 This exercise shows that when the elliptic curve factorization method is applied to the singular curve $y^2 = x^2(x+a)$ where a is not a square mod a prime p, then we obtain a method equivalent to the $p + 1$ factoring method [110]. We first describe a version of the $p + 1$ method. Let p be an odd prime factor of the integer n that we want to factor. Let $t_0 = 2$ and choose a random integer $t_1 \bmod n$. Define t_m by the recurrence relation $t_{m+2} = t_1 t_{m+1} - t_m$ for $m \geq 0$. Let β, γ be the two roots of $f(X) = X^2 - t_1 X + 1$ in $\mathbf{F}_{p^2}$. Assume that $t_1^2 - 4$ is not a square in $\mathbf{F}_p$, so $\beta, \gamma \notin \mathbf{F}_p$. Let $s_m = \beta^m + \gamma^m$ for $m \geq 0$.

(a) Show that $\beta^{m+2} = t_1 \beta^{m+1} - \beta^m$ for $m \geq 0$, and similarly for γ.

(b) Show that $s_{m+2} = t_1 s_{m+1} - s_m$ for all $m \geq 0$.

(c) Show that $t_m \equiv s_m \pmod{p}$ for all $m \geq 0$.

(d) Show that β^p is a root of $f(X)$ (mod p), and that $\beta^p \neq \beta$. Therefore, $\gamma = \beta^p$.

(e) Show that $\beta^{p+1} = 1$ and $\gamma^{p+1} = 1$.

(f) Show that $t_{p+1} - 2 \equiv 0$ (mod p).

(g) Show that if $p+1 | B!$ for some bound B (so $p+1$ is B-smooth) then $\gcd(t_{B!} - 2, n)$ is a multiple of p. Since there are ways to compute $t_{B!}$ mod n quickly, this gives a factorization method.

We now show the relation with the elliptic curve factorization method. Consider a curve E given by $y^2 = x^2(x + a)$ mod n, where a is not a square mod p. Choose a random point P on E. To factor n by the elliptic curve method, we compute $B!P$. By Theorem 2.30, P mod p corresponds to an element $\beta = u + v\sqrt{a} \in \mathbf{F}_{p^2}$ with $u^2 - v^2 a = 1$.

(h) Show that β is a root of $X^2 - 2uX + 1$.

(i) Show that $B!P = \infty$ mod p if and only if $\beta^{B!} = 1$ in $\mathbf{F}_{p^2}$.

(j) Let $t_1 = 2u$ and define the sequence t_m as above. Show that $B!P = \infty$ mod p if and only if p divides $\gcd(t_{B!} - 2, n)$. Therefore, the elliptic curve method factors n exactly when the $p+1$ method factors n.

7.2 (a) Show that if n is prime and g is a primitive root mod n, then $a_\ell = g$ satisfies the hypotheses of Proposition 7.1 for all ℓ.

   (b) Suppose we take $r = n - 1$ and $s = 1$ in Proposition 7.1, and suppose that there is some number g such that $a_\ell = g$ satisfies the conditions on a_ℓ for each ℓ. Show that g is a primitive root mod n. (*Hint:* What power of ℓ divides the order of g mod n?)

7.3 The proof of Theorem 7.3 works for singular curves given by a Weierstrass equation where the cubic has a double root, as in Theorem 2.30. This yields a theorem that uses $\mathbf{Z}_n^\times$, rather than $E(\mathbf{Z}_n)$, to prove that n is prime. State Theorem 7.3 in this case in terms of $\mathbf{Z}_n^\times$. (*Remark:* The analogue of Theorem 7.3 for $\mathbf{Z}_n$ is rather trivial. The condition that P_i is a finite point becomes the condition that P_i is a number mod n such that $\gcd(P_i, n) = 1$ (that is, it is not the identity for the group law mod any prime factor of n). Therefore $\ell_i P_i = \infty$ translates to $\ell_i P_i \equiv 0$ (mod n), which implies that $\ell_i \equiv 0$ (mod n). Since ℓ_i is prime, we must have $n = \ell_i$. Hence n is prime.)

Chapter 8

Elliptic Curves over Q

As we saw in Chapter 1, elliptic curves over $\mathbf{Q}$ represent an interesting class of Diophantine equations. In the present chapter, we study the group structure of the set of rational points of an elliptic curve E defined over $\mathbf{Q}$. First, we show how the torsion points can be found quite easily. Then we prove the Mordell-Weil theorem, which says that $E(\mathbf{Q})$ is a finitely generated abelian group. As we'll see in Section 8.6, the method of proof has its origins in Fermat's method of infinite descent. Finally, we reinterpret the descent calculations in terms of Galois cohomology and define the Shafarevich-Tate group.

8.1 The Torsion Subgroup. The Lutz-Nagell Theorem

The torsion subgroup of $E(\mathbf{Q})$ is easy to calculate. In this section we'll give examples of how this can be done. The crucial step is the following theorem, which was used in Chapter 5 to study anomalous curves. For convenience, we repeat some of the notation introduced there.

Let $a/b \neq 0$ be a rational number, where a, b are relatively prime integers. Write $a/b = p^r a_1/b_1$ with $p \nmid a_1 b_1$. Define the *p*-**adic valuation** to be

$$v_p(a/b) = r.$$

For example, $v_2(7/40) = -3$, $v_5(50/3) = 2$, and $v_7(1/2) = 0$. Define $v_p(0) = +\infty$ (so $v_p(0) > n$ for every integer n).

Let E be an elliptic curve over $\mathbf{Z}$ given by $y^2 = x^3 + Ax + B$. Let $r \geq 1$ be an integer. Define

$$E_r = \{(x, y) \in E(\mathbf{Q}) \mid v_p(x) \leq -2r, \quad v_p(y) \leq -3r\} \cup \{\infty\}.$$

These are the points such that x has at least p^{2r} in its denominator and y has at least p^{3r} in its denominator. These should be thought of as the points that are close to ∞ mod powers of p (that is, p-adically close to ∞).

THEOREM 8.1

Let E be given by $y^2 = x^3 + Ax + B$ with $A, B \in \mathbf{Z}$. Let p be a prime and let r be a positive integer. Then

1. E_r is a subgroup of $E(\mathbf{Q})$.

2. If $(x, y) \in E(\mathbf{Q})$, then $v_p(x) < 0$ if and only if $v_p(y) < 0$. In this case, there exists an integer $r \geq 1$ such that $v_p(x) = -2r$ and $v_p(y) = -3r$.

3. The map

$$\lambda_r : E_r/E_{5r} \to \mathbf{Z}_{p^{4r}}$$
$$(x, y) \mapsto p^{-r} x/y \quad (\mathrm{mod}\ p^{4r})$$
$$\infty \mapsto 0$$

   is an injective homomorphism (where $\mathbf{Z}_{p^{4r}}$ is a group under addition).

4. If $(x, y) \in E_r$ but $(x, y) \notin E_{r+1}$, then $\lambda_r(x, y) \not\equiv 0 \ (\mathrm{mod}\ p)$.

REMARK 8.2   The map λ_r should be regarded as a logarithm for the group E_r/E_{r+1} since it changes the law of composition in the group to addition in $\mathbf{Z}_{p^{4r}}$, just as the classical logarithm changes the composition law in the multiplicative group of positive real numbers to addition in $\mathbf{R}$.   ∎

PROOF   The denominator of $x^3 + Ax + B$ equals the denominator of y^2. It is easy to see that the denominator of y is divisible by p if and only if the denominator of x is divisible by p. If p^j, with $j > 0$, is the exact power of p dividing the denominator of y, then p^{2j} is the exact power of p in the denominator of y^2. Similarly, if p^k, with $k > 0$, is the exact power of p dividing the denominator of x, then denominator of $x^3 + Ax + B$ is exactly divisible by p^{3k}. Therefore, $2j = 3k$. It follows that there exists r with $j = 3r$ and $k = 2r$. This proves (2). Also, we see that

$$\{(x, y) \in E_r \,|\, v_p(x) = -2r,\ v_p(y) = -3r\} = \{(x, y) \in E_r \,|\, v_p(x/y) = r\}$$

is the set of points in E_r not in E_{r+1}. This proves (4). Moreover, if $\lambda_r(x, y) \equiv 0 \ (\mathrm{mod}\ p^{4r})$, then $v_p(x/y) \geq 5r$, so $(x, y) \in E_{5r}$. This proves that λ_r is injective (as soon as we prove it is a homomorphism).
   Let

$$t = \frac{x}{y}, \quad s = \frac{1}{y}.$$

Dividing the equation $y^2 = x^3 + Ax + B$ by y^3 yields

$$\frac{1}{y} = \left(\frac{x}{y}\right)^3 + A\left(\frac{x}{y}\right)\left(\frac{1}{y}\right)^2 + B\left(\frac{1}{y}\right)^3,$$

which can be written as

$$s = t^3 + Ast^2 + Bs^3.$$

In the following, it will be convenient to write $p^j | z$ for a rational number z when p^j divides the numerator of z. Similarly, we'll write $z \equiv 0 \pmod{p^j}$ in this case. These extended notions of divisibility and congruence satisfy properties similar to those for the usual notions.

LEMMA 8.3
$(x, y) \in E_r$ if and only if $p^{3r} | s$. If $p^{3r} | s$, then $p^r | t$.

PROOF   If $(x, y) \in E_r$, then p^{3r} divides the denominator of y, so p^{3r} divides the numerator of $s = 1/y$. Conversely, suppose $p^{3r} | s$. Then p^{3r} divides the denominator of y. Part (2) of the theorem shows that p^{2r} divides the denominator of x. Therefore, $(x, y) \in E_r$.

If $p^{3r} | s$, then the exact power of p dividing the denominator of y is p^{3k}, with $k \geq r$. Part (2) of the theorem implies that the exact power of p dividing $t = x/y$ is p^k. Since $k \geq r$, we have $p^r | t$. $\blacksquare$

We now continue with the proof of Theorem 8.1. Let λ_r be as in the statement of the theorem. Note that

$$\lambda_r(-(x, y)) = \lambda_r(x, -y) = -p^{-r} x/y = -\lambda_r(x, y).$$

We now claim that if $P_1 + P_2 + P_3 = \infty$ then

$$\lambda_r(P_1) + \lambda_r(P_2) + \lambda_r(P_3) \equiv 0 \pmod{p^{4r}}.$$

The proof will also show that if $P_1, P_2 \in E_r$, then $P_3 \in E_r$. Therefore,

$$\lambda_r(P_1 + P_2) = \lambda_r(-P_3) = -\lambda_r(P_3) = \lambda_r(P_1) + \lambda_r(P_2),$$

so λ_r is a homomorphism.

Recall that three points add to ∞ if and only if they are collinear (Exercise 2.3). To prove the claim, let P_1, P_2, P_3 lie on the line

$$ax + by + d = 0$$

and assume that $P_1, P_2 \in E_r$. Dividing by y yields the s, t line

$$at + b + ds = 0.$$

Let P_i' denote the point P_i written in terms of the s, t coordinates. In other words, if

$$P_i = (x_i, y_i),$$

then

$$P_i' = (s_i, t_i)$$

with

$$s_i = 1/y_i, \quad t_i = x_i/y_i.$$

The points P_1', P_2', P_3' lie on the line $at + b + ds = 0$.

Since $P_1, P_2 \in E_r$, Lemma 8.3 implies that

$$p^{3r} | s_i, \quad p^r | t_i, \quad \text{for } i = 1, 2.$$

As discussed in Section 2.4, at a finite point (x, y), the order of intersection of the line $ax + by + d = 0$ and the curve $y^2 = x^3 + Ax + B$ can be calculated by using projective coordinates and considering the line $aX + bY + dZ = 0$ and the curve $ZY^2 = X^3 + AXZ^2 + BZ^3$. In this case, $x = X/Z$ and $y = Y/Z$.

If we start with a line $at + b + ds = 0$ and the curve $s = t^3 + Ats^2 + Bs^3$, we can homogenize to get $aT + bU + dS = 0$ and $SU^2 = T^3 + ATS^2 + BS^3$. In this case, we have $t = T/U$ and $s = S/U$. If we let $Z = S$, $Y = U$, $X = T$, we find that we are working with the same line and curve as above. A point (x, y) corresponds to

$$t = T/U = X/Y = x/y \text{ and } s = S/U = Z/Y = 1/y.$$

Since orders of intersection can be calculated using the projective models, it follows that the order of intersection of the line $ax + by + d = 0$ with the curve $y^2 = x^3 + Ax + B$ at (x, y) is the same as the order of intersection of the line $at + b + ds = 0$ with the curve $s = t^3 + Ats^2 = Bs^3$ at $(t, s) = (x/y, 1/y)$. For example, the line and curve are tangent in the variables x, y if and only if they are tangent in the variables t, s. This allows us to do the elliptic curve group calculations using t, s instead of x, y.

LEMMA 8.4

A line $t = c$, where $c \in \mathbf{Q}$ is a constant with $c \equiv 0 \pmod{p}$, intersects the curve $s = t^3 + As^2t + Bs^3$ in at most one point (s, t) with $s \equiv 0 \pmod{p}$. This line is not tangent at such a point of intersection.

PROOF   Suppose we have two values of s, call them s_1, s_2 with $s_1 \equiv s_2 \equiv 0$ $\pmod{p}$. Suppose $s_1 \equiv s_2 \pmod{p^k}$ for some $k \geq 1$. Write $s_i = ps_i'$. Then $s_1' \equiv s_2' \pmod{p^{k-1}}$, so ${s_1'}^2 \equiv {s_2'}^2 \pmod{p^{k-1}}$, so $s_1^2 = p^2{s_1'}^2 \equiv p^2{s_2'}^2 = s_2^2$ $\pmod{p^{k+1}}$. Similarly, $s_1^3 \equiv s_2^3 \pmod{p^{k+2}}$. Therefore,

$$s_1 = c^3 + Acs_1^2 + Bs_1^3 \equiv c^3 + Acs_2^2 + Bs_2^3 = s_2 \pmod{p^{k+1}}.$$

By induction, we have $s_1 \equiv s_2 \pmod{p^k}$ for all k. It follows that $s_1 = s_2$, so there is at most one point of intersection with $s \equiv 0 \pmod{p}$.

The slope of the tangent line to the curve can be found by implicit differentiation:

$$\frac{ds}{dt} = 3t^2 + As^2 + 2Ast\frac{ds}{dt} + 3s^2\frac{ds}{dt},$$

so

$$\frac{ds}{dt} = \frac{3t^2 + As^2}{1 - 2Ast - 3s^2}.$$

If the line $t = c$ is tangent to the curve at (t, s), then $1 - 2Ast - 3s^2 = 0$. But $s \equiv t \equiv 0 \pmod{p}$ implies that

$$1 - 2Ast - 3s^2 \equiv 1 \not\equiv 0 \pmod{p}.$$

Therefore, $t = c$ is not tangent to the curve.   ∎

If $d = 0$, then our line is of the form in the lemma. But it passes through the points P_1' and P_2', so we must have $P_1' = P_2'$, and the line is tangent to the curve. Changing back to x, y coordinates, we obtain $P_1 = P_2$. The definition of the group law says that since the points P_1 and P_2 are equal, the line $ax + by + d = 0$ is tangent at (x, y). As pointed out above, this means that $at + b + ds = 0$ is tangent at (s, t). The lemma says that this cannot happen. Therefore, $d \neq 0$.

Dividing by d, we obtain

$$s = \alpha t + \beta$$

for some $\alpha, \beta \in \mathbf{Q}$. Then P_1', P_2', P_3' lie on the line $s = \alpha t + \beta$.

LEMMA 8.5

$$\alpha = \frac{t_2^2 + t_1 t_2 + t_1^2 + As_2^2}{1 - A(s_1 + s_2)t_1 - B(s_2^2 + s_1 s_2 + s_1^2)}.$$

PROOF   If $t_1 \neq t_2$, then $\alpha = (s_2 - s_1)/(t_2 - t_1)$. Since $s_i = t_i^3 + As_i^2 t_i + Bs_i^3$, we have

$$(s_2 - s_1)\left(1 - A(s_1 + s_2)t_1 - B(s_2^2 + s_1 s_2 + s_1^2)\right)$$
$$= (s_2 - s_1) - A(s_2^2 - s_1^2)t_1 - B(s_2^3 - s_1^3)$$
$$= (s_2 - As_2^2 t_2 - Bs_2^3) - (s_1 - As_1^2 t_1 - Bs_1^3) + As_2^2(t_2 - t_1)$$
$$= t_2^3 - t_1^3 + As_2^2(t_2 - t_1)$$
$$= (t_2 - t_1)(t_2^2 + t_1 t_2 + t_1^2 + As_2^2).$$

This proves that $(s_2 - s_1)/(t_2 - t_1)$ equals the expression in the lemma.

Now suppose that $t_1 = t_2$. Since a line $t = c$ with $c \equiv 0 \pmod{p}$ intersects the curve $s = t^3 + As^2 t + Bs^3$ in only one point with $s \equiv 0 \pmod{p}$ by Lemma 8.4, the points (t_1, s_1) and (t_2, s_2) must be equal. The line $s = \alpha t + \beta$

is therefore the tangent line at this point, and the slope is computed by implicit differentiation of $s^2 = t^3 + As^t + Bs^3$:

$$\frac{ds}{dt} = 3t^2 + As^2 + 2Ast\frac{ds}{dt} + 3Bs^2\frac{ds}{dt}.$$

Solving for ds/dt yields the expression in the statement of the lemma when $t_1 = t_2 = t$ and $s_1 = s_2 = s$.   ∎

Since $s_1 \equiv s_2 \equiv 0 \pmod{p}$, we find that the denominator

$$1 - A(s_1 + s_2)t_1 - B(s_2^2 + s_1 s_2 + s_1^2) \equiv 1 \pmod{p}.$$

Since $p^r | t_i$, we have

$$t_2^2 + t_1 t_2 + t_1^2 + As_2^2 \equiv 0 \pmod{p^{2r}}.$$

Therefore, $\alpha \equiv 0 \pmod{p^{2r}}$. Since $p^{3r} | s_i$, we have

$$\beta = s_i - \alpha t_i \equiv 0 \pmod{p^{3r}}.$$

The point P_3' is the third point of intersection of the line $s = \alpha t + \beta$ with $s = t^3 + As^2 t + Bs^3$. Therefore, we need to solve for t:

$$\alpha t + \beta = t^3 + A(\alpha t + \beta)^2 t + B(\alpha t + \beta)^3.$$

This can be rearranged to obtain

$$0 = t^3 + \frac{2A\alpha\beta + 3B\alpha^2\beta}{1 + B\alpha^3 + A\alpha^2}t^2 + \cdots .$$

The sum of the three roots is the negative of the coefficient of t^2, so

$$t_1 + t_2 + t_3 = -\frac{2A\alpha\beta + 3B\alpha^2\beta}{1 + B\alpha^3 + A\alpha^2}$$
$$\equiv 0 \pmod{p^{5r}}.$$

The last congruence holds because $p^{2r} | \alpha$ and $p^{3r} | \beta$. Since $t_1 \equiv t_2 \equiv 0 \pmod{p^r}$, we have $t_3 \equiv 0 \pmod{p^r}$. Therefore, $s_3 = \alpha t_3 + \beta \equiv 0 \pmod{p^{3r}}$. By Lemma 8.3, $P_3 \in E_r$. Moreover,

$$\lambda_r(P_1) + \lambda_r(P_2) + \lambda_r(P_3) \equiv p^{-r}(t_1 + t_2 + t_3) \equiv 0 \pmod{p^{4r}}.$$

Therefore, λ_r is a homomorphism. This completes the proof of Theorem 8.1. ∎

COROLLARY 8.6
Let the notations be as in Theorem 8.1. If $n > 1$ and n is not a power of p, then E_1 contains no points of exact order n.

PROOF   Suppose $P \in E_1$ has order n. Since n is not a power of p, we may multiply P by the largest power of p dividing n and obtain a point, not equal to ∞, of order prime to p. Therefore, we may assume that P has order n with $p \nmid n$. Let r be the largest integer such that $P \in E_r$. Then

$$n\lambda_r(P) = \lambda_r(nP) - \lambda_r(\infty) \equiv 0 \pmod{p^{4r}}.$$

Since $p \nmid n$, we have $\lambda_r(P) \equiv 0 \pmod{p^{4r}}$, so $P \in E_{4r}$. Since $4r > r$, this contradicts the choice of r. Therefore, P does not exist.   ∎

The following theorem was proved independently by Lutz and Nagell in the 1930s. Quite often it allows a quick determination of the torsion points on an elliptic curve over $\mathbf{Q}$.

THEOREM 8.7 (Lutz-Nagell)
Let E be given by $y^2 = x^3 + Ax + B$ with $A, B \in \mathbf{Z}$. Let $P = (x, y) \in E(\mathbf{Q})$. Suppose P has finite order. Then $x, y \in \mathbf{Z}$. If $y \neq 0$ then

$$y^2 \mid 4A^3 + 27B^2.$$

PROOF   Suppose x or y is not in $\mathbf{Z}$. Then there is some prime p dividing the denominator of one of them. By part (2) of Theorem 8.1, $P \in E_r$ for some $r \geq 1$. Let ℓ be a prime dividing the order n of P. Then $Q = (n/\ell)P$ has order ℓ. By Corollary 8.6, $\ell = p$. Choose j such that $Q \in E_j$, $Q \notin E_{j+1}$. Then $\lambda_j(Q) \not\equiv 0 \pmod{p}$, and

$$p\lambda_j(Q) = \lambda_j(pQ) \equiv 0 \pmod{p^{4j}}.$$

Therefore,

$$\lambda_j(Q) \equiv 0 \pmod{p^{4j-1}}.$$

This contradicts the fact that $\lambda_j(Q) \not\equiv 0 \pmod{p}$. It follows that $x, y \in \mathbf{Z}$.

Assume $y \neq 0$. Then $2P = (x_2, y_2) \neq \infty$. Since $2P$ has finite order, $x_2, y_2 \in \mathbf{Z}$. By Theorem 3.6,

$$x_2 = \frac{x^4 - 2Ax^2 - 8Bx + A^2}{4y^2}.$$

Since $x_2 \in \mathbf{Z}$, this implies that

$$y^2 \mid x^4 - 2Ax^2 - 8Bx + A^2.$$

A straightforward calculation shows that

$$(3x^2 + 4A)(x^4 - 2Ax^2 - 8Bx + A^2) - (3x^3 - 5Ax - 27B)(x^3 + Ax + B)$$
$$= 4A^3 + 27B^2.$$

Since $y^2 = x^3 + Ax + B$, we see that y^2 divides both terms on the left. Therefore, $y^2 | 4A^3 + 27B^2$.   ∎

COROLLARY 8.8

Let E be an elliptic curve over $\mathbf{Q}$. Then the torsion subgroup of $E(\mathbf{Q})$ is finite.

PROOF   A suitable change of variables puts the equation for E into Weierstrass form with integer coefficients. Theorem 8.7 now shows that there are only finitely many possibilities for the torsion points.   ∎

Example 8.1

Let E be given by $y^2 = x^3 + 4$. Then $4A^3 + 27B^2 = 432$. Let $P = (x, y)$ be a point of finite order in $E(\mathbf{Q})$. Since $0 = x^3 + 4$ has no rational solutions, we have $y \neq 0$. Therefore, $y^2 | 432$, so

$$y = \pm 1, \ \pm 2, \ \pm 3, \ \pm 4, \ \pm 6, \ \pm 12.$$

Only $y = \pm 2$ yields a rational value of x, so the only possible torsion points are $(0, 2)$ and $(0, -2)$. A quick calculation shows that $3(0, \pm 2) = \infty$. Therefore, the torsion subgroup of $E(\mathbf{Q})$ is cyclic of order 3.   ☐

Example 8.2

Let E be given by $y^2 = x^3 + 8$. Then $4A^3 + 17B^2 = 1728$. If $y = 0$, then $x = -2$. The point $(-2, 0)$ has order 2. If $y \neq 0$, then $y^2 | 1728$, which means that $y | 24$. Trying the various possibilities, we find the points $(1, \pm 3)$ and $(2, \pm 4)$. However,

$$2(1, 3) = (-7/4, \ -13/8) \ \text{ and } \ 2(2, 4) = (-7/4, \ 13/8).$$

Since these points do not have integer coordinates, they cannot have finite order. Therefore, $(1, 3)$ and $(2, 4)$ cannot have finite order. It follows that the torsion subgroup of $E(\mathbf{Q})$ is $\{\infty, (-2, 0)\}$. (*Remark:* The fact that $2(1, 3) = -2(2, 4)$ leads us to suspect, and easily verify, that $(1, 3) + (2, 4) = (-2, 0)$.)
☐

Suppose we use the Lutz-Nagell theorem and obtain a possible torsion point P. How do we decide whether or not it's a torsion point? In the previous example, we multiplied P by an integer and obtained a non-torsion point. Therefore, P was not a torsion point. In general, the Lutz-Nagell theorem explicitly gives a finite list of possibilities for torsion points. If P is a torsion point, then, for every n, the point nP must either be ∞ or be on that list. Since there are only finitely many points on the list, either we'll have $nP = mP$ for some $m \neq n$, in which case P is torsion and $(n - m)P = \infty$, or some multiple nP is not on the list and P is not torsion. Alternatively, we can use Mazur's theorem (Theorem 8.11 below), which says that the order of a torsion point in $E(\mathbf{Q})$ is at most 12. Therefore, if $nP \neq \infty$ for all $n \leq 12$, then P is not torsion. Consequently, it is not hard to check each possibility in the Lutz-Nagell theorem and see which ones yield torsion points.

Another technique that helps us determine the torsion subgroup involves reduction mod primes. The main result needed is the following.

THEOREM 8.9

Let E be an elliptic curve given by $y^2 = x^3 + Ax + B$ with $A, B \in \mathbf{Z}$. Let p be an odd prime and assume $p \nmid 4A^3 + 27B^2$. Let

$$\rho_p : E(\mathbf{Q}) \to E(\mathbf{F}_p)$$

be the **reduction mod** p map. If $P \in E(\mathbf{Q})$ has finite order and $\rho_p(P) = \infty$, then P has p-power order.

REMARK 8.10   The theorem says that reduction mod p is injective on the prime-to-p torsion in $E(\mathbf{Q})$. This is similar to the situation in algebraic number theory, where reduction mod a prime ideal is injective on roots of unity of order prime to p, where p is the rational prime in the prime ideal (see [106]).   ∎

PROOF   The kernel of ρ_p is E_1 (Theorem 8.1). By Corollary 8.6, E_1 contains only p-power torsion.   ∎

Example 8.3

Let's use Theorem 8.9 to find the torsion on $y^2 = x^3 + 8$. We have $4A^3 + 27B^2 = 1728 = 2^6 \cdot 3^3$, so we cannot use the primes $2, 3$. The reduction mod 5 has 6 points, so Theorem 8.9 implies that the torsion in $E(\mathbf{Q})$ has order dividing 6 times a power of 5. The reduction mod 7 has 12 points, so the torsion has order 12 times a power of 7. At this point, we can conclude that the torsion in $E(\mathbf{Q})$ has order dividing 6. The reduction mod 11 has 12 points, so we get no new information. However, the reduction mod 13 has 16 points, so the torsion group of $E(\mathbf{Q})$ has order dividing 16 times a power of 13. It follows that the torsion group has order dividing 2. Since $(-2, 0)$ is a point of order 2, the torsion has order exactly 2. This is of course the same result that we obtained earlier using the Lutz-Nagell theorem.   ⬜

Example 8.4

In the preceding example, the Lutz-Nagell theorem was perhaps at least as fast as Theorem 8.9 in determining the order of the torsion subgroup. This is not always the case. Let E be given by $y^2 = x^3 + 18x + 72$. Then

$$4A^3 + 27B^2 = 163296 = 2^5 \cdot 3^6 \cdot 7.$$

The Lutz-Nagell theorem would require us to check all y with $y^2 | 163296$, which amounts to checking all $y | 108 = 2^2 \cdot 3^3$. Instead, the reduction mod 5 has 5 points and the reduction mod 11 has 8 points. It follows that the torsion subgroup of $E(\mathbf{Q})$ is trivial.   ⬚

Finally, we mention a deep result of Mazur, which we will not prove (see [62]).

THEOREM 8.11
Let E be an elliptic curve defined over $\mathbf{Q}$. Then the torsion subgroup of $E(\mathbf{Q})$ is one of the following:

$$\mathbf{Z}_n \;\; with \; 1 \leq n \leq 10 \; or \; n = 12,$$
$$\mathbf{Z}_2 \oplus \mathbf{Z}_{2n} \;\; with \; 1 \leq n \leq 4.$$

REMARK 8.12   For each of the groups in the theorem, there are infinitely many elliptic curves E (with distinct j-invariants) having that group as the torsion subgroup of $E(\mathbf{Q})$. See Exercise 8.1 for examples of each possibility. ∎

8.2 Descent and the Weak Mordell-Weil Theorem

We start with an example that has its origins in work of Fermat (see Section 8.6).

Example 8.5
Let's look at rational points on the curve E given by

$$y^2 = x(x - 2)(x + 2).$$

If $y = 0$, we have $x = 0, \pm 2$. Therefore, assume $y \neq 0$. Since the product of x, $x - 2$, and $x + 2$ is a square, intuition suggests that each of these factors should, in some sense, be close to being a square. Write

$$x = au^2$$
$$x - 2 = bv^2$$
$$x + 2 = cw^2$$

with rational numbers a, b, c, u, v, w. Then $y^2 = abc(uvw)^2$, so

$$abc \text{ is a square.}$$

By adjusting u, v, w, we may assume that a, b, c are squarefree integers. In fact, we claim that

$$a, b, c \in \{\pm 1, \pm 2\}.$$

Suppose that p is an odd prime dividing a. Since a is squarefree, $p^2 \nmid a$, so the exact power p^k dividing $x = au^2$ has k odd. If $k < 0$, then p^k is the exact power of p in the denominator of $x \pm 2$, so p^{3k} is the power of p in the denominator of $y^2 = x(x-2)(x+2)$. Since $3k$ is odd and y^2 is a square, this is impossible. If $k > 0$ then $x \equiv 0 \pmod{p}$, so $x \pm 2 \not\equiv 0 \pmod{p}$. Therefore, p^k is the power of p dividing y^2. Since k is odd, this is impossible. Therefore, $p \nmid a$. Similarly, no odd prime divides b or c. Therefore, each of a, b, c is, up to sign, a power of 2. Since they are squarefree, this proves the claim.

The procedure we are following is called **descent**, or, more precisely, a **2-descent**. Suppose x is a rational number with at most N digits in its numerator and denominator. Then u, v, w should have at most $N/2$ digits (approximately) in their numerators and denominators. Therefore, if we are searching for points (x, y), we can instead search for smaller numbers u, v, w. This method was developed by Fermat. See Section 8.6.

We have four choices for a and four choices for b. Since a and b together determine c (because abc is a square), there are 16 possible combinations for a, b, c. We can eliminate some of them quickly. Note that

$$bv^2 = x - 2 < au^2 = x < cw^2 = x + 2.$$

Therefore, if $a < 0$, then $b < 0$. Moreover, if $a > 0$, then $c > 0$, hence $b > 0$, since $abc > 0$. Therefore, a and b have the same sign. We are now down to 8 possible combinations.

Let's consider $(a, b, c) = (1, 2, 2)$. We have

$$x = u^2, \quad x - 2 = 2v^2, \quad x + 2 = 2w^2$$

with rational numbers u, v, w. Therefore,

$$u^2 - 2v^2 = 2, \quad u^2 - 2w^2 = -2.$$

If v has 2 in its denominator, then $2v^2$ has an odd power of 2 in its denominator. But u^2 has an even power of 2 in its denominator, so $u^2 - 2v^2$ cannot be an integer. This contradiction shows that v and u have odd denominators. Therefore, we may consider u, v mod powers of 2. Since $2 | u^2$, we have $2 | u$, hence $4 | u^2$. Therefore, $-2v^2 \equiv 2 \pmod{4}$, which implies that $2 \nmid v$. Similarly, $-2w^2 \equiv -2 \pmod{4}$, so $2 \nmid w$. It follows that $v^2 \equiv w^2 \equiv 1 \pmod{8}$, so

$$2 \equiv u^2 - 2v^2 \equiv u^2 - 2 \equiv u^2 - 2w^2 \equiv -2 \pmod{8},$$

which is a contradiction. It follows that $(a, b, c) = (1, 2, 2)$ is impossible. Similar considerations eliminate the combinations $(-1, -1, 1)$, $(2, 1, 2)$, and $(-2, 2, -1)$ for (a, b, c) (later, we'll see a faster way to eliminate them). Only the combinations

$$(a, b, c) = (1, 1, 1), \ (-1, -2, 2), \ (2, 2, 1), \ (-2, -1, 2)$$

remain. As we'll see below, these four combinations correspond to the four points that we already know about, namely,

$$\infty, \ (0, 0), \ (2, 0), \ (-2, 0)$$

(this requires some explanation, which will be given later). As we'll see later, the fact that we eliminated all combinations except those coming from known points implies that we have found all points, except possibly points of odd order, on the curve. The Lutz-Nagell theorem, or reduction mod 5 and 7 (see Theorem 8.9), shows that there are no nontrivial points of odd order. Therefore, we have found all rational points on E:

$$E(\mathbf{Q}) = \{\infty, \ (0, 0), \ (2, 0), \ (-2, 0)\}.$$

$\Box$

The calculations of the example generalize to elliptic curves E of the form

$$y^2 = (x - e_1)(x - e_2)(x - e_3)$$

with $e_1, e_2, e_3 \in \mathbf{Z}$ and $e_i \neq e_j$ when $i \neq j$. In fact, they extend to even more general situations. If $e_i \in \mathbf{Q}$ but $e_i \notin \mathbf{Z}$, then a change of variables transforms the equation to one with $e_i \in \mathbf{Z}$, so this situation gives nothing new. However, if $e_i \notin \mathbf{Q}$, the method still applies. In order to keep the discussion elementary, we'll not consider this case, though we'll say a few things about it later.

Write

$$x - e_1 = au^2$$
$$x - e_2 = bv^2$$
$$x - e_3 = cw^2$$

with rational numbers a, b, c, u, v, w. Then $y^2 = abc(uvw)^2$, so

$$abc \text{ is a square.}$$

By adjusting u, v, w, we may assume that a, b, c are squarefree integers.

PROPOSITION 8.13

Let

$$S = \{p \,|\, p \text{ is prime and } p|(e_1 - e_2)(e_1 - e_3)(e_2 - e_3)\}.$$

If p is a prime and $p|abc$, then $p \in S$.

PROOF Suppose $p|a$. Then p^k, with k odd, is the exact power of p dividing $x - e_1$. If $k < 0$, then p^k is the power of p in the denominator of $x - e_2$ and $x - e_3$. Therefore, p^{3k} is the power of p in the denominator of y^2, which is impossible. Therefore $k > 0$. This means that $x \equiv e_1 \pmod{p}$. Also, x has no p in its denominator, so the same is true of $bv^2 = x - e_2$ and $cw^2 = x - e_3$. Moreover, $bv^2 \equiv e_1 - e_2$ and $cw^2 \equiv e_1 - e_3 \pmod{p}$. If $p \notin S$, then the power of p in

$$y^2 = (au^2)(bv^2)(cw^2)$$

is $p^k p^0 p^0 = p^k$. Since k is odd, this is impossible. Therefore, $p \in S$. ∎

Since S is a finite set, there are only finitely many combinations (a, b, c) that are possible. The following theorem shows that the set of combinations that actually come from points (x, y) has a group structure modulo squares.

Let $\mathbf{Q}^\times/\mathbf{Q}^{\times 2}$ denote the group of rational numbers modulo squares. This means that we regard two nonzero rational numbers x_1, x_2 as equivalent if the ratio x_1/x_2 is the square of a rational number. Every element of $\mathbf{Q}^\times/\mathbf{Q}^{\times 2}$ can be represented by ± 1 times a (possibly empty) product of distinct primes. Note that if $x - e_1 = av_1^2$, then $x - e_1$ is equivalent to a mod squares. Therefore, the map ϕ in the following theorem maps a point $(x, y) \notin E[2]$ to the corresponding triple (a, b, c).

THEOREM 8.14
Let E be given by $y^2 = (x - e_1)(x - e_2)(x - e_3)$ with $e_1, e_2, e_3 \in \mathbf{Z}$. The map

$$\phi : E(\mathbf{Q}) \to (\mathbf{Q}^\times/\mathbf{Q}^{\times 2}) \oplus (\mathbf{Q}^\times/\mathbf{Q}^{\times 2}) \oplus (\mathbf{Q}^\times/\mathbf{Q}^{\times 2})$$

defined by

$$
\begin{aligned}
(x, y) &\mapsto (x - e_1, \quad x - e_2, \quad x - e_3) \quad \text{when } y \neq 0 \\
\infty &\mapsto (1, \quad 1, \quad 1) \\
(e_1, 0) &\mapsto ((e_1 - e_2)(e_1 - e_3), \quad e_1 - e_2, \quad e_1 - e_3) \\
(e_2, 0) &\mapsto (e_2 - e_1, \quad (e_2 - e_1)(e_2 - e_3), \quad e_2 - e_3) \\
(e_3, 0) &\mapsto (e_3 - e_1, \quad e_3 - e_2, \quad (e_3 - e_1)(e_3 - e_2))
\end{aligned}
$$

is a homomorphism. The kernel of ϕ is $2E(\mathbf{Q})$.

PROOF First, we show that ϕ is a homomorphism. Suppose $P_i = (x_i, y_i)$, $i = 1, 2, 3$, are points lying on the line $y = ax + b$. Assume for the moment that $y_i \neq 0$. The polynomial

$$(x - e_1)(x - e_2)(x - e_3) - (ax + b)^2$$

has leading coefficient 1 and has roots x_1, x_2, x_3 (with the correct multiplicities). Therefore,

$$(x - e_1)(x - e_2)(x - e_3) - (ax + b)^2 = (x - x_1)(x - x_2)(x - x_3).$$

Evaluating at e_i yields

$$(x_1 - e_i)(x_2 - e_i)(x_3 - e_i) = (ae_i + b)^2 \in \mathbf{Q}^{\times 2}.$$

Since this is true for each i,

$$\phi(P_1)\phi(P_2)\phi(P_3) = 1 \in \mathbf{Q}^\times / \mathbf{Q}^{\times 2}$$

(that is, the product is a square, hence is equivalent to 1 mod squares). Since any number z is congruent to its multiplicative inverse mod squares (that is, z equals $1/z$ times a square),

$$\phi(P_3)^{-1} = \phi(P_3) = \phi(-P_3).$$

Therefore,

$$\phi(P_1)\phi(P_2) = \phi(-P_3) = \phi(P_1 + P_2).$$

To show that ϕ is a homomorphism, it remains to check what happens when one or both of P_1, P_2 is a point of order 1 or 2. The case where a point P_i is of order 1 (that is, $P_i = \infty$) is trivial. If both P_1 and P_2 have order 2, a case by case check shows that $\phi(P_1 + P_2) = \phi(P_1)\phi(P_2)$. Finally, suppose that P_1 has order 2 and P_2 has $y_2 \neq 0$. Let's assume $P_1 = (e_1, 0)$. The other possibilities are similar. Since the values of ϕ are triples, let ϕ_1, ϕ_2, ϕ_3 denote the three components of ϕ (so $\phi = (\phi_1, \phi_2, \phi_3)$). The proof given above shows that

$$\phi_i(P_1)\phi_i(P_2) = \phi_i(P_1 + P_2)$$

for $i = 2, 3$. So it remains to consider ϕ_1.

If we try to use the formula $\phi_1(x, y) = x - e_1$, then $\phi_1(e_1, 0)$ would equal 0, which is not allowed. This is why the definition of ϕ treats the points of order 2 separately. Correspondingly, our proof needs to do this, too.

Let $y = ax + b$ be the line through P_1 and P_2, and let P_3 be the third point of intersection of the line. As above, we have (since $x_1 = e_1$)

$$(x - e_1)(x - e_2)(x - e_3) - (ax + b)^2$$
$$= (x - e_1)(x - x_2)(x - x_3). \tag{8.1}$$

Since $y = ax + b$ passes through $(e_1, 0)$, we must have

$$ax + b = a(x - e_1).$$

Therefore, we can divide Equation 8.1 by $x - e_1$ to obtain

$$(x - e_2)(x - e_3) - a^2(x - e_1) = (x - x_2)(x - x_3).$$

Substituting $x = e_1$ yields

$$\phi_1(P_1) = (e_1 - e_2)(e_1 - e_3) = (e_1 - x_2)(e_1 - x_3) = \phi_1(P_2)\phi_1(P_3).$$

Since $\phi_1(P_i) = \phi_1(P_i)^{-1}$, this may be rearranged to obtain

$$\phi_1(P_1)\phi_1(P_2) = \phi_1(P_3) = \phi_1(-P_3) = \phi_1(P_1 + P_2).$$

Therefore, ϕ_1 is a homomorphism. Putting everything together, we see that ϕ is a homomorphism.

To prove the second half of the theorem, we need to show that if $x - e_i$ is a square for all i, then $(x, y) = 2P$ for some point $P \in E(\mathbf{Q})$. Let

$$x - e_i = v_i^2, \quad i = 1, 2, 3.$$

For simplicity, we'll assume that $e_1 + e_2 + e_3 = 0$, which means that the equation for our elliptic curve has the form $y^2 = x^3 + Ax + B$. (If $e_1 + e_2 + e_3 \neq 0$, the coefficient of x^2 is nonzero. A simple change of variables yields the present case.) Let

$$f(T) = u_0 + u_1 T + u_2 T^2$$

satisfy

$$f(e_i) = v_i, \quad i = 1, 2, 3.$$

Such an f exists since there is a unique quadratic polynomial whose graph passes through any three points that have distinct x-coordinates. In fact

$$\begin{aligned}
f(T) = e_1 &\frac{1}{(e_1 - e_2)(e_1 - e_3)}(T - e_2)(T - e_3) \\
+ e_2 &\frac{1}{(e_2 - e_1)(e_2 - e_3)}(T - e_1)(T - e_3) \\
+ e_3 &\frac{1}{(e_3 - e_1)(e_3 - e_2)}(T - e_1)(T - e_2).
\end{aligned}$$

Let $g(T) = x - T - f(T)^2$. Then $g(e_i) = 0$ for all i, so

$$T^3 + AT + B = (T - e_1)(T - e_2)(T - e_3) \text{ divides } g(T).$$

Therefore, $g(T) \equiv 0 \pmod{T^3 + AT + B}$, so

$$x - T \equiv (u_0 + u_1 T + u_2 T^2)^2 \pmod{T^3 + AT + B}.$$

(We say that two polynomials P_1, P_2 are congruent mod P_3 if $P_1 - P_2$ is a multiple of P_3.) This congruence for $x - T$ can be thought of as a way to simultaneously capturing the information that $x - e_i$ is a square for all i. Mod $T^3 + AT + B$, we have

$$T^3 \equiv -AT - B, \quad T^4 \equiv T \cdot T^3 \equiv -AT^2 - BT.$$

Therefore,

$$\begin{aligned}
x - T &\equiv (u_0 + u_1 T + u_2 T^2)^2 \\
&\equiv u_0^2 + 2u_0 u_1 T + (u_1^2 + 2u_0 u_2)T^2 + 2u_1 u_2 T^3 + u_2^2 T^4 \\
&\equiv (u_0^2 - 2Bu_1 u_2) + (2u_0 u_1 - 2Au_1 u_2 - Bu_2^2)T \\
&\quad + (u_1^2 + 2u_0 u_2 - Au_2^2)T^2.
\end{aligned}$$

If two polynomials P_1 and P_2 of degree at most two are congruent mod a polynomial of degree three, then their difference $P_1 - P_2$ is a polynomial of degree at most two that is divisible by a polynomial of degree three. This can only happen if $P_1 = P_2$. In our case, this means that

$$x = u_0^2 - 2Bu_1 u_2 \tag{8.2}$$
$$-1 = 2u_0 u_1 - 2Au_1 u_2 - Bu_2^2 \tag{8.3}$$
$$0 = u_1^2 + 2u_0 u_2 - Au_2^2. \tag{8.4}$$

If $u_2 = 0$ then (8.4) implies that also $u_1 = 0$. Then $f(T)$ is constant, so $v_1 = v_2 = v_3$. This means that $e_1 = e_2 = e_3$, contradiction. Therefore, $u_2 \neq 0$. Multiply (8.4) by u_1/u_2^3 and multiply (8.3) by u_2/u_2^3, then subtract to obtain

$$\left(\frac{1}{u_2}\right)^2 = \left(\frac{u_1}{u_2}\right)^3 + A\left(\frac{u_1}{u_2}\right) + B.$$

Let

$$x_1 = \frac{u_1}{u_2}, \quad y_1 = \frac{1}{u_2},$$

so $(x_1, y_1) \in E(\mathbf{Q})$. We claim that $2(x_1, y_1) = \pm(x, y)$.

Equation 8.4 implies that

$$u_0 = \frac{Au_2^2 - u_1^2}{2u_2} = \frac{A - x_1^2}{2y_1}.$$

Substituting this into (8.2) yields

$$x = \frac{x_1^4 - 2Ax_1^2 - 8Bx_1 + A^2}{4y_1^2}.$$

This is the x-coordinate of $2(x_1, y_1)$ (see Theorem 3.6). The y-coordinate is determined up to sign by the x-coordinate, so $2(x_1, y_1) = (x, \pm y) = \pm(x, y)$. It follows that $(x, y) = 2(x_1, y_1)$ or $2(x_1, -y_1)$. In particular, $(x, y) \in 2E(\mathbf{Q})$.

∎

Example 8.6

We continue with Example 8.5. For the curve $y^2 = x(x-2)(x+2)$, we have

$$\phi(\infty) = (1, 1, 1), \quad \phi(0, 0) = (-1, -2, 2),$$
$$\phi(2, 0) = (2, 2, 1), \quad \phi(-2, 0) = (-2, -1, 2)$$

(we used the fact that 4 and 1 are equivalent mod squares to replace 4 by 1). We eliminated the triple $(a, b, c) = (1, 2, 2)$ by working mod powers of 2. We now show how to eliminate $(-1, -1, 1), (2, 1, 2), (-2, 2, -1)$. Suppose there is a point P with $\phi(P) = (-1, -1, 1)$. Then

$$\phi(P + (0, 0)) = \phi(P)\phi(0, 0) = (-1, -1, 1)(-1, -2, 2) = (1, 2, 2).$$

But we showed that $(1, 2, 2)$ does not come from a point in $E(\mathbf{Q})$. Therefore, P does not exist. The two other triples are eliminated similarly.   ☐

Theorem 8.14 has a very important corollary.

THEOREM 8.15 (Weak Mordell-Weil Theorem)
Let E be an elliptic curve defined over $\mathbf{Q}$. Then

$$E(\mathbf{Q})/2E(\mathbf{Q})$$

is finite.

PROOF   We give the proof in the case that $e_1, e_2, e_3 \in \mathbf{Q}$. As remarked earlier, we may assume that $e_1, e_2, e_3 \in \mathbf{Z}$. The map ϕ in Theorem 8.14 gives an injection

$$E(\mathbf{Q})/2E(\mathbf{Q}) \hookrightarrow (\mathbf{Q}^\times/\mathbf{Q}^{\times 2}) \oplus (\mathbf{Q}^\times/\mathbf{Q}^{\times 2}) \oplus (\mathbf{Q}^\times/\mathbf{Q}^{\times 2}).$$

Proposition 8.13 says that if (a, b, c) (where a, b, c are chosen to be squarefree integers) is in the image of ϕ, then a, b, c are products of primes in the set S of Proposition 8.13. Since S is finite, there are only finitely many such a, b, c mod squares. Therefore, the image of ϕ is finite. This proves the theorem.

∎

REMARK 8.16   (for those who know some algebraic number theory) Let $K/\mathbf{Q}$ be a finite extension. The theorem can be extended to say that if E is an elliptic curve over K then $E(K)/2E(K)$ is finite. If we assume that $x^3 + Ax + B = (x - e_1)(x - e_2)(x - e_3)$ with all $e_i \in K$, then the proof is the same except that the image of ϕ is contained in

$$(K^\times/K^{\times 2}) \oplus (K^\times/K^{\times 2}) \oplus (K^\times/K^{\times 2}).$$

Let $\mathcal{O}_K$ be the ring of algebraic integers of K. To make things simpler, we invert some elements in order to obtain a unique factorization domain. Take a nonzero element from an integral ideal in each ideal class of $\mathcal{O}_K$ and let M be the multiplicative subset generated by these elements. Then $M^{-1}\mathcal{O}_K$ is a principal ideal domain, hence a unique factorization domain. The analogue of Proposition 8.13 says that the primes of $M^{-1}\mathcal{O}_K$ dividing a, b, c also divide

$(e_1 - e_2)(e_1 - e_3)(e_2 - e_3)$. Let $S \subset M^{-1}\mathcal{O}_K$ be the set of prime divisors of $(e_1 - e_2)(e_1 - e_3)(e_2 - e_3)$. Then the image of ϕ is contained in the group generated by S and the units of $M^{-1}\mathcal{O}_K$. Since the class number of K is finite, M is finitely generated. A generalization of the Dirichlet unit theorem (often called the S-unit theorem) says that the units of $M^{-1}\mathcal{O}_K$ are a finitely generated group. Therefore, the image of ϕ is a finitely generated abelian group of exponent 2, hence is finite. This proves that $E(K)/2E(K)$ is finite. ∎

8.3 Heights and the Mordell-Weil Theorem

The purpose of this section is to change the weak Mordell-Weil theorem into the Mordell-Weil theorem. This result was proved by Mordell in 1922 for elliptic curves defined over $\mathbf{Q}$. It was greatly generalized in 1928 by Weil in his thesis, where he proved the result not only for elliptic curves over number fields (that is, finite extensions of $\mathbf{Q}$) but also for abelian varieties (higher-dimensional analogues of elliptic curves).

THEOREM 8.17 (Mordell-Weil)
Let E be an elliptic curve defined over $\mathbf{Q}$. Then $E(\mathbf{Q})$ is a finitely generated abelian group.

The theorem says that there is a finite set of points on E from which all other points can be obtained by repeatedly drawing tangent lines and lines through points, as in the definition of the group law. The proof will be given below. Since we proved the weak Mordell-Weil theorem only in the case that $E[2] \subseteq E(\mathbf{Q})$, we obtain the theorem only for this case. However, the weak Mordell-Weil theorem is true in general, and the proof of the passage from the weak result to the strong result holds in general.

From the weak Mordell-Weil theorem, we know that $E(\mathbf{Q})/2E(\mathbf{Q})$ is finite. This alone is not enough to deduce the stronger result. For example, $\mathbf{R}/2\mathbf{R} = 0$, hence is finite, even though $\mathbf{R}$ is not finitely generated. In our case, suppose we have points $R_1, \ldots, R_n$ representing the finitely many cosets in $E(\mathbf{Q})/2E(\mathbf{Q})$. Let $P \in E(\mathbf{Q})$ be an arbitrary point. We can write

$$P = R_i + 2P_1$$

for some i and some point P_1. Then we write

$$P_1 = R_j + 2P_2,$$

etc. If we can prove the process stops, then we can put things back together and obtain the theorem. The theory of heights will show that the points

$P_1, P_2, \ldots$ are getting smaller, in some sense, so the process will eventually yield a point P_k that lies in some finite set of small points. These points, along with the R_i, yield the generators of $E(\mathbf{Q})$. We make these ideas more precise after Theorem 8.18 below. Note that sometimes the points R_i by themselves do not suffice to generate $E(\mathbf{Q})$. See Exercise 8.7.

Let a/b be a rational number, where a, b are integers with $\gcd(a, b) = 1$. Define

$$H(a/b) = \mathrm{Max}(|a|, |b|)$$

and

$$h(a/b) = \log H(a/b).$$

The function h is called the **(logarithmic) height function**. It is closely related to the number of digits required to write the rational number a/b. Note that, given a constant c, there are only finitely many rational numbers x with $h(x) \leq c$.

Now let E be an elliptic curve over $\mathbf{Q}$ and let $(x, y) \in E(\mathbf{Q})$. Define

$$h(x, y) = h(x), \quad h(\infty) = 0.$$

It might seem strange using only the x-coordinate. Instead, we could use the y-coordinate. Since the square of the denominator of the y-coordinate is the cube of the denominator of the x-coordinate (when the coefficients A, B of E are integers), it can be shown that this would change the function h approximately by a factor of $3/2$. This would cause no substantial change in the theory. In fact, the canonical height $\hat{h}$, which will be introduced shortly, is defined using a limit of values of $\frac{1}{2}h$. It could also be defined as a limit of values of $1/3$ of the height of the y-coordinate. These yield the same canonical height function. See [90, Lemma 6.3]. The numbers 2 and 3 are the orders of the poles of the functions x and y on E (see Section 11.1).

It is convenient to replace h with a function $\hat{h}$ that has slightly better properties. The function $\hat{h}$ is called the **canonical height**.

THEOREM 8.18

Let E be an elliptic curve defined over $\mathbf{Q}$. There is a function

$$\hat{h} : E(\mathbf{Q}) \to \mathbf{R}_{\geq 0}$$

with the following properties:

1. *$\hat{h}(P) \geq 0$ for all $P \in E(\mathbf{Q})$.*

2. *There is a constant c_0 such that $|\frac{1}{2}h(P) - \hat{h}(P)| \leq c_0$ for all P.*

3. *Given a constant c, there are only finitely many points $P \in E(\mathbf{Q})$ with $\hat{h}(P) \leq c$.*

4. $\hat{h}(mP) = m^2\hat{h}(P)$ *for all integers m and all P.*

5. $\hat{h}(P + Q) + \hat{h}(P - Q) = 2\hat{h}(P) + 2\hat{h}(Q)$ *for all P, Q.*

6. $\hat{h}(P) = 0$ *if and only if P is a torsion point.*

Property (5) is often called the **parallelogram law** because if the origin 0 and vectors $P, Q, P + Q$ (ordinary vector addition) are the vertices of a parallelogram, then the sum of the squares of the lengths of the diagonals equals the sum of the squares of the lengths of the four sides:

$$||P + Q||^2 + ||P - Q||^2 = 2||P||^2 + 2||Q||^2.$$

The proof of Theorem 8.18 will occupy most of the rest of this section. First, let's use the theorem to deduce the Mordell-Weil theorem.

Proof of the Mordell-Weil theorem: Let $R_1, \ldots, R_n$ be representatives for $E(\mathbf{Q})/2E(\mathbf{Q})$. Let

$$c = \text{Max}_i\{\hat{h}(R_i)\}$$

and let $Q_1, \ldots, Q_m$ be the set of points with $\hat{h}(Q_i) \le c$. This is a finite set by Theorem 8.18. Let G be the subgroup of $E(\mathbf{Q})$ generated by

$$R_1, \ldots, R_n, Q_1, \ldots, Q_m.$$

We claim that $G = E(\mathbf{Q})$. Suppose not. Let $P \in E(\mathbf{Q})$ be an element not in G. Since, for a point P, there are only finitely many points of height less than P, we may change P to one of these, if necessary, and assume P has the smallest height among points not in G. We may write

$$P - R_i = 2P_1$$

for some i and some P_1. By Theorem 8.18,

$$
\begin{aligned}
4\hat{h}(P_1) = \hat{h}(2P_1) &= \hat{h}(P - R_i) \\
&= 2\hat{h}(P) + 2\hat{h}(R_i) - \hat{h}(P + R_i) \\
&\le 2\hat{h}(P) + 2c + 0 \\
&< 2\hat{h}(P) + 2\hat{h}(P) = 4\hat{h}(P)
\end{aligned}
$$

(since $c < \hat{h}(P)$, because $P \ne Q_j$). Therefore,

$$\hat{h}(P_1) < \hat{h}(P).$$

Since P had the smallest height for points not in G, we must have $P_1 \in G$. Therefore,

$$P = R_i + 2P_1 \in G.$$

This contradiction proves that $E(\mathbf{Q}) = G$. This completes the proof of the Mordell-Weil theorem.   ∎

It remains to prove Theorem 8.18. The key step is the following.

PROPOSITION 8.19

There exists a constant c_1 such that

$$|h(P+Q) - h(P-Q) - 2h(P) - 2h(Q)| \leq c_1$$

for all $P, Q \in E(\mathbf{Q})$.

The proof is rather technical, so we postpone it in order to complete the proof of Theorem 8.18.

Proof of Theorem 8.18:

Proof of parts (1) and (2): Letting $Q = P$ in Proposition 8.19, we obtain

$$|h(2P) - 4h(P)| \leq c_1 \tag{8.5}$$

for all P. Define

$$\hat{h}(P) = \frac{1}{2} \lim_{n \to \infty} \frac{1}{4^n} h(2^n P).$$

We need to prove the limit exists. We have

$$\lim_{n \to \infty} \frac{1}{4^n} h(2^n P) = h(P) + \sum_{j=1}^{\infty} \frac{1}{4^j} (h(2^j P) - 4h(2^{j-1} P)). \tag{8.6}$$

By (8.5),

$$\left| \frac{1}{4^j} (h(2^j P) - 4h(2^{j-1} P)) \right| \leq \frac{c_1}{4^j},$$

so the infinite sum converges. Therefore, $\hat{h}(P)$ exists. Since

$$\sum_{j=1}^{\infty} \frac{c_1}{4^j} = \frac{c_1}{3},$$

we obtain $|\hat{h}(P) - \frac{1}{2} h(P)| \leq c_1/6$. It is clear from the definitions that $\hat{h}(P) \geq 0$ for all P.

Proof of part (3): If $\hat{h}(P) \leq c$, then $h(P) \leq 2c + \frac{c_1}{3}$. There are only finitely many P satisfying this inequality.

Proof of part (5): We have

$$\frac{1}{4^n} |h(2^n P) + h(2^n Q) - 2h(2^n P + 2^n Q) - 2h(2^n P + 2^n Q)| \leq \frac{c_1}{4^n}.$$

Letting $n \to \infty$ yields the result.

Proof of part (4): Since the height depends only on the x-coordinate, $\hat{h}(-P) = \hat{h}(P)$. Therefore, we may assume $m \geq 0$. The cases $m = 0, 1$

are trivial. Letting $Q = P$ in part (5) yields the case $m = 2$. Assume that we know the result for $m - 1$ and m. Then

$$
\begin{aligned}
\hat{h}((m+1)P) &= -\hat{h}((m-1)P) + 2\hat{h}(mP) + 2\hat{h}(P) \text{ (by part (5))} \\
&= \left(-(m-1)^2 + 2m^2 + 2\right)\hat{h}(P) \\
&= (m+1)^2\hat{h}(P).
\end{aligned}
$$

By induction, the result is true for all m.

Proof of part (6): If $mP = \infty$, then $m^2\hat{h}(P) = \hat{h}(mP) = \hat{h}(\infty) = 0$, so $\hat{h}(P) = 0$. Conversely, if $\hat{h}(P) = 0$, then $\hat{h}(mP) = m^2\hat{h}(P) = 0$ for all m. Since there are only finitely many points of height 0, the set of multiples of P is finite. Therefore, P is a torsion point. This completes the proof of Theorem 8.18. ∎

Proof of Proposition 8.19. It remains to prove Proposition 8.19. It can be restated as saying that there exist constants c', c'' such that

$$2h(P) + 2h(Q) - c' \le h(P+Q) + h(P-Q) \tag{8.7}$$

$$h(P+Q) + h(P-Q) \le 2h(P) + 2h(Q) + c'' \tag{8.8}$$

for all P, Q. These two inequalities will be proved separately. We'll start with the second one.

Let the elliptic curve E be given by $y^2 = x^3 + Ax + B$ with $A, B \in \mathbf{Z}$. Let

$$P = (\frac{a_1}{b_1}, y_1), \qquad Q = (\frac{a_2}{b_2}, y_2)$$

$$P + Q = (\frac{a_3}{b_3}, y_3), \qquad P - Q = (\frac{a_4}{b_4}, y_4),$$

be points on E, where $y_i \in \mathbf{Q}$ and a_i, b_i are integers with $\gcd(a_i, b_i) = 1$. Let

$$
\begin{aligned}
g_1 &= 2(a_1b_2 + a_2b_1)(Ab_1b_2 + a_1a_2) + 4Bb_1^2b_2^2 \\
g_2 &= (a_1a_2 - Ab_1b_2)^2 - 4B(a_1b_2 + a_2b_1)b_1b_2 \\
g_3 &= (a_1b_2 - a_2b_1)^2.
\end{aligned}
$$

Then a short calculation shows that

$$\frac{a_3}{b_3} + \frac{a_4}{b_4} = \frac{g_1}{g_3}, \qquad \frac{a_3a_4}{b_3b_4} = \frac{g_2}{g_3}.$$

LEMMA 8.20

Let $c_1, c_2, d_1, d_2 \in \mathbf{Z}$ with $\gcd(c_i, d_i) = 1$ for $i = 1, 2$. Then

$$\mathrm{Max}(|c_1|, |d_1|) \cdot \mathrm{Max}(|c_2|, |d_2|) \le 2\mathrm{Max}(|c_1c_2|, |c_1d_2 + c_2d_1|, |d_1d_2|).$$

PROOF   Without loss of generality, we may assume that $|c_1| \le |d_1|$ (otherwise, switch c_1, d_1). Let L denote the left side of the inequality of the lemma and let R denote the right side. There are three cases to consider.

1. If $|c_2| \leq |d_2|$, then $L = |d_1 d_2|$ and $2|d_1 d_2| \leq R$, so $L \leq R$.

2. If $|c_2| \geq |d_2| \geq (1/2)|c_2|$, then $L = |d_1 c_2|$ and

$$R \geq 2|d_1 d_2| \geq |d_1 c_2| \geq L.$$

3. If $|d_2| \leq (1/2)|c_2|$, then $L = |d_1 c_2|$ and

$$\begin{aligned} R &\geq 2|c_1 d_2 + c_2 d_1| \\ &\geq 2(|c_2 d_1| - |c_1 d_2|) \\ &\geq 2(|c_2 d_1| - |d_1|(1/2)|c_2|) \\ &= |c_1 d_2| = L. \end{aligned}$$

This completes the proof of the lemma. ∎

LEMMA 8.21
  Let $c_1, c_2, d_1, d_2 \in \mathbf{Z}$ with $\gcd(c_i, d_i) = 1$ for $i = 1, 2$. Then

$$\gcd(c_1 c_2,\ c_1 d_2 + c_2 d_1,\ d_1 d_2) = 1.$$

PROOF   Let $d = \gcd(c_1 d_2 + c_2 d_1,\ d_1 d_2)$. Suppose p is a prime such that $p|c_1$ and $p|d$. Then $p \nmid d_1$ since $\gcd(c_1, d_1) = 1$. Since $p|d_1 d_2$, we have $p|d_2$. Therefore, $p \nmid c_2$. Therefore, $p|c_1 d_2$ and $p \nmid c_2 d_1$, so $p \nmid c_1 d_2 + c_2 d_1$. Therefore $p \nmid d$, contradiction. Similarly, there is no prime dividing both c_2 and d. It follows that there is no prime dividing $c_1 c_2$ and d, so the gcd in the lemma is 1. ∎

We can apply the lemmas to a_3, a_4, b_3, b_4. Since $\gcd(a_3, b_3) = 1$ and $\gcd(a_4, b_4) = 1$, we have

$$\gcd(a_3 a_4,\ a_3 b_4 + a_4 b_3,\ b_3 b_4) = 1.$$

Therefore, there exist integers x, y, z such that

$$a_3 a_4 x + (a_3 b_4 + a_4 b_3)y + b_3 b_4 z = 1.$$

Since

$$g_3(a_3 b_4 + a_4 b_3) = g_1(b_3 b_4) \quad \text{and} \quad g_3(a_3 a_4) = g_2(b_3 b_4), \qquad (8.9)$$

we have

$$\begin{aligned} g_3 &= g_3(a_3 a_4)x + g_3(a_3 b_4 + a_4 b_3)y + g_3(b_3 b_4)z \\ &= g_2(b_3 b_4)x + g_1(b_3 b_4)y + g_3(b_3 b_4)z. \end{aligned}$$

Therefore, $b_3 b_4 | g_3$, so

$$|b_3 b_4| \leq |g_3|.$$

Similarly,

$$|a_3 a_4| \leq |g_2|.$$

Equation 8.9 and the fact that $|b_3 b_4| \leq |g_3|$ imply that

$$|a_3 b_4 + a_4 b_3| \leq |g_1|.$$

In terms of the non-logarithmic height H, these inequalities say that

$$
\begin{aligned}
H(P+Q) \cdot H(P-Q) &= \mathrm{Max}(|a_3|, |b_3|) \cdot \mathrm{Max}(|a_4|, |b_4|) \\
&\leq 2\mathrm{Max}(|a_3 a_4|, |a_3 b_4 + a_4 b_3|, |b_3 b_3|) \\
&\leq 2\mathrm{Max}(|g_2|, |g_1|, |g_3|).
\end{aligned}
$$

Let $H_1 = \mathrm{Max}(|a_1|, |b_1|)$ and $H_2 = \mathrm{Max}(|a_2|, |b_2|)$. Then

$$
\begin{aligned}
|g_1| &= |2(a_1 b_2 + a_2 b_1)(A b_1 b_2 + a_1 a_2) + 4B b_1^2 b_2^2| \\
&\leq 2(H_1 H_2 + H_2 H_1)(|A| H_1 H_2 + H_1 H_2) + 4|B| H_1^2 H_2^2 \\
&\leq 4(|A| + 1 + |B|) H_1^2 H_2^2.
\end{aligned}
$$

Similarly,

$$|g_2| \leq ((1 + |A|)^2 + 8|B|) H_1^2 H_2^2, \qquad |g_3| \leq 4 H_1^2 H_2^2.$$

Therefore,

$$H(P+Q) \cdot H(P-Q) \leq C H_1^2 H_2^2 = C H(P)^2 H(Q)^2$$

for some constant C. Taking logs yields

$$h(P+Q) + h(P-Q) \leq 2h(P) + 2h(Q) + c'' \tag{8.10}$$

for some constant c''.

We now need to prove the inequality in Equation 8.7. First we'll prove an inequality between $h(R)$ and $h(2R)$ for points R.

LEMMA 8.22
Let $R \in E(\mathbf{Q})$. There exists a constant C_2, independent of R, such that

$$4h(R) \leq h(2R) + C_2.$$

PROOF   Let

$$R = (\frac{a}{b}, y)$$

with $y \in \mathbf{Q}$ and $a, b \in \mathbf{Z}$ with $\gcd(a, b) = 1$. Let

$$h_1 = a^4 - 2Aa^2b^2 - 8Bab^3 + A^2b^4$$
$$h_2 = (4b)(a^3 + Aab^2 + Bb^3)$$
$$\Delta = 4A^3 + 27B^2.$$

By Lemma 3.8, there exist homogeneous polynomials $r_1, r_2, s_1, s_2 \in \mathbf{Z}[a, b]$ of degree 3 (the coefficients depend on A, B) such that

$$4\Delta b^7 = r_1 h_1 + r_2 h_2 \qquad\qquad (8.11)$$
$$4\Delta a^7 = s_1 h_1 + s_2 h_2. \qquad\qquad (8.12)$$

For a homogeneous polynomial

$$p(x, y) = c_0 x^3 + c_1 x^2 y + c_2 x y^2 + c_3 y^3,$$

we have

$$|p(a, b)| \le (|c_0| + |c_1| + |c_2| + |c_3|)\mathrm{Max}(|a|, |b|)^3.$$

Suppose $|b| \ge |a|$. It follows that

$$|4\Delta||b|^7 \le |r_1(a, b)||h_1| + |r_2(a, b)||h_2|$$
$$\le C_1|b|^3 \mathrm{Max}(|h_1|, |h_2|),$$

for some constant C_1 independent of R. Therefore,

$$|4\Delta||b|^4 \le C_1 \mathrm{Max}(|h_1|, |h_2|).$$

Let $d = \gcd(h_1, h_2)$. Then Equations 8.11 and 8.12 imply that

$$d \,|\, 4\Delta b^7 \quad \text{and} \quad d \,|\, 4\Delta a^7.$$

Since $\gcd(a, b) = 1$, we have $d|4\Delta$, so $d \le |4\Delta|$. Since

$$H(2R) = \mathrm{Max}\left(\frac{|h_1|}{d}, \frac{|h_2|}{d}\right),$$

we have

$$|4\Delta|H(R)^4 = |4\Delta||b|^4$$
$$\le C_1 \mathrm{Max}(|h_1|, |h_2|)$$
$$\le C_1|4\Delta| \,\mathrm{Max}(\frac{|h_1|}{d}, \frac{|h_2|}{d})$$
$$\le C_1|4\Delta|H(2R).$$

Dividing by $|4\Delta|$ and taking logs yields

$$4h(R) \le h(2R) + C_2$$

for some constant C_2, independent of R.

The case where $|a| \geq |b|$ is similar. This completes the proof of Lemma 8.22.

∎

Changing P to $P + Q$ and Q to $P - Q$ in Equation 8.10 yields

$$h(2P) + h(2Q) \leq 2h(P + Q) + 2h(P - Q) + c''.$$

By Lemma 8.22,

$$4h(P) + 4h(Q) - 2C_2 \leq h(2P) + h(2Q).$$

Therefore,

$$2h(P) + 2h(Q) - c' \leq h(P + Q) + h(P - Q)$$

for some constant c'. This completes the proof of Proposition 8.19.   ∎

8.4 Examples

The Mordell-Weil theorem says that if E is an elliptic curve defined over $\mathbf{Q}$, then $E(\mathbf{Q})$ is a finitely generated abelian group. The structure theorem for such groups (see Appendix B) says that

$$E(\mathbf{Q}) \simeq T \oplus \mathbf{Z}^r,$$

where T is a finite group (the **torsion subgroup**) and $r \geq 0$ is an integer, called the **rank** of $E(\mathbf{Q})$. In Section 8.1, we showed how to compute T. The integer r is harder to compute. In this section, we show how to use the methods of the previous sections to compute r in some cases. In Section 8.8, we'll give an example that shows why the computation of r is sometimes difficult.

Example 8.7

Let E be the curve

$$y^2 = x^3 - 4x.$$

In Section 8.2, we showed that

$$E(\mathbf{Q})/2E(\mathbf{Q}) = \{\infty, (0,0), (2,0), (-2,0)\}$$

(more precisely, the points on the right are representatives for the cosets on the left). Moreover, an easy calculation using the Lutz-Nagell theorem shows that the torsion subgroup of $E(\mathbf{Q})$ is

$$T = E[2].$$

From Theorem 8.15, we have $E(\mathbf{Q}) \simeq T \oplus \mathbf{Z}^r$, so

$$E(\mathbf{Q})/2E(\mathbf{Q}) \simeq (T/2T) \oplus \mathbf{Z}_2^r = T \oplus \mathbf{Z}_2^r.$$

Since $E(\mathbf{Q})/2E(\mathbf{Q})$ has order 4, we must have $r = 0$. Therefore,

$$E(\mathbf{Q}) = E[2] = \{\infty, (0,0), (2,0), (-2,0)\}.$$

$\square$

Example 8.8

Let E be the curve

$$y^2 = x^3 - 25x.$$

This curve E appeared in Chapter 1, where we found the points

$$(0,0), (5,0), (-5,0), (-4,6).$$

We also calculated the point

$$2(-4,6) = (\frac{41^2}{12^2}, \frac{-62279}{1728}).$$

Since $2(-4,6)$ does not have integer coordinates, $(-4,6)$ cannot be a torsion point, by Theorem 8.7. In fact, a calculation using the Lutz-Nagell theorem shows that the torsion subgroup is

$$T = \{\infty, (0,0), (5,0), (-5,0)\} \simeq \mathbf{Z}_2 \oplus \mathbf{Z}_2.$$

We claim that

$$E(\mathbf{Q}) \simeq \mathbf{Z}_2 \oplus \mathbf{Z}_2 \oplus \mathbf{Z}.$$

We know that the rank r is at least 1, because there is a point $(-4,6)$ of infinite order. The problem is to show that the rank is exactly 1.

Consider the map

$$\phi : E(\mathbf{Q}) \to (\mathbf{Q}^\times/\mathbf{Q}^{\times 2}) \oplus (\mathbf{Q}^\times/\mathbf{Q}^{\times 2}) \oplus (\mathbf{Q}^\times/\mathbf{Q}^{\times 2})$$

of Theorem 8.14 defined by

$$(x, y) \mapsto (x, \; x - 5, \; x + 5)$$

when $y \neq 0$. Therefore,

$$\phi(-4,6) = (-1, -1, 1),$$

where we have used the fact that -4 and -9 are equivalent to -1 mod squares. Also, from Theorem 8.14,

$$\phi(\infty) = (1, 1, 1)$$
$$\phi(0,0) = (-1, -5, 5)$$
$$\phi(5,0) = (5, 2, 10)$$
$$\phi(-5,0) = (-5, -10, 2).$$

Since ϕ is a homomorphism, we immediately find that $\phi(-4,6)$ times any of these triples is in the image of ϕ, so

$$(1,5,5), (-5,-2,10), (5,10,2)$$

correspond to points.

If we write

$$x = au^2$$
$$x - 5 = bv^2$$
$$x + 5 = cw^2,$$

we have $\phi(x,y) = (a,b,c)$. From Proposition 8.13, we may assume

$$a, b, c \in \{\pm 1, \pm 2, \pm 5, \pm 10\}.$$

Also, abc is a square, so c is determined by a, b. Therefore, we'll often ignore c and concentrate on the possibilities for a, b. There are 64 possible pairs a, b. So far, we have 8 pairs that correspond to points. Let's record them in a list, which we'll refer to as L in the following:

$$L = \{(1,1), (1,5), (-1,-1), (-1,-5), (5,2), (5,10), (-5,-2), (-5,-10)\}.$$

Our job is to eliminate the remaining 56 possibilities.

Observe that

$$x - 5 = bv^2 < x = au^2 < x + 5 = cw^2.$$

If $a < 0$, then $b < 0$. If $a > 0$ then $c > 0$, hence $b > 0$ since abc is a square. Therefore, a and b have the same sign. This leaves 32 possible pairs a, b.

We now consider, and eliminate, three special pairs a, b. The fact that ϕ is a homomorphism will then suffice to eliminate all but the eight pairs corresponding to known points.

(a,b)=(2,1). We have

$$x = 2u^2$$
$$x - 5 = v^2$$
$$x + 5 = 2w^2.$$

Therefore,

$$2u^2 - v^2 = 5, \quad 2w^2 - 2u^2 = 5.$$

If one of u or v has an even denominator, then so does the other. However, $2u^2$ has an odd power of 2 in its denominator, while v^2 has an even power of 2 in its denominator. Therefore, $2u^2 - v^2$ is not an integer, contradiction. It follows that u, v have odd denominators, so we may work with them mod powers of 2. Since $v^2 \equiv -5 \pmod 2$, we must have v odd. Therefore, $v^2 \equiv 1$ (mod 8), so

$$2u^2 \equiv 6 \pmod 8.$$

This implies that $u^2 \equiv 3 \pmod 4$, which is impossible. Therefore, the pair $(a, b) = (2, 1)$ is eliminated.

(a,b)=(5,1). We have

$$x = 5u^2$$
$$x - 5 = v^2$$
$$x + 5 = 5w^2.$$

Therefore,

$$5u^2 - v^2 = 5, \quad 5w^2 - 5u^2 = 5.$$

If the denominator of one of u or v is divisible by 5, then so is the other. But $5u^2$ then has an odd power of 5 in its denominator, while v^2 has an even power of 5 in its denominator. This is impossible, so the denominators of both u and v are not divisible by 5. Since $w^2 - u^2 = 1$, the same holds for w. Therefore, we can work with u, v, w mod 5. We have $v \equiv 0 \pmod 5$, so we can write $v = 5v_1$. Then

$$u^2 - 5v_1^2 = 1,$$

so $u^2 \equiv 1 \pmod 5$. Therefore, $w^2 = 1 + u^2 \equiv 2 \pmod 5$. This is impossible. Therefore, the pair $(a, b) = (5, 1)$ is eliminated.

(a,b)=(10, 1). We have

$$x = 10u^2$$
$$x - 5 = v^2$$
$$x + 5 = 10w^2.$$

Therefore,

$$10u^2 - v^2 = 5, \quad 10w^2 - 10u^2 = 5.$$

As before, the denominators of u, v, w are not divisible by 5. Write $v = 5v_1$. Then $2u^2 - 5v_1^2 = 1$, so $2u^2 \equiv 1 \pmod 5$. This is impossible, so the pair $(a, b) = (10, 1)$ is eliminated.

The pairs $(a, 1)$ with $a < 0$ are eliminated since a, b must have the same sign. Therefore, $(1, 1) = \phi(\infty)$ is the only pair of the form $(a, 1)$ corresponding to a point.

Let (a, b) be any pair with $a = \pm 1$ or ± 5. There is a point P with $\phi(P) = (a, b')$ on the list L for some b'. If there is a point Q with $\phi(Q) = (a, b)$, then

$$\phi(P - Q) = (a, b)(a, b')^{-1} = (1, b'')$$

for some b''. If $b'' \neq 1$, then we showed above that $(1, b'')$ is not in the image of ϕ. Therefore, $b'' = 1$, so $b = b'$. Therefore, the only pairs with $a = \pm 1$ or ± 5 that are in the image of ϕ are those on the list L.

Suppose (a, b) is a pair that has been eliminated and $(a', b') = \phi(P)$ for some point P. If $(aa', bb') = \phi(Q)$ for some point Q, then $\phi(Q - P) = (a, b)$, contradiction. This observation allows us to eliminate the remaining pairs, as follows.

Since the pair $(2, 1)$ has been eliminated, if we multiply the elements of the list L by $(2, 1)$, we eliminate the pairs

$$(2, 1), (2, 5), (-2, -1), (-2, -5), (10, 2), (10, 10), (-10, -2), (-10, -10).$$

Multiplying the list L by $(5, 1)$ eliminates the pairs

$$(5, 1), (5, 5), (-5, -1), (-5, -5), (1, 2), (1, 10), (-1, -2), (-1, -10).$$

Multiplying the list L by $(10, 1)$ eliminates the pairs

$$(10, 1), (10, 5), (-10, -1), (-10, -5), (2, 2), (2, 10), (-2, -2), (-2, -10).$$

The only remaining pairs are those on the list L, which correspond to known points.

As stated above, the torsion subgroup of $E(\mathbf{Q})$ is $E[2]$, so

$$E(\mathbf{Q})/2E(\mathbf{Q}) \simeq \mathbf{Z}_2 \oplus \mathbf{Z}_2 \oplus \mathbf{Z}_2^r$$

for some r. Since the image of ϕ has order 8 and the kernel of ϕ is $2E(\mathbf{Q})$, the order of $E(\mathbf{Q})/2E(\mathbf{Q})$ is 8. Therefore, $r = 1$. This implies that

$$E(\mathbf{Q}) \simeq \mathbf{Z}_2 \oplus \mathbf{Z}_2 \oplus \mathbf{Z}.$$

Note that we have also proved that $E[2]$ and $(-4, 6)$ generate a subgroup of $E(\mathbf{Q})$ of odd index. It can be shown that they actually generate the whole group. This would require making the constants in the proof of Theorem 8.17 more explicit, then finding all points with heights less than an explicit bound to obtain a generating set.   ∎

Silverman [91] proved the following.

THEOREM 8.23
Let E be defined over $\mathbf{Q}$ by the equation

$$y^2 = x^3 + Ax + B$$

with $A, B \in \mathbf{Z}$. Then

$$-\frac{1}{8}h(j) - \frac{1}{12}h(\Delta) - 0.973 \le \hat{h}(P) - \frac{1}{2}h(P)$$

$$\le \frac{1}{12}h(j) + \frac{1}{12}h(\Delta) + 1.07$$

for all $P \in E(\mathbf{Q})$. Here $\Delta = -16(4A^3 + 27B^2)$ and $j = -1728(4A)^3/\Delta$.

For the curve $y^2 = x^3 - 25x$, we have $\Delta = 10^6$ and $j = 1728$. Therefore,

$$-3.057 < \hat{h}(P) - \frac{1}{2}h(P) < 2.843$$

for all $P \in E(\mathbf{Q})$. The points $(0,0), (5,0), (-5,0), (-4,6)$ generate the group $E(\mathbf{Q})/2E(\mathbf{Q})$. The first three of these points have canonical height 0 since they are torsion points. The point $(-4,6)$ has canonical height $0.94974\ldots$ (this can be calculated using the series (8.6)). The proof of Theorem 8.17 shows that the points with canonical height at most $0.94974\ldots$ generate $E(\mathbf{Q})$. Theorem 8.23 says that such points have non-canonical height $h(P) <$ 8.02. Since $e^{8.02} \approx 3041$, the non-logarithmic height of the x-coordinate is at most 3041. Therefore, we need to find all points $(x,y) \in E(\mathbf{Q})$ such that

$$x = \frac{a}{b} \quad \text{with} \quad \text{Max}(|a|, |b|) \le 3041.$$

It is possible to find all such points using a computer. The fact that the denominator of x must be a perfect square can be used to speed up the search. We find the points

$$(0,0), \quad (-5,0), \quad (5,0), \quad (-4,6)$$
$$(45, -300) = (-5,0) + (-4,6)$$
$$(25/4, 75/8) - (0,0) + (-4,6)$$
$$(-5/9, -100/27) = (5,0) + (-4,6)$$
$$(1681/144, -62279/1728) = 2(-4,6)$$

and the negatives of these points. Since these points generate $E(\mathbf{Q})$, we conclude that $(0,0), (5,0), (-5,0), (-4,6)$ generate $E(\mathbf{Q})$.

REMARK 8.24   In Chapter 1, we needed to find an x such that x, $x - 5$, and $x + 5$ were all squares. We did this by starting with the point $(-4, 6)$ and finding the other point of intersection of the tangent line with the curve. In effect, we computed

$$2(-4, 6) = (\frac{41^2}{12^2}, \frac{-62279}{1728})$$

and miraculously obtained $x = 41^2/12^2$ with the desired property. We now see that this can be explained by the fact that ϕ is a homomorphism. Since $\phi(2P) = (1,1,1)$ for any point P, we always obtain an x such that x, $x - 5$, and $x + 5$ are squares when we double a point on the curve $y^2 = x(x-5)(x+5)$.

Example 8.9

One use of descent is to find points on elliptic curves. The idea is that in the equations

$$x - e_1 = au^2$$
$$x - e_2 = bv^2$$
$$x - e_3 = cw^2,$$

the numerators and denominators of u, v, w are generally smaller than those of x. Therefore, an exhaustive search for u, v, w is faster than searching for x directly. For example, suppose we are looking for points on

$$y^2 = x^3 - 36x.$$

One of the triples that we encounter is $(a, b, c) = (3, 6, 2)$. This gives the equations

$$x = 3u^2$$
$$x - 6 = 6v^2$$
$$x + 6 = 2w^2.$$

These can be written as

$$3u^2 - 6v^2 = 6, \quad 2w^2 - 3u^2 = 6,$$

which simplify to

$$u^2 - 2v^2 = 2, \quad 2w^2 - 3u^2 = 6.$$

A quick search through small values of u yields $(u, v, w) = (2, 1, 3)$. This gives

$$(x, y) = (12, 36).$$

Note that the value of u is smaller than x. Of course, we are lucky in this example since the value of u turned out to be integral. Otherwise, we would have had to search through values of u with small numerator and small denominator.

The curve $y^2 = x^3 - 36x$ can be transformed to the curve $y^2 = x(x+1)(2x+1)/6$ that we met in Chapter 1 (see Exercise 1.5). The point $(1/2, 1/2)$ on that curve corresponds to the point $(12, 36)$ that we found here.   ☐

Example 8.10

The elliptic curves that we have seen up to now have had small generators for their Mordell-Weil groups. However, frequently the generators of Mordell-Weil groups have very large heights. For example, the Mordell-Weil group of the elliptic curve (see [61])

$$y^2 = x^3 - 59643$$

over $\mathbf{Q}$ is infinite cyclic, generated by

$$\left(\frac{62511752209}{9922500}, \frac{15629405421521177}{31255875000} \right)$$

(there are much larger examples, but the margin is not large enough to contain them). This curve can be transformed to the curve $u^3 + v^3 = 94$ by the techniques of Section 2.5.2. $\square$

8.5 The Height Pairing

Suppose we have points $P_1, \ldots, P_r$ that we want to prove are independent. How do we do it?

THEOREM 8.25

*Let E be an elliptic curve defined over $\mathbf{Q}$ and let $\hat{h}$ be the canonical height function. For $P, Q \in E(\mathbf{Q})$, define the **height pairing***

$$\langle P, Q \rangle = \hat{h}(P + Q) - \hat{h}(P) - \hat{h}(Q).$$

Then $\langle \, , \, \rangle$ is bilinear in each variable. If $P_1, \ldots, P_r$ are points in $E(\mathbf{Q})$, and the $r \times r$ determinant

$$\det(\langle P_i, P_j \rangle) \neq 0,$$

then $P_1, \ldots, P_r$ are independent (that is, if there are integers a_i such that $a_1 P_1 + \cdots + a_r P_r = \infty$, then $a_i = 0$ for all i).

PROOF   The second part of the theorem is true for any bilinear pairing. Let's assume for the moment that the pairing is bilinear and prove the second part. Suppose $a_1 P_1 + \cdots + a_r P_r = \infty$, and $a_r \neq 0$, for example. Then a_r times the last row of the matrix $\langle P_i, P_j \rangle$ is a linear combination of the first $r - 1$ rows. Therefore, the determinant vanishes. This contradiction proves that the points must be independent.

The proof of bilinearity is harder. Since the pairing is symmetric (that is, $\langle P, Q \rangle = \langle Q, P \rangle$), it suffices to prove bilinearity in the first variable:

$$\langle P + Q, R \rangle = \langle P, R \rangle + \langle Q, R \rangle.$$

Recall the parallelogram law:

$$\hat{h}(S + T) + \hat{h}(S - T) = 2\hat{h}(S) + 2\hat{h}(T).$$

Successively letting $(S, T) = (P + Q, R)$, $(P, Q - R)$, $(P + R, Q)$, and (Q, R) yields the following equations:

$$\hat{h}(P + Q + R) + \hat{h}(P + Q - R) = 2\hat{h}(P + Q) + 2\hat{h}(R)$$
$$2\hat{h}(P) + 2\hat{h}(Q - R) = \hat{h}(P + Q - R) + \hat{h}(P - Q + R)$$
$$\hat{h}(P + R + Q) + \hat{h}(P + R - Q) = 2\hat{h}(P + R) + 2\hat{h}(Q)$$
$$4\hat{h}(Q) + 4\hat{h}(R) = 2\hat{h}(Q + R) + 2\hat{h}(Q - R).$$

Adding together all of these equations yields

$$2\left(\hat{h}(P + Q + R) - \hat{h}(P + Q) - \hat{h}(R)\right)$$
$$= 2\left(\hat{h}(P + R) - \hat{h}(P) - \hat{h}(R) + \hat{h}(Q + R) - \hat{h}(Q) - \hat{h}(R)\right).$$

Dividing by 2 and using the definition of the pairing yields the result. ∎

Example 8.11
Let E be given by $y^2 = x^3 + 73$. Let $P = (2, 9)$ and $Q = (3, 10)$. Then

$$\langle P, P \rangle = 0.9239\ldots$$
$$\langle P, Q \rangle = -0.9770\ldots$$
$$\langle Q, Q \rangle = 1.9927\ldots.$$

Since

$$\det\begin{pmatrix} 0.9239 & -0.9770 \\ -0.9770 & 1.9927 \end{pmatrix} = 0.8865\cdots \neq 0,$$

the points P and Q are independent on E. ☐

8.6   Fermat's Infinite Descent

The methods in this chapter have their origins in Fermat's **method of infinite descent**. In the present section, we'll give an example of Fermat's method and show how it relates to the calculations we have been doing.

Consider the equation

$$a^4 + b^4 = c^2. \tag{8.13}$$

The goal is to show that it has no solutions in nonzero integers a, b, c. Recall the parameterization of **Pythagorean triples**:

PROPOSITION 8.26

Suppose x, y, z are relatively prime positive integers such that

$$x^2 + y^2 = z^2.$$

Then one of x, y is even. Suppose it is x. Then there exist positive integers m, n such that

$$x = 2mn, \quad y = m^2 - n^2, \quad z = m^2 + n^2.$$

Moreover, $\gcd(m, n) = 1$ and $m \not\equiv n$ (mod 2).

This result is proved in most elementary number theory texts. Alternatively, see Exercise 2.16.

Suppose now that there are nonzero integers a, b, c satisfying Equation 8.13. We may assume a, b, c are positive and relatively prime. Proposition 8.26 implies we may assume that a is even and that there exist integers m, n with

$$a^2 = 2mn, \quad b^2 = m^2 - n^2, \quad c = m^2 + n^2.$$

If n is odd, then m is even, which implies that $b^2 \equiv -1$ (mod 4). This is impossible, so n is even and m is odd. Write $n = 2q$ for some integer q. We then have

$$(a/2)^2 = mq.$$

Since $\gcd(m, n) = 1$, we also have $\gcd(m, q) = 1$. Since m, q are relatively prime and their product is a square, it follows easily from looking at the prime factorizations of m, q that both m and q must be squares:

$$m = t^2, \quad q = u^2$$

for some positive integers t, u. Therefore, we have

$$b^2 = m^2 - n^2 = t^4 - 4u^4.$$

This may be rewritten as

$$(2u^2)^2 + b^2 = t^4.$$

Since m is odd, t is odd. Since $\gcd(m, q) = 1$, we also have $\gcd(t, u) = 1$. Therefore, $\gcd(t, 2u^2) = 1$. Proposition 8.26 implies that

$$2u^2 = 2vw, \quad b = v^2 - w^2, \quad t^2 = v^2 + w^2$$

with $\gcd(v, w) = 1$. Since the product vw is a square, it follows that both v and w are squares:

$$v = r^2, \quad w = s^2.$$

Therefore, $t^2 = v^2 + w^2$ becomes

$$t^2 = r^4 + s^4.$$

This is the same equation we started with. Since

$$0 < t \le t^4 = m^2 < c, \tag{8.14}$$

we have proved that for every triple (a, b, c) with $a^4 + b^4 = c^2$, there is another solution (r, s, t) with $0 < t < c$. We therefore have an infinitely descending sequence $c > t > \ldots$ of positive integers. This is impossible. Therefore, there is no solution (a, b, c).

Observe that $m^2 > n^2$, so $c < 2m^2 = 2t^4$. Combining this with Equation 8.14 yields

$$t^4 < c < 2t^4.$$

This implies that the logarithmic height of t is approximately one fourth the logarithmic height of c. Recall that the canonical height of $2P$ is four times the height of P. Therefore, we suspect that Fermat's procedure amounts to halving a point on an elliptic curve. We'll show that this is the case.

We showed in Section 2.5.3 that the transformation

$$x = \frac{2(z+1)}{w^2}, \quad y = \frac{4(z+1)}{w^3}$$

maps the curve

$$C : w^2 = z^4 + 1$$

to the curve

$$E : y^2 = x^3 - 4x.$$

If we start with

$$a^4 + b^4 = c^2,$$

then the point

$$(z, w) = (\frac{a}{b}, \frac{c}{b^2})$$

lies on C. It maps to a point (x, y) on E, with

$$x = \frac{2(\frac{c}{b^2} + 1)}{(a/b)^2} = \frac{2(c + b^2)}{a^2}$$

$$= \frac{2(t^4 + 4r^4 s^4 + (r^4 - s^4)^2)}{(2rst)^2}$$

$$= \left(\frac{t}{rs}\right)^2.$$

This implies that

$$x - 2 = \frac{t^2 - 2r^2 s^2}{(rs)^2} = \left(\frac{r^2 - s^2}{rs}\right)^2$$

$$x + 2 = \frac{t^2 + 2r^2 s^2}{(rs)^2} = \left(\frac{r^2 + s^2}{rs}\right)^2.$$

Let ϕ be the map in Theorem 8.14. Since x, $x-2$, $x+2$ are squares, $\phi(x, y) = 1$. Theorem 8.14 implies that

$$(x, y) = 2P$$

for some point $P \in E(\mathbf{Q})$.

Let's find P. We follow the procedure used to prove Theorem 8.14. In the notation of the proof of Theorem 8.14, the polynomial

$$f(T) = \frac{t}{rs} - \frac{s}{rs}T + \frac{r^2 - t}{4rs}T^2$$

satisfies

$$f(0) = \frac{t}{rs}, \quad f(2) = \frac{r^2 - s^2}{rs}, \quad f(-2) = \frac{r^2 + s^2}{rs}.$$

The formulas from the proof of Theorem 8.14 say that the point (x_1, y_1) with

$$x_1 = \frac{-s/2r}{(r^2 - t)/4rs} = \frac{-2s^2}{r^2 - t}$$

$$y_1 = \frac{4rs}{r^2 - t}$$

satisfies $2(x_1, y_1) = (x, y)$.

The transformation

$$z = \frac{2x}{y}, \quad w = -1 + \frac{2x^3}{y^2}$$

maps E to C. The point (x_1, y_1) maps to

$$z_1 = \frac{2x_1}{y_1} = -\frac{s}{r}$$

$$w_1 = -1 + \frac{2x_1^3}{y_1^2} = -1 - \frac{s^4}{r^2(r^2 - t)}$$

$$= -\frac{r^4 + s^4 - r^2 t}{r^2(r^2 - t)} = -\frac{t^2 - r^2 t}{r^2(r^2 - t)}$$

$$= \frac{t}{r^2}.$$

We have

$$\left(\frac{t}{r^2}\right)^2 = \left(\frac{-s}{r}\right)^4 + 1.$$

Therefore, the solution $(r, -s, t)$ corresponds to a point P on E such that $2P$ corresponds to (a, b, c). Fermat's procedure, therefore, can be interpreted as starting with a point on an elliptic curve and halving it. The height decreases by a factor of 4. The procedure cannot continue forever, so we must conclude that there are no nontrivial solutions to start with.

On $y^2 = x^3 - 4x$, the points of order 2 played a role in the descent procedure in Section 8.2. We showed that the image of the map ϕ was equal to the image of $E[2]$ under ϕ. If we start with a possible point $P \in E(\mathbf{Q})$, then $\phi(P) = \phi(T)$ for some $T \in E[2]$. Therefore, $P - T = 2Q$ for some $Q \in E(\mathbf{Q})$. In Fermat's method, the points of order 2 appear more subtly. If (x, y) on E corresponds to the solution a, b, c of $a^4 + b^4 = c^2$, then a calculation shows that

$$(x, y) + (0, 0) \longleftrightarrow -a, b, -c$$
$$(x, y) + (2, 0) \longleftrightarrow -b, a, c$$
$$(x, y) + (-2, 0) \longleftrightarrow b, a, -c.$$

Since we assumed that a was even and b was odd, we removed the solutions $\pm b, a, \mp c$ from consideration. The solution $-a, b, -c$ was implicitly removed by the equation $c = m^2 + n^2$, which required c to be positive. Therefore, the choices that were made, which seemed fairly natural and innocent, were exactly those that caused $\phi(P)$ to be trivial and thus allowed us to halve the point.

Finally, we note that in the descent procedure for E in Section 8.2, we eliminated many possibilities by congruences mod powers of 2. The considerations also appear in Fermat's method, for example, in the argument that n is even.

In Fermat's descent, the equation

$$b^2 = t^4 - 4u^4$$

appears in an intermediate stage. This means we are working with the point $(w, z) = (u/t, b/t^2)$ on the curve

$$C' : w^2 = -4z^4 + 1.$$

The transformation (see Theorem 2.17)

$$x' = \frac{2(z + 1)}{w^2}, \quad y' = \frac{4(z + 1)}{w^3}$$

maps C' to the elliptic curve

$$E' : y'^2 = x'^3 + 16x'.$$

There is a map $\psi : E \to E'$ given by

$$(x', y') = \psi(x, y) = \left(\frac{y^2}{x^2}, \frac{y(x^2 + 4)}{x^2} \right).$$

There is also a map $\psi' : E' \to E$ given by

$$(x, y) = \psi'(x', y') = \left(\frac{y'^2}{4x'^2}, \frac{y'(x'^2 - 16)}{8x'^2} \right).$$

It can be shown that $\psi' \circ \psi$ is multiplication by 2 on E. Fermat's descent procedure can be analyzed in terms of the maps ψ and ψ'.

More generally, if E is an elliptic curve given by $y^2 = x^3 + Cx^2 + Ax$ and E' is given by $y'^2 = x'^3 - 2Ax'^2 + (A^2 - 4C)x'$, then there are maps $\psi : E \to E'$ given by

$$(x', y') = \psi(x, y) = \left(\frac{y^2}{x^2}, \frac{y(x^2 - C)}{x^2} \right), \quad \psi(0,0) = \psi(\infty) = \infty,$$

and $\psi' : E' \to E$ given by

$$(x, y) = \psi'(x', y') = \left(\frac{y'^2}{4x'^2}, \frac{y'(x'^2 - A^2 + 4C)}{8x'^2} \right), \quad \psi'(0,0) = \psi'(\infty) = \infty.$$

The composition $\psi' \circ \psi$ is multiplication by 2 on E. It is possible to do descent and prove the Mordell-Weil theorem using the maps ψ and ψ'. This is a more powerful method than the one we have used since it requires only one two-torsion to be rational, rather than all three. For details, see [95], [90].

The maps ψ and ψ' can be shown to be homomorphisms between $E(\mathbf{Q})$ and $E'(\mathbf{Q})$ and are described by rational functions. In general, for elliptic curves E_1 and E_2 over a field K, a homomorphism from $E_1(\overline{K})$ to $E_2(\overline{K})$ that is given by rational functions is called an **isogeny**.

8.7   2-Selmer Groups; Shafarevich-Tate Groups

Let's return to the basic descent procedure of Section 8.2. We start with an elliptic curve E defined over $\mathbf{Q}$ by

$$y^2 = (x - e_1)(x - e_2)(x - e_3)$$

with all $e_i \in \mathbf{Z}$. This leads to equations

$$x - e_1 = au^2$$
$$x - e_2 = bv^2$$
$$x - e_3 = cw^2.$$

These lead to the equations

$$au^2 - bv^2 = e_2 - e_1, \quad au^2 - cw^2 = e_3 - e_1.$$

This defines a curve $C_{a,b,c}$ in u, v, w. In fact, it is the intersection of two quadratic surfaces. If it has a rational point, then it can be changed to an elliptic curve, as in Section 2.5.4. A lengthy calculation, using the formulas of

Theorem 2.17, shows that this elliptic curve is the original curve E. If $C_{a,b,c}$ does not have any rational points, then the triple (a, b, c) is eliminated.

The problem is how to decide which curves $C_{a,b,c}$ have rational points. In the examples of Section 8.2, we used considerations of sign and congruences mod powers of 2 and 5. These can be interpreted as showing that the curves $C_{a,b,c}$ that are being eliminated have no real points, no 2-adic points, or no 5-adic points (for a summary of the relevant properties of p-adic numbers, see Appendix A). For example, when we used inequalities to eliminate the triple $(a, b, c) = (-1, 1, -1)$ for the curve $y^2 = x(x-2)(x+2)$, we were showing that the curve

$$C_{-1,1,-1} : -u^2 - v^2 = 2, \quad -u^2 + w^2 = -2$$

has no real points. When we eliminated $(a, b, c) = (1, 2, 2)$, we used congruences mod powers of 2. This meant that

$$C_{1,2,2} : u^2 - 2v^2 = 2, \quad u^2 - 2w^2 = -2$$

has no 2-adic points.

The **2-Selmer group** S_2 is defined to be the set of (a, b, c) such that $C_{a,b,c}$ has a real point and has p-adic points for all p. For notational convenience, the real numbers are sometimes called the ∞-adics $\mathbf{Q}_\infty$. Instead of saying that something holds for the reals and for all the p-adics $\mathbf{Q}_p$, we say that it holds for $\mathbf{Q}_p$ for all $p \leq \infty$. Therefore,

$$S_2 = \{(a, b, c) \mid C_{a,b,c}(\mathbf{Q}_p) \text{ is non-empty for all } p \leq \infty\}.$$

Therefore, S_2 is the set of (a, b, c) that cannot be eliminated by sign or congruence considerations. It is a group under multiplication mod squares. Namely, we regard

$$S_2 \subset (\mathbf{Q}^\times / \mathbf{Q}^{\times 2}) \oplus (\mathbf{Q}^\times / \mathbf{Q}^{\times 2}) \oplus (\mathbf{Q}^\times / \mathbf{Q}^{\times 2}).$$

The prime divisors of a, b, c divide $(e_1 - e_2)(e_1 - e_3)(e_2 - e_3)$, which implies that S_2 is a finite group.

The descent map ϕ gives a map

$$\phi : E(\mathbf{Q})/2E(\mathbf{Q}) \hookrightarrow S_2.$$

The 2-torsion in the **Shafarevich-Tate group** is the cokernel of this map:

$$\text{Ш}_2 = S_2/\text{Im } \phi.$$

The symbol Ш is the Cyrillic letter "sha," which is the first letter of "Shafarevich" (in Cyrillic). We'll define the full group Ш in Section 8.9. The group Ш_2 represents those triples (a, b, c) such that $C_{a,b,c}$ has a p-adic point for all $p \leq \infty$, but has no rational point. If $\text{Ш}_2 \neq 1$, then it is much more difficult to find the points on the elliptic curve E. If (a, b, c) represents a nontrivial element of Ш, then it is usually difficult to show that $C_{a,b,c}$ does not have rational points.

Suppose we have an elliptic curve on which we want to find rational points. If we do a 2-descent, then we encounter curves $C_{a,b,c}$. If we search for points on a curve $C_{a,b,c}$ and also try congruence conditions, both with no success, then perhaps (a, b, c) represents a nontrivial element of $Ш_2$. Or we might need to search longer for points. It is difficult to decide which is the case. Fortunately for Fermat, the curves on which he did 2-descents had trivial $Ш_2$.

The possible nontriviality of the group $Ш_2$ means that we do not have a general procedure for finding the rank of the group $E(\mathbf{Q})$. The group S_2 can be computed exactly and allows us to obtain an upper bound for the rank. But we do not know how much of S_2 is the image of ϕ and how much consists of triples (a, b, c) representing elements of a possibly nontrivial $Ш_2$. Since the generators of $E(\mathbf{Q})$ can sometimes have very large height, it is sometimes quite difficult to find points representing elements of the image of ϕ. Without this information, we don't know that the triple is actually in the image.

The Shafarevich-Tate group is often called the **Tate-Shafarevich group** in English and the Shafarevich-Tate group in Russian. Since $Ш$ comes after T in the Cyrillic alphabet, these names for the group, in each language, are the reverse of the standard practice in mathematics, which is to put names in alphabetic order. The symbol $Ш$ was given to the group by Cassels (see [17, p. 109]).

REMARK 8.27   The Hasse-Minkowski theorem (see [86]) states that a quadratic form over the rationals represents 0 nontrivially over $\mathbf{Q}$ if and only if it represents 0 nontrivially in $\mathbf{Q}_p$ for all $p \leq \infty$. This is an example of a **local-global principle**. For a general algebraic variety over $\mathbf{Q}$ (for example, an algebraic curve), we can ask whether the local-global principle holds. Namely, if the variety has a p-adic point for all $p \leq \infty$, does it have a rational point? Since it is fairly easy to determine when a variety has p-adic points, and most varieties fail to have p-adic points for at most a finite set of p, this would make it easy to decide when a variety has rational points. However, the local-global principle fails in many cases. In Section 8.8, we'll give an example of a curve, one that arises in a descent on an elliptic curve, for which the local-global principle fails.   ∎

8.8   A Nontrivial Shafarevich-Tate Group

Let E be the elliptic curve over $\mathbf{Q}$ given by

$$y^2 = x(x - 2p)(x + 2p),$$

where p is a prime. If we do a 2-descent on E, we encounter the equations

$$x = u^2$$
$$x - 2p = pv^2$$
$$x + 2p = pw^2.$$

These yield the curve defined by the intersection of two quadratic surfaces:

$$C_{1,p,p} : u^2 - pv^2 = 2p, \quad u^2 - pw^2 = -2p. \tag{8.15}$$

THEOREM 8.28

If $p \equiv 9 \pmod{16}$, then $C_{1,p,p}$ has p-adic points for all $p \le \infty$, but has no rational points.

PROOF   First, we'll show that there are no rational points. Suppose there is a rational point (u, v, w). We may assume that $u, v, w > 0$. If p divides the denominator of v, then an odd power of p is in the denominator of pv^2 and an even power of p is in the denominator of u^2, so $u^2 - pv^2$ cannot be an integer, contradiction. Therefore, u, v, and hence also w have no p in their denominators. It follows easily that the denominators of u, v, w are equal. Since $u^2 = 2p + pv^2$, we have $u \equiv 0 \pmod{p}$. Write

$$u = \frac{pr}{e}, \quad v = \frac{s}{e}, \quad w = \frac{t}{e},$$

with positive integers r, s, t, e and with

$$\gcd(r, e) = \gcd(s, e) = \gcd(t, e) = 1.$$

The equations for $C_{1,p,p}$ become

$$pr^2 - s^2 = 2e^2, \quad pr^2 - t^2 = -2e^2.$$

Subtracting yields

$$s^2 + 4e^2 = t^2.$$

If s is even, then $pr^2 = s^2 + 2e^2$ is even, so r is even. Then $2e^2 = pr^2 - s^2 \equiv 0 \pmod{4}$, which implies that e is even. This contradicts the fact that $\gcd(s, e) = 1$. Therefore, s is odd, so

$$\gcd(s, 2e) = 1.$$

By Proposition 8.26, there exist integers m, n with $\gcd(m, n) = 1$ such that

$$2e = 2mn, \quad s = m^2 - n^2, \quad t = m^2 + n^2.$$

Therefore,

$$pr^2 = s^2 + 2e^2 = (m^2 - n^2)^2 + 2(mn)^2 = m^4 + n^4.$$

Let q be a prime dividing r. Proposition 8.26 says that $m \not\equiv n \pmod 2$, which implies that pr^2 must be odd. Therefore, $q \neq 2$. Since $\gcd(m, n) = 1$, at least one of m, n is not divisible by q. It follows that both m, n are not divisible by q, since $m^4 + n^4 \equiv 0 \pmod q$. Therefore,

$$(m/n)^4 \equiv -1 \pmod q.$$

It follows that m/n has order 8 in $\mathbf{F}_q^\times$, so $q \equiv 1 \pmod 8$. Since r is a positive integer and all prime factors of r are 1 mod 8, we obtain

$$r \equiv 1 \pmod 8.$$

Therefore, $r^2 = 1 \pmod{16}$, so

$$m^4 + n^4 = pr^2 \equiv 9 \pmod{16}.$$

But, for an arbitrary integer j, we have $j^4 \equiv 0, 1 \pmod{16}$. Therefore,

$$m^4 + n^4 \equiv 0, 1, 2 \pmod{16},$$

so $pr^2 \neq m^4 + n^4$. This contradiction proves that $C_{1,p,p}$ has no rational points.

We now need to show that $C_{1,p,p}$ has q-adic points for all primes $q \leq \infty$. The proof breaks into four cases: $q = \infty$, $q = 2$, $q = p$, and all other q.

The case of the reals is easy. Let u be large enough that $u^2 > 2p$. Then choose v, w satisfying Equation 8.15.

For $q - 2$, write

$$u = 1/2, \quad v = v_1/2, \quad w = w_1/2.$$

The equations for $C_{1,p,p}$ become

$$1 - pv_1^2 = 8p, \quad 1 - pw_1^2 = -8p.$$

We need to solve

$$v_1^2 = (1 - 8p)/p, \quad w_1^2 = (1 + 8p)/p$$

in the 2-adics. Since

$$(1 \pm 8p)/p \equiv 1 \pmod 8,$$

and since any number congruent to 1 mod 8 has a 2-adic square root (see Appendix A), v_1, w_1 exist. Therefore, $C_{1,p,p}$ has a 2-adic point.

Now let's consider $q = p$. Since $p \equiv 1 \pmod 4$, there is a square root of -1 mod p. Since $p \equiv 1 \pmod 8$, there is a square root of -2 mod p. Therefore, both 2 and -2 have square roots mod p. Hensel's lemma (see Appendix A) implies that both 2 and -2 have square roots in the p-adics. Let

$$u = 0, \quad v = \sqrt{-2}, \quad w = \sqrt{2}.$$

Then u, v, w is a p-adic point on $C_{1,p,p}$.

Finally, we need to consider $q \neq 2, p, \infty$. From a more advanced standpoint, we could say that the curve $C_{1,p,p}$ is a curve of genus 1 and that Hasse's theorem holds for such curves. If we use the estimates from Hasse's theorem, then we immediately find that $C_{1,p,p}$ has points mod q for all q (except maybe for a few small q, since we are not looking at the points at infinity on $C_{1,p,p}$). However, we have only proved Hasse's theorem for elliptic curves, rather than for arbitrary genus 1 curves. In the following, we'll use Hasse's theorem only for elliptic curves and show that $C_{1,p,p}$ has points mod q. Hensel's lemma then will imply that there is a q-adic point.

Subtracting the two equations defining $C_{1,p,p}$ allows us to put the equations into a more convenient form:

$$w^2 - v^2 = 4, \qquad u^2 - pv^2 = 2p. \tag{8.16}$$

Suppose we have a solution (u_0, v_0, w_0) mod q. It is impossible for both u_0 and w_0 to be 0 mod q.

Suppose $u_0 \equiv 0 \pmod{q}$. Then $w_0 \not\equiv 0 \pmod{q}$. Also, $v_0 \not\equiv 0 \pmod{q}$. Let $u = 0$. Since $-pv_0^2 \equiv 2p \pmod{q}$, Hensel's lemma says that there exists $v \equiv v_0 \pmod{q}$ in the q-adics such that $-pv^2 = 2p$. Applying Hensel's lemma again gives the existence of $w \equiv w_0$ satisfying $w^2 - v^2 = 4$. Therefore, we have found a q-adic point. Similarly, if $w_0 \equiv 0 \pmod{q}$, there is a q-adic point. Finally, suppose $u_0 \not\equiv 0 \pmod{q}$ and $w_0 \not\equiv 0 \pmod{q}$. Choose any $v \equiv v_0 \pmod{q}$. Now use Hensel's lemma to find u, w. This yields a q-adic point.

It remains to show that there is a point mod q. Let n be a quadratic nonresidue mod q. Then every element of $\mathbf{F}_q^\times$ is either of the form u^2 or nu^2. Consider the curve

$$C' : w^2 - v^2 = 4, \quad nu^2 - pv^2 = 2p.$$

Let N be the number of points mod q on $C_{1,p,p}$ and let N' be the number of points mod q on C'. (We are not counting points at infinity.)

LEMMA 8.29
$N + N' = 2(q - 1)$.

PROOF   Let $x \not\equiv 0 \pmod{q}$. Solving

$$w + v \equiv x, \quad w - v \equiv 4/x \pmod{q}$$

yields a pair (v, w) for each x. There are $q - 1$ choices for x, hence there are $q - 1$ pairs (v, w) satisfying $w^2 - v^2 = 4$. Let (v, w) be such a pair. Consider the congruences

$$u^2 \equiv 2p + pv^2 \pmod{q} \text{ and } nu^2 \equiv 2p + pv^2 \pmod{q}.$$

If $2p + pv^2 \not\equiv 0 \pmod{q}$, then exactly one of these has a solution, and it has 2 solutions. If $2p + pv^2 \equiv 0 \pmod{q}$, then both congruences have 1 solution. Therefore, each of the $q - 1$ pairs (v, w) contributes 2 to the sum $N + N'$, so $N + N' = 2(q - 1)$. ∎

The strategy now is the following. If $N > 0$, we're done. If $N' > 0$, then C' can be transformed into an elliptic curve with approximately N' points. Hasse's theorem then gives a bound on N', which will show that $N = 2(q - 1) - N' > 0$, so there must be points on $C_{1,p,p}$.

LEMMA 8.30
If $q \geq 11$, then $N > 0$.

PROOF   If $N = 0$ then $N' = 2(q-1) > 0$, by Lemma 8.29. In Section 2.5.4, we showed how to start with the intersection of two quadratic surfaces and a point and obtain an elliptic curve. Therefore, we can transform C' to an elliptic curve E'. By Hasse's theorem, E' has less than $q + 1 + 2\sqrt{q}$ points. We need to check that every point on C' gives a point on E'. In the parameterization

$$v = \frac{4t}{1 - t^2}, \quad w = \frac{2 + 2t^2}{1 - t^2} \tag{8.17}$$

of $w^2 - v^2 = 4$, the value $t = \infty$ corresponds to $(v, w) = (0, -2)$. All of the other points (v, w) correspond to finite values of t. No (finite) pair (v, w) corresponds to $t = \pm 1$ (the lines through $(0, 2)$ of slope $t = \pm 1$ are parallel to the asymptotes of the hyperbola). Substituting the parameterization (8.17) into $nu^2 - pv^2 = 2p$ yields the curve

$$Q' : \quad u_1^2 = \frac{2p}{n}(t^4 + 6t^2 + 1),$$

where $u_1 = (1 - t^2)u$. A point on C' with $(v, w) \neq (0, -2)$ yields a finite point on the quartic curve Q'. Since C' has $2(q - 1) > 1$ points mod q, there is at least one finite point on Q'. Section 2.5.3 describes how to change Q' to an elliptic curve E' (the case where Q' is singular does not occur since Q' is easily shown to be nonsingular mod q when $q \neq 2, p$). Every point mod q on Q' (including those at infinity, if they are defined over $\mathbf{F}_q$) yields a point (possibly ∞) on E' (points at infinity on Q' yield points of order 2 on E'). Therefore, the number of points on C' is less than or equal to the number of points on E'. By Hasse's theorem,

$$2(q - 1) = N' \leq q + 1 + 2\sqrt{q}.$$

This may be rearranged to obtain

$$(\sqrt{q} - 1)^2 \leq 4,$$

which yields $q \leq 9$. Therefore, if $q \geq 11$, we must have $N \neq 0$.   ∎

It remains to treat the cases $q = 3, 5, 7$. First, suppose p is a square mod q. There are no points on $C_{1,p,p}$ with coordinates in $\mathbf{F}_3$, for example, so we introduce denominators. Let's try

$$u = u_1/q, \quad v = 1/q, \quad w = w_1/q.$$

Then we want to solve

$$w_1^2 = 1 + 4q^2, \quad u_1^2 = p + 2pq^2.$$

Since p is assumed to be a square mod q, Hensel's lemma implies that there are q-adic solutions u_1, w_1.

Now suppose that p is not a square mod q. Divide the second equation in (8.16) by p to obtain

$$w^2 - v^2 = 4, \quad \frac{1}{p}u^2 - v^2 = 2.$$

Let n be any fixed quadratic nonresidue mod q, and write $1/p \equiv nx^2 \pmod{q}$. Letting $u_1 = xu$, we obtain

$$w^2 - v^2 = 4, \quad nu_1^2 - v^2 = 2.$$

For $q = 3$ and $q = 5$, we may take $n = 2$ and obtain

$$w^2 - v^2 \equiv 4, \quad 2u_1^2 - v^2 \equiv 2 \pmod{q}.$$

This has the solution $(u_1, v, w) = (1, 0, 2)$. As above, Hensel's lemma yields a q-adic solution.

For $q = 7$, take $n = 3$ to obtain

$$w^2 - v^2 \equiv 4, \quad 3u_1^2 - v^2 \equiv 2 \pmod{7}.$$

This has the solution $(u_1, v, w) = (3, 2, 1)$, which yields a 7-adic solution.

Therefore, we have shown that there is a q-adic solution for all $q \leq \infty$. This completes the proof of Theorem 8.28.   ∎

8.9   Galois Cohomology

In this section, we give the definition of the full Shafarevich-Tate group. This requires reinterpreting and generalizing the descent calculations in terms of Galois cohomology. Fortunately, we only need the first two cohomology groups, and they can be defined in concrete terms.

Let G be a group and let M be an additive abelian group on which G acts. This means that each $g \in G$ gives a automorphism $g : M \to M$. Moreover,

$$(g_1 g_2)(m) = g_1(g_2(m))$$

for all $m \in M$ and all $g_1, g_2 \in G$. We call such an M a **G-module**. One possibility is that g is the identity map for all $g \in G$. In this case, we say that the action of G is **trivial**.

If G is a topological group, and M has a topology, then we require that the action of G on M be continuous. We also require all maps to be continuous. In the cases below where the groups have topologies, this will always be the case, so we will not discuss this point further.

A **homomorphism** $\phi : M_1 \to M_2$ **of G-modules** is a homomorphism of abelian groups that is compatible with the action of G:

$$\phi(g m_1) = g \, \phi(m_1)$$

for all $g \in G$ and all $m_1 \in M_1$. Note that $\phi(m_1)$ is an element of M_2, so $g \, \phi(m_1)$ is the action of g on an element of M_2. An **exact sequence**

$$0 \to M_1 \to M_2 \to M_3 \to 0$$

is a short way of writing that the map from M_1 to M_2 is injective, the map from M_2 to M_3 is surjective, and the image of $M_1 \to M_2$ is the kernel of $M_2 \to M_3$. The most common situation is when $M_1 \subseteq M_2$ and $M_3 = M_2/M_1$.

More generally, a sequence of abelian groups and homomorphisms

$$\cdots \to A \to B \to C \to \cdots$$

is said to be **exact at** B if the image of $A \to B$ is the kernel of $B \to C$. Such a sequence is said to be **exact** if it is exact at each group in the sequence.

Define the **zeroth cohomology group** to be

$$H^0(G, M) = M^G = \{ m \in M \mid gm = m \text{ for all } g \in G \}.$$

For example, if G acts trivially, then $H^0(G, M) = M$.

Define the **cocycles**

$$Z(G, M) =$$
$$\{ \text{ maps } f : G \to M \mid f(g_1 g_2) = f(g_1) + g_1 \, f(g_2) \text{ for all } g_1, g_2 \in G \}.$$

The maps f are (continuous) maps of sets that are required to satisfy the given condition. Note that $g_1 \, f(g_2)$ means that we evaluate $f(g_2)$ and obtain an element of M, then act on this element of M by the automorphism g_1. The set Z is sometimes called the set of **twisted homomorphisms** from G to M. It is a group under addition of maps.

We note one important case. If G acts trivially on M, then

$$Z(G, M) = \operatorname{Hom}(G, M)$$

is the set of group homomorphisms from G to M.

There is an easy way to construct elements of $Z(G, M)$. Let m be a fixed element of M and define

$$f_m(g) = gm - m.$$

Then f_m gives a map from G to M. Since

$$\begin{aligned}
f_m(g_1 g_2) &= g_1(g_2 m) - m \\
&= g_1 m - m + g_1(g_2 m - m) \\
&= f_m(g_1) + g_1 f_m(g_2),
\end{aligned}$$

we have $f_m \in Z(G, M)$. Let

$$B(G, M) = \{ f_m \,|\, m \in M \}.$$

Then $B(G, M) \subseteq Z(G, M)$ is called the set of **coboundaries**. Define the **first cohomology group**

$$H^1(G, M) = Z/B.$$

In the important special case where G acts trivially, $B(G, M) = 0$ since $gm - m = 0$ for all g, m. Therefore

$$H^1(G, M) = \operatorname{Hom}(G, M)$$

is simply the set of group homomorphisms from G to M.

A homomorphism $\phi : M_1 \to M_2$ of G-modules induces a map

$$\phi_* : H^j(G, M_1) \to H^j(G, M_2)$$

of cohomology groups for $j = 0, 1$. For H^0, this is simply the restriction of ϕ to M_1^G. Note that if $g m_1 = m_1$, then $g\,\phi(m_1) = \phi(g m_1) = \phi(m_1)$, so ϕ maps M_1^G into M_2^G. For H^1, we obtain ϕ_* by taking an element $f \in Z$ and defining

$$(\phi_*(f))(g) = \phi(f(g)).$$

It is easy to see that this induces a map on cohomology groups.

The main property we need is the following.

PROPOSITION 8.31

An exact sequence

$$0 \to M_1 \to M_2 \to M_3 \to 0$$

of G-modules induces a long exact sequence

$$0 \to H^0(G, M_1) \to H^0(G, M_2) \to H^0(G, M_3)$$
$$\to H^1(G, M_1) \to H^1(G, M_2) \to H^1(G, M_3)$$

of cohomology groups.

For a proof, see any book on group cohomology, for example [108], [16], or [3]. The hardest part of the proposition is the existence of the map from $H^0(G, M_3)$ to $H^1(G, M_1)$.

Suppose now that we have an elliptic curve defined over $\mathbf{Q}$. Let n be a positive integer. Multiplication by n gives an endomorphism of E. By Theorem 2.21, it is surjective from $E(\overline{\mathbf{Q}}) \to E(\overline{\mathbf{Q}})$, since $\overline{\mathbf{Q}}$ is algebraically closed. Therefore, we have an exact sequence

$$0 \to E[n] \to E(\overline{\mathbf{Q}}) \xrightarrow{n} E(\overline{\mathbf{Q}}) \to 0. \tag{8.18}$$

Let

$$G = \mathrm{Gal}(\overline{\mathbf{Q}}/\mathbf{Q})$$

be the Galois group of $\overline{\mathbf{Q}}/\mathbf{Q}$. The reader who doesn't know what this group looks like should not worry. No one does. Much of modern number theory can be interpreted as trying to understand the structure of this group. The one property we need at the moment is that

$$H^0(G, E(\overline{\mathbf{Q}})) = E(\overline{\mathbf{Q}})^G = E(\mathbf{Q}).$$

Applying Proposition 8.31 to the exact sequence (8.18) yields the long exact sequence

$$0 \to E(\mathbf{Q})[n] \to E(\mathbf{Q}) \xrightarrow{n} E(\mathbf{Q})$$
$$\to H^1(G, E[n]) \to H^1(G, E(\overline{\mathbf{Q}})) \xrightarrow{n} H^1(G, E(\overline{\mathbf{Q}})).$$

This induces the short exact sequence

$$0 \to E(\mathbf{Q})/nE(\mathbf{Q}) \to H^1(G, E[n]) \to H^1(G, E(\overline{\mathbf{Q}}))[n] \to 0, \tag{8.19}$$

where we have written $A[n]$ for the n-torsion in an abelian group A. This sequence is similar to the sequence

$$0 \to E(\mathbf{Q})/2E(\mathbf{Q}) \to S_2 \to \mathrm{III}_2 \to 0$$

that we met in Section 8.7. In the remainder of this section, we'll show how the two sequences relate when $n = 2$ and also consider the situation for arbitrary n.

First, we give a way to construct elements of $H^1(G, E(\overline{\mathbf{Q}}))$. Let C be a curve defined over $\mathbf{Q}$ such that C is isomorphic to E over $\overline{\mathbf{Q}}$. This means that there is a map $\phi : E \to C$ given by rational functions with coefficients in $\overline{\mathbf{Q}}$ and an inverse function $\phi^{-1} : C \to E$ also given by rational functions with coefficients in $\overline{\mathbf{Q}}$. Let $g \in G$, and let ϕ^g denote the map obtained by applying g to the coefficients of the rational functions defining ϕ. Since C is defined over $\mathbf{Q}$, the map ϕ^g maps E to $gC = C$. Note that

$$g(\phi(P)) = (\phi^g)(gP) \tag{8.20}$$

for all $P \in E(\overline{\mathbf{Q}})$, since the expression $g(\phi(P))$ means we apply g to everything, while ϕ^g means applying g to the coefficients of ϕ and gP means applying g to P.

We have to be a little careful when applying $g_1 g_2$. The rule is

$$\phi^{g_1 g_2} = (\phi^{g_2})^{g_1},$$

since applying $g_1 g_2$ to the coefficients of ϕ means first applying g_2, then applying g_1 to the result.

We say that a map ϕ is **defined over Q** if $\phi^g(P) = \phi(P)$ for all $P \in E(\overline{\mathbf{Q}})$ and all $g \in G$ (this is equivalent to saying that the coefficients of the rational functions defining ϕ can be taken to be in $\mathbf{Q}$, though proving this requires results such as Hilbert's Theorem 90).

The map $\phi^{-1}\phi^g$ gives a map from E to E. We assume the following:

Assumption: Assume that there is a point $T_g \in E(\overline{\mathbf{Q}})$ such that

$$\phi^{-1}(\phi^g(P)) = P + T_g \tag{8.21}$$

for all $P \in E(\mathbf{Q})$. Equation (8.21) can be rewritten as

$$\phi^g(P) = \phi(P + T_0) \tag{8.22}$$

for all $P \in E(\mathbf{Q})$. If we let $P = (\phi^g)^{-1}(Q)$ for a point $Q \in C(\overline{\mathbf{Q}})$, then the assumption becomes

$$\phi^{-1}(Q) = (\phi^g)^{-1}(Q) + T_g, \tag{8.23}$$

which says that ϕ^{-1} and $(\phi^g)^{-1}$ differ by a translation. We'll give an example of such a map ϕ below.

LEMMA 8.32
Define $\tau_\phi : G \to E(\overline{\mathbf{Q}})$ by $\tau_\phi(g) = T_g$. Then $\tau_\phi \in Z(G, E(\overline{\mathbf{Q}}))$.

PROOF

$$
\begin{aligned}
g_1^{-1}\phi(P + T_{g_1 g_2}) &= g_1^{-1}\phi^{g_1 g_2}(P) \\
&= \phi^{g_2}(g_1^{-1}P) \quad \text{(by (8.20))} \\
&= \phi(g_1^{-1}P + T_{g_2}) \quad \text{(by (8.22))} \\
&= g_1^{-1}\phi^{g_1}(P + g_1 T_{g_2}) \quad \text{(by (8.20))} \\
&= g_1^{-1}\phi(P + g_1 T_{g_2} + T_{g_1}) \quad \text{(by (8.22))}.
\end{aligned}
$$

Applying g_1 then ϕ^{-1} yields

$$T_{g_1 g_2} = g_1 T_{g_2} + T_{g_1}.$$

This is the desired relation. ∎

Suppose we have curves C_i and maps $\phi_i : E \to C_i$, for $i = 1, 2$, as above. We say that the pairs (C_1, ϕ_1) and (C_2, ϕ_2) are **equivalent** if there is a map $\theta : C_1 \to C_2$ defined over $\mathbf{Q}$ and a point $P_0 \in E(\overline{\mathbf{Q}})$ such that

$$\phi_2^{-1} \theta \phi_1(P) = P + P_0 \tag{8.24}$$

for all $P \in E(\overline{\mathbf{Q}})$. In other words, if we identify C_1 and C_2 with E via ϕ_1 and ϕ_2, then θ is simply translation by P_0.

PROPOSITION 8.33

The pairs (C_1, ϕ_1) and (C_2, ϕ_2) are equivalent if and only if the cocycles τ_{ϕ_1} and τ_{ϕ_2} differ by a coboundary. This means that there is a point $P_1 \in E(\overline{\mathbf{Q}})$ such that

$$\tau_{\phi_1}(g) - \tau_{\phi_2}(g) = gP_1 - P_1$$

for all $g \in G$.

PROOF For $i = 1, 2$, denote $\tau_{phi_i}(g) = T_g^i$, so

$$\psi_i^g(P) = \phi_i(P + T_g^i) \tag{8.25}$$

for all $P \in E(\overline{\mathbf{Q}})$. Suppose the pairs (C_1, ϕ_1) and (C_2, ϕ_2) are equivalent, so there exists $\theta : C_1 \to C_2$ and P_0 as above. For any $P \in E(\overline{\mathbf{Q}})$, we have

$$
\begin{aligned}
P + T_g^1 + P_0 &= \phi_2^{-1} \theta \phi_1(P + T_g^1) \quad \text{(by (8.24))} \\
&= \phi_2^{-1} \theta \phi_1^g(P) \quad \text{(by (8.25))} \\
&= \phi_2^{-1} \phi_2^g (\phi_2^{-1} \theta \phi_1)^g(P) \quad \text{(since } \theta^g = \theta\text{)} \\
&= (\phi_2^{-1} \theta \phi_1)^g(P) + T_g^2 \quad \text{(by (8.21))} \\
&= g(\phi_2^{-1} \theta \phi_1)(g^{-1}P) + T_g^2 \quad \text{(by (8.20))} \\
&= g(g^{-1}P + P_0) + T_g^2 \quad \text{(by (8.24))} \\
&= P + gP_0 + T_g^2.
\end{aligned}
$$

Therefore,

$$T_g^1 - T_g^2 = \tau_{\phi_1}(g) - \tau_{\phi_2}(g) = gP_0 - P_0.$$

Conversely, suppose there exists P_1 such that

$$\tau_{\phi_1}(g) - \tau_{\phi_2}(g) = gP_1 - P_1. \tag{8.26}$$

Define $\theta : C_1 \to C_2$ by

$$\theta(Q) = \phi_2(\phi_1^{-1}(Q) + P_1).$$

Clearly, θ satisfies Equation 8.24. We need to show that θ is defined over $\mathbf{Q}$. If $Q \in C(\overline{\mathbf{Q}})$, then

$$
\begin{aligned}
\theta^g(Q) &= g\theta(g^{-1}Q) \quad \text{(by (8.20))}\\
&= g\phi_2\left(\phi_1^{-1}(g^{-1}Q) + P_1\right)\\
&= \phi_2^g((\phi_1^g)^{-1}(Q) + gP_1)\\
&= \phi_2(\phi_2^{-1}\phi_2^g)((\phi_1^g)^{-1}(Q) + gP_1)\\
&= \phi_2\left((\phi_1^g)^{-1}(Q) + gP_1 + T_g^2\right) \quad \text{(by (8.25))}\\
&= \phi_2\left(\phi_1^{-1}(Q) - T_g^1(g) + gP_1 + T_g^2\right) \quad \text{(by (8.23))}\\
&= \phi_2(\phi_1^{-1}(Q) + P_1) \quad \text{(by (8.26))}\\
&= \theta(Q).
\end{aligned}
$$

Therefore, θ is defined over $\mathbf{Q}$, so the pairs (C_1, ϕ_1) and (C_2, ϕ_2) are equivalent.
∎

Proposition 8.33 says that we have a map

$$\text{equivalence classes of pairs } (C, \phi) \hookrightarrow H^1(G, E(\overline{\mathbf{Q}})).$$

It can be shown that this is a bijection (see [90]). The most important property for us is the following.

PROPOSITION 8.34
Let τ_ϕ correspond to the pair (C, ϕ). Then $\tau_\phi \in B(G, E(\overline{\mathbf{Q}}))$ (= coboundaries) if and only if C has a rational point (that is, a point with coordinates in $\mathbf{Q}$).

PROOF   Let $P \in E(\overline{\mathbf{Q}})$. Then

$$gP + T_g = \phi^{-1}\phi^g(gP) = \phi^{-1}(g\phi(P))$$

and

$$P = \phi^{-1}(\phi(P)).$$

Therefore,

$$T_g = P - gP \iff g\phi(P) = \phi(P).$$

If C has a rational point Q, choose P such that $\phi(P) = Q$. Then $gQ = Q$ for all g implies that

$$T_g = g(-P) - (-P)$$

for all $g \in G$. Conversely, if $T_g = g(-P) - (-P)$ for all g then $g\phi(P) = \phi(P)$ for all $g \in G$, so $\phi(P)$ is a rational point.   ∎

Propositions 8.33 and 8.34 give us a reinterpretation in terms of cohomology groups of the fundamental question of when certain curves have rational points.

Example 8.12

Consider the curve $C_{1,p,p}$ from Section 8.8. It was given by the equations

$$x = u^2$$
$$x - 2p = pv^2$$
$$x + 2p = pw^2.$$

These were rewritten as

$$w^2 - v^2 = 4, \quad u^2 - pv^2 = 2p.$$

The method of Section 2.5.4 changes this to

$$C : s^2 = 2p(t^4 + 6t^2 + 1).$$

Finally, the transformation

$$t = \frac{\sqrt{2p}\,(x + 2p)}{y}, \quad s = -\sqrt{2p} + \frac{2t^2(x - p)}{\sqrt{2p}} = \sqrt{2p}\,\frac{x^2 + 4px - 4p^2}{x(x - 2p)}$$

(use the formulas of Section 2.5.3, plus a minor change of variables) changes the equation to

$$E : y^2 = x(x - 2p)(x + 2p).$$

We want to relate the curve $C_{1,p,p}$ from Section 8.8 to a cohomology class in $H^1(G, E(\overline{\mathbf{Q}}))$. The map

$$\phi : E \to C$$
$$(x, y) \mapsto (t, s)$$

gives a map from E to C. Since the equations for E and C have coefficients in $\mathbf{Q}$, these curves are defined over $\mathbf{Q}$. However, ϕ is not defined over $\mathbf{Q}$.

A short computation shows that

$$(x, y) + (-2p, 0) = (x_1, y_1)$$

on E, where

$$x_1 = 2p\,\frac{2p - x}{2p + x}, \quad y_1 = \frac{-8p^2 y}{(x + 2p)^2}.$$

Another calculation shows that

$$\phi(x_1, y_1) = (-t, -s).$$

Let $g \in G$ be such that $g(\sqrt{2p}) = -\sqrt{2p}$. Then ϕ^g is the transformation obtained by changing $\sqrt{2p}$ to $-\sqrt{2p}$ in the formulas for ϕ. Therefore,

$$\phi^g(x, y) = (-t, -s) = \phi(x_1, y_1).$$

We obtain

$$\phi^{-1}\phi^g(x, y) = (x, y) + (-2p, 0).$$

Now suppose $g \in G$ satisfies $g\sqrt{2p} = +\sqrt{2p}$. Then $\phi^g = \phi$, so

$$\phi^{-1}\phi^g(x, y) = (x, y).$$

Putting everything together, we see that the pair (C, ϕ) is of the type considered above. We obtain an element of $H^1(G, E[2])$ that can be regarded as an element of $H^1(G, E(\overline{\mathbf{Q}}))$. The cocycle τ_ϕ is given by

$$\tau_\phi(g) = T_g = \begin{cases} \infty & \text{if } g\sqrt{2p} = +\sqrt{2p} \\ (-2p, 0) & \text{if } g\sqrt{2p} = -\sqrt{2p} \end{cases}$$

The cohomology class of τ_ϕ is nontrivial in $H^1(G, E(\overline{\mathbf{Q}}))$, and hence also in $H^1(G, E[2])$, because C has no rational points. Note that τ_ϕ is a homomorphism from G to $E[2]$. This corresponds to the fact that G acts trivially on $E[2]$ in the present case, so $H^1(G, E[2]) = \text{Hom}(G, E[2])$. The kernel of τ is the subgroup of G of index 2 that fixes $\mathbf{Q}(\sqrt{2p})$. □

In general, if E is given by $y^2 = (x - e_1)(x - e_2)(x - e_3)$ with $e_1, e_2, e_3 \in \mathbf{Q}$, then a 2-descent yields curves $C_{a,b,c}$, as in Section 8.2. These curves yield elements of $H^1(G, E[2])$. The curves that have rational points give cocycles in $Z(G, E(\overline{\mathbf{Q}}))$ that are coboundaries. We also saw in the descent procedure that a rational point on a curve $C_{a,b,c}$ comes from a rational point on E. This discussion is summarized by the exact sequence

$$0 \to E(\mathbf{Q})/2E(\mathbf{Q}) \to H^1(G, E[2]) \to H^1(G, E(\overline{\mathbf{Q}}))[2] \to 0.$$

All of the preceding applies when $\mathbf{Q}$ is replaced by a p-adic field $\mathbf{Q}_p$ with $p \le \infty$. We have an exact sequence

$$0 \to E(\mathbf{Q}_p)/2E(\mathbf{Q}_p) \to H^1(G_p, E[2]) \to H^1(G_p, E(\overline{\mathbf{Q}_p}))[2] \to 0,$$

where

$$G_p = \text{Gal}(\overline{\mathbf{Q}}_p/\mathbf{Q}_p).$$

The group G_p can be regarded as a subgroup of G. Recall that cocycles in $Z(G, E[2])$ are maps from G to $E[2]$ with certain properties. Such maps may be restricted to G_p to obtain elements of $Z(G_p, E[2])$. A curve $C_{a,b,c}$ yields an element of $H^1(G, E[2])$. This yields an element of $H^1(G_p, E[2])$ that becomes trivial in $H^1(G_p, E(\overline{\mathbf{Q}}_p))$ if and only if $C_{a,b,c}$ has a p-adic point.

In Section 8.7, we defined S_2 to be those triples (a, b, c) such that $C_{a,b,c}$ has a p-adic point for all $p \leq \infty$. This means that S_2 is the set of triples (a, b, c) such that the corresponding cohomology class in $H^1(G, E[2])$ becomes trivial in $H^1(G_p, E(\overline{\mathbf{Q}}_p))$ for all $p \leq \infty$. Moreover, III_2 is S_2 modulo those triples coming from points in $E(\mathbf{Q})$. All of this can be expressed in terms of cohomology. We can also replace 2 by an arbitrary $n \geq 1$. Define the **Shafarevich-Tate group** to be

$$\text{III} = \text{Ker}\left(H^1(G, E(\overline{\mathbf{Q}})) \to \prod_{p \leq \infty} H^1(G_p, E(\overline{\mathbf{Q}}_p)) \right)$$

and define the n-**Selmer group** to be

$$S_n = \text{Ker}\left(H^1(G_p, E[n]) \to \prod_{p \leq \infty} H^1(G_p, E(\overline{\mathbf{Q}}_p)) \right).$$

The Shafarevich-Tate group can be thought of as consisting of equivalence classes of pairs (C, ϕ) such that C has a p-adic point for all $p \leq \infty$. This group is nontrivial if there exists such a C that has no rational points. In Section 8.8, we gave an example of such a curve. The n-Selmer group S_n can be regarded as the generalization to n-descents of the curves $C_{a,b,c}$ that arise in 2-descents. It is straightforward to use the definitions to deduce the basic descent sequence

$$0 \to E(\mathbf{Q})/nE(\mathbf{Q}) \to S_n \to \text{III}[n] \to 0,$$

where $\text{III}[n]$ is the n-torsion in III. When one is doing descent, the goal is to obtain information about $E(\mathbf{Q})/nE(\mathbf{Q})$. However, the calculations take place in S_n. The group $\text{III}[n]$ is the obstruction to transferring information back to $E(\mathbf{Q})/nE(\mathbf{Q})$.

The group S_n depends on n. It is finite (we proved this in the case where $n = 2$ and $E[2] \subseteq E(\mathbf{Q})$). The group III is independent of n. Its n-torsion $\text{III}[n]$ is finite since it is the quotient of the finite group S_n. It was conjectured by Tate and Shafarevich in the early 1960s that III is finite; this is still unproved in general. The first examples where III was proved finite were given by Rubin in 1986 (for all CM curves over $\mathbf{Q}$ with analytic rank 0; see Section 12.2) and by Kolyvagin in 1987 (for all elliptic curves over $\mathbf{Q}$ with analytic rank 0 or 1). No other examples over $\mathbf{Q}$ are known.

Exercises

8.1 Show that each of the following elliptic curves has the the the stated torsion group.

(a) $y^2 = x^3 - 2$; 0

(b) $y^2 = x^3 + 8$; $\mathbf{Z}_2$

(c) $y^2 = x^3 + 4$; $\mathbf{Z}_3$

(d) $y^2 = x^3 + 4x$; $\mathbf{Z}_4$

(e) $y^2 = x^3 - 432x + 8208$; $\mathbf{Z}_5$

(f) $y^2 = x^3 + 1$; $\mathbf{Z}_6$

(g) $y^2 = x^3 - 1323x + 6395814$; $\mathbf{Z}_7$

(h) $y^2 = x^3 - 44091x + 3304854$; $\mathbf{Z}_8$

(i) $y^2 = x^3 - 219x + 1654$; $\mathbf{Z}_9$

(j) $y^2 = x^3 - 58347x + 3954150$; $\mathbf{Z}_{10}$

(k) $y^2 = x^3 - 33339627x + 73697852646$; $\mathbf{Z}_{12}$

(l) $y^2 = x^3 - x$; $\mathbf{Z}_2 \oplus \mathbf{Z}_2$

(m) $y^2 = x^3 - 12987x - 263466$; $\mathbf{Z}_4 \oplus \mathbf{Z}_2$

(n) $y^2 = x^3 - 24003x + 1296702$; $\mathbf{Z}_6 \oplus \mathbf{Z}_2$

(o) $y^2 = x^3 - 1386747x + 368636886$; $\mathbf{Z}_8 \oplus \mathbf{Z}_2$

Parameterizations of elliptic curves with given torsion groups can be found in [52].

8.2 Let E be an elliptic curve over $\mathbf{Q}$ given by an equation of the form $y^2 = x^3 + Cx^2 + Ax + B$, with $A, B, C \in \mathbf{Z}$.

(a) Modify the proof of Theorem 8.1 to obtain a homomorphism

$$\lambda_r : E_r/E_{3r} \longrightarrow \mathbf{Z}_{p^{2r}}$$

(see [53, pp. 51-52]).

(b) Show that $(x, y) \in E(\mathbf{Q})$ is a torsion point, then $x, y \in \mathbf{Z}$.

8.3 (a) Show that the map λ_r, applied to the curve $y^2 = x^3$, is the map of Theorem 2.29 divided by p^r and reduced mod p^{4r}.

(b) Consider the map λ_r of Exercise 8.2, applied to the curve $E : y^2 = x^3 + ax^2$. Let ϕ be as in Theorem 2.30. The map $\lambda_r \phi^{-1}$ gives a map

$$\frac{y + \alpha x}{y - \alpha x} \mapsto p^{-r} \frac{x}{y} \pmod{p^{2r}}.$$

Use the Taylor series for $\log((1 + t)/(1 - t))$ to show that the map $(2\alpha)\lambda_r \phi^{-1}$ is p^{-r} times the logarithm map, reduced mod p^{2r}.

8.4 Let E be given by $y^2 = x^3 + Ax + B$ with $A, B \in \mathbf{Z}$. Let $P = (x, y)$ be a point on E.

(a) Let $2P = (x_2, y_2)$. Show that

$$y^2 \left(4x_2(3x^2 + 4A) - 3x^2 + 5Ax + 27B\right) = 4A^3 + 27B^2.$$

(b) Show that if both P and $2P$ have coordinates in $\mathbf{Z}$, then y^2 divides $4A^3 + 27B^2$. This gives another way to finish the proof of the Lutz-Nagell theorem.

8.5 Let E be the elliptic curve over $\mathbf{Q}$ given by $y^2 + xy = x^3 + x^2 - 11x$. Show that the point

$$P = \left(\frac{11}{4}, -\frac{11}{8}\right)$$

is a point of order 2. This shows that the integrality part of Theorem 8.7 (see also Exercise 8.2), which is stated for Weierstrass equations, does not hold for generalized Weierstrass equations. However, since changing from generalized Weierstrass form to the form in Exercise 8.2 affects only powers of 2 in the denominators, only the prime 2 can occur in the denominators of torsion points in generalized Weierstrass form.

8.6 Show that the Mordell-Weil group $E(\mathbf{Q})$ of the elliptic curve $y^2 = x^3 - x$ is isomorphic to $\mathbf{Z}_2 \oplus \mathbf{Z}_2$.

8.7 Suppose $E(\mathbf{Q})$ is generated by one point Q of infinite order. Suppose we take $R_1 = 3Q$, which generates $E(\mathbf{Q})/2E(\mathbf{Q})$. Show that the process with $P_0 = Q$ and

$$P_i = R_{j_i} + 2P_{i+1},$$

as in Section 8.3, never terminates. This shows that a set of representatives of $E(\mathbf{Q})/2E(\mathbf{Q})$ does not necessarily generate $E(\mathbf{Q})$.

8.8 Show that there is a set of representatives of $E(\mathbf{Q})/2E(\mathbf{Q})$ that generates $E(\mathbf{Q})$. (*Hint:* This mostly follows from the Mordell-Weil theorem. However, it does not handle the odd order torsion. Use Corollary 3.13 to show that the odd order torsion in $E(\mathbf{Q})$ is cyclic. In the set of representatives, use a generator of this cyclic group for the representative of the trivial coset.)

8.9 Let E be an elliptic curve defined over $\mathbf{Q}$ and let n be a positive integer. Assume that $E[n] \subseteq E(\mathbf{Q})$. Let $P \in E(\mathbf{Q})$ and let $Q \in E(\overline{\mathbf{Q}})$ be such that $nQ = P$. Define a map $\delta_P : \mathrm{Gal}(\overline{\mathbf{Q}}/\mathbf{Q}) \to E[n]$ by $\delta_P(\sigma) = \sigma Q - Q$.

(a) Let $\sigma \in \mathrm{Gal}(\overline{\mathbf{Q}}/\mathbf{Q})$. Show that $\sigma Q - Q \in E[n]$.

(b) Show that δ_P is a cocycle in $Z(G, E[n])$.

(c) Suppose we choose Q' with $nQ' = P$, and thus obtain a cocycle δ'_P. Show that $\delta_P - \delta'_P$ is a coboundary.

(d) Suppose that $\delta_P(\sigma)$ is a coboundary. Show that there exists $Q \in E(\mathbf{Q})$ such that $nQ = P$.

This shows that we have an injection $E(\mathbf{Q})/nE(\mathbf{Q}) \rightarrow H^1(G, E[n])$. This is the map of Proposition 8.19.

Chapter 9

Elliptic Curves over C

The goal of this chapter is to show that an elliptic curve over the complex numbers is the same thing as a torus. First, we show that a torus is isomorphic to an elliptic curve. To do this, we need to study functions on a torus, which amounts to studying doubly periodic functions on $\mathbf{C}$, especially the Weierstrass $\wp$-function. We then introduce the j-function and use its properties to show that every elliptic curve over $\mathbf{C}$ comes from a torus. Since most of the fields of characteristic 0 that we meet can be embedded in $\mathbf{C}$, many properties of elliptic curves over fields of characteristic 0 can be deduced from properties of a torus. For example, the n-torsion on a torus is easily seen to be isomorphic to $\mathbf{Z}_n \oplus \mathbf{Z}_n$, so we can deduce that this holds for all elliptic curves over algebraically closed fields of characteristic 0 (see Corollary 9.20).

9.1 Doubly Periodic Functions

Let ω_1, ω_2 be complex numbers that are linearly independent over $\mathbf{R}$. Then

$$L = \mathbf{Z}\omega_1 + \mathbf{Z}\omega_2 = \{n_1\omega_1 + n_2\omega_2 \,|\, n_1, n_2 \in \mathbf{Z}\}$$

is called a **lattice**. The main reason we are interested in lattices is that $\mathbf{C}/L$ is a **torus**, and we want to show that a torus gives us an elliptic curve.

The set

$$F = \{a_1\omega_1 + a_2\omega_2 \,|\, 0 \le a_i < 1, \, i = 1, 2\}$$

(see Figure 9.1) is called a **fundamental parallelogram** for L. A different choice of basis ω_1, ω_2 for L will of course give a different fundamental parallelogram. Since it will occur several times, we denote

$$\omega_3 = \omega_1 + \omega_2.$$

A function on $\mathbf{C}/L$ can be regarded as a function f on $\mathbf{C}$ such that $f(z+\omega) = f(z)$ for all $z \in \mathbf{C}$ and all $\omega \in L$. We are only interested in meromorphic

Figure 9.1

The Fundamental Parallelogram

functions, so we define a **doubly periodic function** to be a meromorphic function

$$f : \mathbf{C} \to \mathbf{C} \cup \infty$$

such that

$$f(z + \omega) = f(z)$$

for all $z \in \mathbf{C}$ and all $\omega \in L$. Equivalently,

$$f(z + \omega_i) = f(z), \quad i = 1, 2.$$

The numbers $\omega \in L$ are called the **periods** of f.

If f is a (not identically 0) meromorphic function and $w \in \mathbf{C}$, then we can write

$$f(z) = a_r(z - w)^r + a_{r+1}(z - w)^{r+1} + \cdots,$$

with $a_r \neq 0$. The integer r can be either positive, negative, or zero. Define the **order** and the **residue** of f at w to be

$$r = \operatorname{ord}_w f$$
$$a_{-1} = \operatorname{Res}_w f.$$

Therefore, $\operatorname{ord}_w f$ is the order of vanishing of f at w, or negative the order of a pole. The order is 0 if and only if the function is finite and nonvanishing at w. It is not hard to see that if f is doubly periodic, then $\operatorname{ord}_{w+\omega} f = \operatorname{ord}_w f$ and $\operatorname{Res}_{w+\omega} f = \operatorname{Res}_w f$ for all $\omega \in L$.

A **divisor** D is a formal sum of points:

$$D = n_1[w_1] + n_2[w_2] + \cdots + n_k[w_k],$$

where $n_i \in \mathbf{Z}$ and $w_i \in F$. In other words, we have a symbol $[w]$ for each $w \in F$, and the divisors are linear combinations with integer coefficients of these symbols. The **degree** of a divisor is

$$\deg(D) = \sum n_i.$$

Define the **divisor of a function** f to be

$$\text{div}(f) = \sum_{w \in F} (\text{ord}_w f)[w].$$

THEOREM 9.1

Let f be a doubly periodic function for the lattice L and let F be a fundamental parallelogram for L.

1. *If f has no poles, then f is constant.*

2. $\sum_{w \in F} \text{Res}_w f = 0.$

3. *If f is not identically 0,*

$$\deg(\text{div}(f)) = \sum_{w \in F} \text{ord}_w f = 0.$$

4. *If f is not identically 0,*

$$\sum_{w \in F} w \cdot \text{ord}_w f \in L.$$

5. *If f is not constant, then $f : \mathbf{C} \to \mathbf{C} \cup \infty$ is surjective. If n is the sum of the orders of the poles of f in F and $z_0 \in \mathbf{C}$, then $f(z) = z_0$ has n solutions (counting multiplicities).*

6. *If f has only one pole in F, then this pole cannot be a simple pole.*

All of the above sums over $w \in F$ have only finitely many nonzero terms.

PROOF   Because f is a meromorphic function, it can have only finitely many zeros and poles in any compact set, for example, the closure of F. Therefore, the above sums have only finitely many nonzero terms.

If f has no poles, then it is bounded in the closure of F, which is a compact set. Therefore, f is bounded in all of $\mathbf{C}$. Liouville's theorem says that a bounded entire function is constant. This proves (1).

Recall Cauchy's theorem, which says that

$$\int_{\partial F} f(z)dz = 2\pi i \sum_{w \in F} \text{Res}_w f,$$

where ∂F is the boundary of F and the line integral is taken in the counterclockwise direction. Write (assuming ω_1, ω_2 are oriented as in Figure 9.1; otherwise, switch them in the following)

$$\int_{\partial F} f(z)dz = \int_0^{\omega_2} f(z)dz + \int_{\omega_2}^{\omega_2+\omega_1} f(z)dz + \int_{\omega_1+\omega_2}^{\omega_1} f(z)dz + \int_{\omega_1}^0 f(z)dz.$$

Since $f(z + \omega_1) = f(z)$, we have

$$\int_{\omega_1+\omega_2}^{\omega_1} f(z)dz = \int_{\omega_2}^{0} f(z)dz = -\int_{0}^{\omega_2} f(z)dz.$$

Similarly,

$$\int_{\omega_2}^{\omega_1+\omega_2} f(z)dz = -\int_{\omega_1}^{0} f(z)dz.$$

Therefore, the sum of the four integrals is 0. There is a small technicality that we have passed over. The function f is not allowed to have any poles on the path of integration. If it does, adjust the path with a small detour around such points as in Figure 9.2. The integrals cancel, just as in the above. This proves (2).

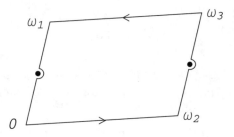

Figure 9.2

Suppose $r = \operatorname{ord}_w f$. Then $f = (z - w)^r g(z)$, where $g(w)$ is finite and nonzero. Then

$$\frac{f'(z)}{f(z)} = \frac{r}{z - w} + \frac{g'(z)}{g(z)},$$

so

$$\operatorname{Res}_w\left(\frac{f'}{f}\right) = r.$$

If f is doubly periodic, then f' is doubly periodic. Therefore, (2) applied to f'/f yields

$$2\pi i \sum_{w \in F} \operatorname{ord}_w f = 2\pi i \sum_{w \in F} \operatorname{Res}_w\left(\frac{f'}{f}\right) = 0.$$

This proves (3).

For (4), we have

$$2\pi i \sum_{w \in F} w \cdot \operatorname{ord}_w f = 2\pi i \sum_{w \in F} \operatorname{Res}_w z\left(\frac{f'}{f}\right) = \int_{\partial F} z\frac{f'}{f}dz.$$

However, in this case, the function zf'/f is not doubly periodic. The integral may be written as a sum of four integrals, as in the proof of (2). The double periodicity of f and f' yield

$$\int_{\omega_1+\omega_2}^{\omega_1} z\frac{f'(z)}{f(z)}\,dz = \int_{\omega_2}^{0}(z+\omega_1)\frac{f'(z)}{f(z)}\,dz$$
$$= -\int_{0}^{\omega_2} z\frac{f'(z)}{f(z)}\,dz - \omega_1\int_{0}^{\omega_2}\frac{f'(z)}{f(z)}\,dz.$$

But

$$\frac{1}{2\pi i}\int_{0}^{\omega_2}\frac{f'(z)}{f(z)}\,dz$$

is the winding number around 0 of the path

$$z = f(t\omega_2),\quad 0 \le t \le 1.$$

Since $f(0) = f(\omega_2)$, this is a closed path. The winding number is an integer, so

$$\int_{0}^{\omega_2} z\frac{f'(z)}{f(z)}\,dz + \int_{\omega_1+\omega_2}^{\omega_1} z\frac{f'(z)}{f(z)}\,dz$$
$$= -\omega_1\int_{0}^{\omega_2}\frac{f'(z)}{f(z)}\,dz \in 2\pi i\mathbf{Z}\omega_1.$$

Similarly,

$$\int_{\omega_2}^{\omega_1+\omega_2} z\frac{f'(z)}{f(z)}\,dz + \int_{\omega_1}^{0} z\frac{f'(z)}{f(z)}\,dz \in 2\pi i\mathbf{Z}\omega_2.$$

Therefore,

$$2\pi i\sum_{w\in F} w\cdot\mathrm{ord}_w f \in 2\pi iL.$$

This proves (4).

To prove (5), let $z_0 \in \mathbf{C}$. Then $h(z) = f(z) - z_0$ is a doubly periodic function whose poles are the same as the poles of f. By (3), the number of zeros of $h(z)$ in F (counting multiplicities) equals the number of poles (counting multiplicities) of h, which is n. This proves (5).

For (6), suppose f has only a simple pole, say at w, and no others. Then $\mathrm{Res}_w f \ne 0$ (otherwise, the pole doesn't exist). The sum in (2) has only one term, and it is nonzero. This is impossible, so we conclude that either the pole cannot be simple or there must be other poles. ∎

REMARK 9.2   As we saw in the proof of (5), part (3) says that the number of zeros of a doubly periodic function equals the number of poles. This is a general fact for compact Riemann surfaces (such as a torus) and for projective algebraic curves (see [31, Ch. 8, Prop. 1] or [36, II, Cor. 6.10]).

If (6) were false for a function f, then f would give a bijective (by (5)) map from the torus to the Riemann sphere (= $\mathbf{C} \cup \infty$). This is impossible for many topological reasons (the torus has a hole but the sphere doesn't).   ∎

So far, we do not have any examples of nonconstant doubly periodic functions. This situation is remedied by the **Weierstrass $\wp$-function**.

THEOREM 9.3
Given a lattice L, define the Weierstrass $\wp$-function by

$$\wp(z) = \wp(z; L) = \frac{1}{z^2} + \sum_{\substack{w \in L \\ w \neq 0}} \left(\frac{1}{(z-w)^2} - \frac{1}{w^2} \right). \tag{9.1}$$

Then

1. *The sum defining $\wp(z)$ converges absolutely and uniformly on compact sets not containing elements of L.*

2. *$\wp(z)$ is meromorphic in $\mathbf{C}$ and has a double pole at each $w \in L$.*

3. *$\wp(-z) = \wp(z)$ for all $z \in \mathbf{C}$.*

4. *$\wp(z + w) = \wp(z)$ for all $w \in L$*

5. *The set of doubly periodic functions for L is $\mathbf{C}(\wp, \wp')$. In other words, every doubly periodic function is a rational function of $\wp$ and its derivative $\wp'$.*

PROOF   Let C be a compact set, and let $M = \text{Max}\{|z| \mid z \in C\}$. If $z \in C$ and $|w| \geq 2M$, then $|z - w| \geq |w|/2$ and $|2w - z| \leq 5|w|/2$, so

$$\left| \frac{1}{(z-w)^2} - \frac{1}{w^2} \right| = \left| \frac{z(2w-z)}{(z-w)^2 w^2} \right|$$
$$\leq \frac{M(5|w|/2)}{|w|^4/4} = \frac{10M}{|w|^3}. \tag{9.2}$$

The preceding calculation explains why the terms $1/w^2$ are included. Without them, the terms in the sum would be comparable to $1/w^2$. Subtracting this $1/w^2$ makes the terms comparable to $1/w^3$. This causes the sum to converge, as the following lemma shows.

LEMMA 9.4
If $k > 2$ then

$$\sum_{\substack{w \in L \\ w \neq 0}} \frac{1}{|w|^k}$$

converges.

PROOF   Let F be a fundamental parallelogram for L and let D be the length of the longer diagonal of F. Then $|z| \leq D$ for all $z \in F$. Let $\omega = m_1\omega_1 + m_2\omega_2 \in L$ with $|\omega| \geq 2D$. If x_1, x_2 are real numbers with $m_i \leq x_i < m_i + 1$, then ω and $x_1\omega_1 + x_2\omega_2$ differ by an element of F, so

$$|m_1\omega_1 + m_2\omega_2| \geq |x_1\omega_1 + x_2\omega_2| - D \geq |x_1\omega_1 + x_2\omega_2| - \frac{1}{2}|m_1\omega_1 + m_2\omega_2|,$$

since $|m_1\omega_1 + m_2\omega_2| \geq 2D$. Therefore,

$$|m_1\omega_1 + m_2\omega_2| \geq \frac{2}{3}|x_1\omega_1 + x_2\omega_2|.$$

Similarly,

$$|x_1\omega_1 + x_2\omega_2| \geq D.$$

Comparing the sum to an integral yields

$$\sum_{|\omega| \geq 2D} \frac{1}{|\omega|^k} \leq \iint_{|x_1\omega_1 + x_2\omega_2| \geq D} \frac{(3/2)^k}{|x_1\omega_1 + x_2\omega_2|^k} dx_1 dx_2.$$

The change of variables defined by $u + iv = x_1\omega_1 + x_2\omega_2$ changes the integral to

$$C \iint_{|u+iv| \geq D} \frac{1}{(u^2 + v^2)^{k/2}} du\, dv = C \int_{\theta=0}^{2\pi} \int_{r=D}^{\infty} \frac{1}{r^k} r\, dr\, d\theta < \infty,$$

where $C = (3/2)^k/(\text{area of } F)$. Therefore, the sum for $|\omega| \geq 2D$ converges. Since there are only finitely many ω with $|\omega| < 2D$, we have shown that the sum converges. ∎

Lemma 9.4 and Equation 9.2 imply that the sum of the terms in Equation 9.1 with $|\omega| \geq 2M$ converges absolutely and uniformly for $z \in C$. Since only finitely many terms have been omitted, we obtain (1). Since a uniform limit of analytic functions is analytic, $\wp(z)$ is analytic for $z \notin L$. If $z \in L$, then the sum of the terms for $\omega \neq z$ is analytic near z, so the term $1/(z-\omega)^2$ causes $\wp$ to have a double pole at z. This proves (2).

To prove (3), note that $\omega \in L$ if and only if $-\omega \in L$. Therefore, in the sum for $\wp(-z)$, we can take the sum over $-\omega \in L$. The terms of this sum are of the form

$$\frac{1}{(-z+\omega)^2} - \frac{1}{(-\omega)^2} = \frac{1}{(z-\omega)^2} - \frac{1}{\omega^2}.$$

Therefore the sum for $\wp(-z)$ equals the sum for $\wp(z)$.

The proof of (4) would be easy if we could ignore the terms $1/\omega^2$, since changing z to $z + \omega$ would simply shift the summands. However, these terms are needed for convergence. With some care, one could justify rearranging the sum, but it is easier to do the following. Differentiating $\wp(z)$ term by term yields

$$\wp'(z) = -2 \sum_{w \in L} \frac{1}{(z-w)^3}.$$

Note that $\omega = 0$ is included in the sum. This sum converges absolutely (by comparison with the case $k = 3$ in Lemma 9.4) when $z \notin L$, and changing z to $z + \omega$ shifts the terms in the sum. Therefore,

$$\wp'(z + \omega) = \wp'(z).$$

This implies that there is a constant c_ω such that

$$\wp(z + \omega) - \wp(z) = c_\omega,$$

for all $z \notin L$. Setting $z = \omega/2$ yields

$$c_\omega = \wp(-\omega/2) - \wp(\omega/2) = 0,$$

by (3). Therefore $\wp(z + \omega) = \wp(z)$. This proves (4).

Let $f(z)$ be any doubly periodic function. Then

$$f(z) = \frac{f(z) + f(-z)}{2} + \frac{f(z) - f(-z)}{2}$$

expresses $f(z)$ as the sum of an even function and an odd function. Therefore, it suffices to prove (5) for even functions and for odd functions. Since $\wp(-z) = \wp(z)$, it follows that $\wp'(-z) = -\wp'(z)$, so $\wp'(z)$ is an odd function. If $f(z)$ is odd, then $f(z)/\wp'(z)$ is even. Therefore, it suffices to show that an even doubly periodic function is a rational function of $\wp(z)$.

Let $f(z)$ be an even doubly periodic function. We may assume that f is not identically zero; otherwise, we're done. By adding a changing f, if necessary, to

$$\frac{af + b}{cf + d}$$

for suitable a, b, c, d with $ad - bc \neq 0$, we may arrange that $f(z)$ does not have a zero or a pole whenever $2z \in L$ (this means that we want $f(0) \neq 0, \infty$ and $f(\omega_i/2) \neq 0$ for $i = 1, 2, 3$). If we prove $(af + b)/(cf + d)$ is a rational function of $\wp$, then we can solve for f and obtain the result for f.

Since $f(z)$ is even and doubly periodic, $f(\omega_3 - z) = f(z)$, so

$$\mathrm{ord}_w f = \mathrm{ord}_{\omega_3 - w} f.$$

We can therefore put the finitely many elements in F where $f(z) = 0$ or where $f(z)$ has a pole into pairs $(w, \omega_3 - w)$. Since we have arranged that $w \neq \omega_3/2$, the two elements of each pair are distinct. There is a slight problem if w lies on a side of F. Suppose $w = x\omega_1$ with $0 < x < 1$. Then $\omega_3 - w = (1 - x)\omega_1 + \omega_2 \notin F$. In this case, we translate by ω_2 to get $(1 - x)\omega_1 \in F$. Since $w \neq \omega_1/2$, we have $x \neq 1/2$, hence $x\omega_1 \neq (1 - x)\omega_1$, and again the two elements of the pair are distinct. The case $w = x\omega_2$ is handled similarly.

For a fixed w, the function $\wp(z) - \wp(w)$ has zeros at $z = w$ and $z = \omega_3 - w$. By Theorem 9.1(5), these are the only two zeros in F and they are simple zeros. Therefore, the function

$$h(z) = \prod_{(w, \omega_3 - w)} (\wp(z) - \wp(w))^{\operatorname{ord}_w f}$$

(the product is over pairs $(w, \omega_3 - w)$) has a zero of order $\operatorname{ord}_w f$ at w and at $\omega_3 - w$ when $\operatorname{ord}_w f > 0$ and has a pole of the same order as f when $\operatorname{ord}_w f < 0$. Since $\sum \operatorname{ord}_w f = 0$ by Theorem 9.1, the poles at $z \in L$ of the factors in the product cancel. Therefore, $f(z)/h(z)$ has no zeros or poles in F. By Theorem 9.1(1), $f(z)/h(z)$ is constant. Since $h(z)$ is a rational function of $\wp(z)$, so is $\wp(z)$. This completes the proof of Theorem 9.3.   ∎

In order to construct functions with prescribed properties, it is convenient to introduce the Weierstrass σ-function. It is not doubly periodic, but it satisfies a simple transformation law for translation by elements of L.

PROPOSITION 9.5

  Let

$$\sigma(z) = \sigma(z; L) = z \prod_{\substack{w \in L \\ w \neq 0}} \left(1 - \frac{z}{\omega}\right) e^{(z/w) + \frac{1}{2}(z/w)^2}.$$

Then

  1. *$\sigma(z)$ is analytic for all $z \in \mathbf{C}$*

  2. *$\sigma(z)$ has simple zeros at each $\omega \in L$ and has no other zeros*

  3. *$\frac{d^2}{dz^2} \log \sigma(z) = -\wp(z)$*

  4. *given $\omega \in L$, there exist $a = a_\omega$ and $b = b_\omega$ such that*

$$\sigma(z + \omega) = e^{az + b} \sigma(z)$$

  for all $z \in \mathbf{C}$.

PROOF   The exponential factor is included to make the product converge. A short calculation yields the power series expansion

$$(1 - u)e^{u + \frac{1}{2}u^2} = 1 + c_3 u^3 + c_4 u^4 + \cdots.$$

Therefore, there is a constant C such that

$$|(1-u)e^{u+\frac{1}{2}u^2} - 1| \le C|u|^3$$

for u near 0. In particular, this inequality holds when $u = z/\omega$ for $|\omega|$ sufficiently large and z in a compact set. Recall that if a sum $\sum |a_n|$ converges, then the product $\prod(1+a_n)$ converges. Moreover, if $(1+a_n) \ne 0$ for all n, then the product is nonzero. Since $\sum |z/\omega_3|^3$ converges by Lemma 9.4 with $k = 3$, the product defining $\sigma(z)$ converges uniformly on compact sets. Therefore, $\sigma(z)$ is analytic. This proves (1). Part (2) follows since the product of the factors, omitting one ω, is nonzero at $z = \omega$.

To prove (3), differentiate the logarithm of the product for $\sigma(z)$ to obtain

$$\frac{d}{dz} \log \sigma(z) = \frac{1}{z} + \sum_{\substack{\omega \in L \\ \omega \ne 0}} \left(\frac{1}{z-\omega} + \frac{1}{\omega} + \frac{z}{\omega^2} \right).$$

Taking one more derivative yields the sum for $-\wp(z)$. This proves (3).

Let $\omega \in L$. Since

$$\frac{d^2}{dz^2} \log \frac{\sigma(z+\omega)}{\sigma(z)} = 0,$$

there are constants $a = a_\omega$ and $b = b_\omega$ such that

$$\log \frac{\sigma(z+\omega)}{\sigma(z)} = az + b.$$

Exponentiating yields (4). We can restrict z in the above to lie in a small region in order to avoid potential complications with branches of the logarithm. Then (4) holds in this small region, and therefore for all $z \in \mathbf{C}$, by uniqueness of analytic continuation. ∎

We can now state exactly when a divisor is a divisor of a function. The following is a special case of what is known as the **Abel-Jacobi theorem**, which states when a divisor on a Riemann surface, or on an algebraic curve, is the divisor of a function.

THEOREM 9.6
Let $D = \sum n_i [w_i]$ be a divisor. Then D is the divisor of a function if and only if $\deg(D) = 0$ and $\sum n_i w_i \in L$.

PROOF   Parts (3) and (4) of Theorem 9.1 are precisely the statements that if D is the divisor of a function then $\deg(D) = 0$ and $\sum n_i w_i \in L$.

Conversely, suppose $\deg(D) = 0$ and $\sum n_i w_i = \ell \in L$. Let

$$f(z) = \frac{\sigma(z)}{\sigma(z-\ell)} \prod_i \sigma(z-w_i)^{n_i}.$$

If $\omega \in L$, then

$$\frac{f(z+\omega)}{f(z)} = e^{a_\omega z + b_\omega - a_\omega(z-\ell) - b_\omega} \, e^{\sum n_i(a_\omega(z-w_i)+b_\omega)} = 1,$$

since $\sum n_i = 0$ and $\sum n_i w_i = \ell$. Therefore, $f(z)$ is doubly periodic. The divisor of f is easily seen to be D, so D is the divisor of a function. ∎

Doubly periodic functions can be regarded as functions on the torus $\mathbf{C}/L$, and divisors can be regarded as divisors for $\mathbf{C}/L$. If we let $\mathbf{C}(L)^\times$ denote the doubly periodic functions that do not vanish identically and let $\mathrm{Div}^0(\mathbf{C}/L)$ denote the divisors of degree 0, then much of the preceding discussion can be expressed by the exactness of the sequence

$$0 \longrightarrow \mathbf{C}^\times \longrightarrow \mathbf{C}(L)^\times \xrightarrow{\mathrm{div}} \mathrm{Div}^0(\mathbf{C}/L) \xrightarrow{\mathrm{sum}} \mathbf{C}/L \longrightarrow 0. \qquad (9.3)$$

The "sum" function adds up the complex numbers representing the points in the divisor mod L. The exactness at $\mathbf{C}(L)^\times$ expresses the fact that a function with no zeros and no poles, hence whose divisor is 0, is a constant. The exactness at $\mathrm{Div}^0(\mathbf{C}/L)$ is Theorem 9.6. The surjectivity of the sum function is easy. If $w \in \mathbf{C}$, then $\mathrm{sum}([w] - [0]) = w \mod L$.

9.2   Tori are Elliptic Curves

The goal of this section is to show that a complex torus $\mathbf{C}/L$ is naturally isomorphic to the complex points on an elliptic curve.

Let L be a lattice, as in the previous section. For integers $k \geq 3$, define the **Eisenstein series**

$$G_k = G_k(L) = \sum_{\substack{\omega \in L \\ \omega \neq 0}} \omega^{-k}. \qquad (9.4)$$

By Lemma 9.4, the sum converges. When k is odd, the terms for ω and $-\omega$ cancel, so $G_k = 0$.

PROPOSITION 9.7
For $0 < |z| < \mathrm{Min}_{0 \neq \omega \in L}(|\omega|)$,

$$\wp(z) = \frac{1}{z^2} + \sum_{j=1}^{\infty} (2j+1) G_{2j+2} z^{2j}.$$

PROOF   When $|z| < |\omega|$,

$$\frac{1}{(z-\omega)^2} - \frac{1}{\omega^2} = \omega^{-2}\left(\frac{1}{(1-(z/\omega))^2} - 1\right)$$

$$= \omega^{-2}\left(\sum_{n=1}^{\infty}(n+1)\frac{z^n}{\omega^n}\right).$$

Therefore,

$$\wp(z) = \frac{1}{z^2} + \sum_{\omega \neq 0}\sum_{n=1}^{\infty}(n+1)\frac{z^n}{\omega^{n+2}}.$$

Summing over ω first, then over n, yields the result.   ∎

THEOREM 9.8

Let $\wp(z)$ be the Weierstrass $\wp$-function for a lattice L. Then

$$\wp'(z)^2 = 4\wp(z)^3 - 60G_4\wp(z) - 140G_6.$$

PROOF   From Proposition 9.7,

$$\wp(z) = z^{-2} + 3G_4 z^2 + 5G_6 z^4 + \cdots$$
$$\wp'(z) = -2z^{-3} + 6G_4 z + 20G_6 z^3 + \cdots.$$

Cubing and squaring these two relations yields

$$\wp(z)^3 = z^{-6} + 9G_4 z^{-2} + 15G_6 + \cdots$$
$$\wp'(z)^2 = 4z^{-6} - 24G_4 z^{-2} - 80G_6 + \cdots.$$

Therefore,

$$f(z) = \wp'(z)^2 - 4\wp(z)^3 + 60G_4\wp(z) + 140G_6 = c_1 z + c_2 z^2 + \cdots$$

is a power series with no constant term and with no negative powers of z. But the only possible poles of $f(z)$ are at the poles of $\wp(z)$ and $\wp'(z)$, namely, the elements of L. Since $f(z)$ is doubly periodic and, as we have just shown, has no pole at 0, $f(z)$ has no poles. By 9.1, $f(z)$ is constant. Since the power series for $f(z)$ has no constant term, $f(0) = 0$. Therefore, $f(z)$ is identically 0.   ∎

It is customary to set

$$g_2 = 60G_4$$
$$g_3 = 140G_6.$$

The theorem then says that

$$\wp'(z)^2 = 4\wp(z)^3 - g_2\wp(z) - g_3. \tag{9.5}$$

Therefore, the points $(\wp(z), \wp'(z))$ lie on the curve

$$y^2 = 4x^3 - g_2 x - g_3.$$

It is traditional to leave the 4 as the coefficient of x^3, rather than performing a change of variables to make the coefficient of x^3 equal to 1. The discriminant of the cubic polynomial is $16(g_2^3 - 27g_3^2)$.

PROPOSITION 9.9
$\Delta = g_2^3 - 27g_3^2 \neq 0.$

PROOF   Since $\wp'(z)$ is doubly periodic, $\wp'(\omega_i/2) = \wp'(-\omega_i/2)$. Since $\wp'(-z) = -\wp'(z)$, it follows that

$$\wp'(\omega_i/2) = 0, \quad i = 1, 2, 3. \tag{9.6}$$

Therefore, each $\wp(\omega_i/2)$ is a root of $4x^3 - g_2 x - g_3$, by (9.5). If we can show that these roots are distinct, then the cubic polynomial has three distinct roots, which means that its discriminant is nonzero. Let

$$h_i(z) = \wp(z) - \wp(\omega_i/2).$$

Then $h_i(\omega_i/2) = 0 = h_i'(\omega_i/2)$, so h_i vanishes to order at least 2 at $\omega_i/2$. Since $h_i(z)$ has only one pole in F, namely the double pole at $z = 0$, Theorem 9.1(5) implies that $\omega_i/2$ is the only zero of $h_i(z)$. In particular,

$$h_i(\omega_j/2) \neq 0, \quad \text{when } j \neq i.$$

Therefore, the values $\wp(\omega_i/2)$ are distinct.   ■

The proposition implies that

$$E : y^2 = 4x^3 - g_2 x - g_3$$

is the equation of an elliptic curve, so we have a map from $z \in \mathbf{C}$ to the points with complex coordinates $(\wp(z), \wp'(z))$ on an elliptic curve. Since $\wp(z)$ and $\wp'(z)$ depend only on $z \mod L$ (that is, if we change z by an element of L, the values of the functions do not change), we have a function from $\mathbf{C}/L$ to $E(\mathbf{C})$. The group $\mathbf{C}/L$ is a group, with the group law being addition of complex numbers mod L. In concrete terms, we can regard elements of $\mathbf{C}/L$ as elements of F. When we add two points, we move the result back into F by subtracting a suitable element of L. For example, $(.7\omega_1 + .8\omega_2) + (.4\omega_1 + .9\omega_2)$ yields $.1\omega_1 + .7\omega_2$.

THEOREM 9.10

Let L be a lattice and let E be the elliptic curve $y^2 = 4x^3 - g_2 x - g_3$. The map

$$\Phi : \mathbf{C}/L \longrightarrow E(\mathbf{C})$$
$$z \longmapsto (\wp(z), \wp'(z))$$
$$0 \longmapsto \infty$$

is an isomorphism of groups.

PROOF   The surjectivity is easy. Let $(x, y) \in E(\mathbf{C})$. Since the function $\wp(z) - x$ has a double pole, Theorem 9.1 implies that it has zeros, so there exists $z \in \mathbf{C}$ such that $\wp(z) = x$. Theorem 9.8 implies that

$$\wp'(z)^2 = y^2,$$

so $\wp'(z) = \pm y$. If $\wp'(z) = y$, we're done. If $\wp'(z) = -y$, then $\wp'(-z) = y$ and $\wp(-z) = x$, so $-z \mapsto (x, y)$.

Suppose $\wp(z_1) = \wp(z_2)$ and $\wp'(z_1) = \wp'(z_2)$, and $z_1 \not\equiv z_2 \mod L$. The only poles of $\wp(z)$ are for $z \in L$. Therefore, if z_1 is a pole of $\wp$, then $z_1 \in L$ and $z_2 \in L$, so $z_1 \equiv z_2 \mod L$. Now assume z_1 is not a pole of $\wp$, so z_1 is not in L. The function

$$h(z) = \wp(z) - \wp(z_1)$$

has a double pole at $z = 0$ and no other poles in F. By Theorem 9.1, it has exactly two zeros. Suppose $z_1 = \omega_i/2$ for some i. From Equation 9.6, we know that $\wp'(\omega_i/2) = 0$, so z_1 is a double root of $h(z)$, and hence is the only root. Therefore $z_2 = z_1$. Finally, suppose z_1 is not of the form $\omega_i/2$. Since $h(-z_1) = h(z_1) = 0$, and since $z_1 \not\equiv -z_1 \mod L$, the two zeros of h are z_1 and $-z_1 \mod L$. Therefore, $z_2 \equiv -z_1 \mod L$. But

$$y = \wp'(z_2) = \wp'(-z_1) = -\wp'(z_1) = -y.$$

This means that $\wp'(z_1) = y = 0$. But $\wp'(z)$ has only a triple pole, so has only three zeros in F. From Equation 9.6, we know that these zeros occur at $\omega_i/2$. This is a contradiction, since $z \neq \omega_i/2$. Therefore, $z_1 \equiv z_2 \mod L$, so Φ is injective.

Finally, we need to show that Φ is a group homomorphism. Let $z_1, z_2 \in \mathbf{C}$ and let

$$\Phi(z_i) = P_i = (x_i, y_i).$$

Assume that both P_1, P_2 are finite and that the line through P_1, P_2 intersects E in three distinct finite points (this means that $P_1 \neq \pm P_2$, that $2P_1 + P_2 \neq \infty$, and that $P_1 + 2P_2 \neq \infty$). For a fixed z_1, this excludes finitely many values of z_2. There are two reasons for these exclusions. The first is that the addition law on E has a different formula when the points are equal. The second is that we do not need to worry about the connection between double roots in the algebraic calculations and double roots of the corresponding analytic functions.

Let $y = ax + b$ be the line through P_1, P_2. Let $P_3 = (x_3, y_3)$ be the third point of intersection of this line with E and let $(x_3, y_3) = P_3 = \Phi(z_3)$ with $z_3 \in \mathbf{C}$. The formulas for the group law on E show that

$$x_3 = \frac{1}{4}\left(\frac{y_2 - y_1}{x_2 - x_1}\right)^2 - x_1 - x_2$$

$$= \frac{1}{4}\left(\frac{\wp'(z_2) - \wp'(z_1)}{\wp(z_2) - \wp(z_1)}\right)^2 - \wp(z_1) - \wp(z_2).$$

The function

$$\ell(z) = \wp'(z) - a\wp(z) - b$$

has zeros at $z = z_1, z_2, z_3$. Since $\ell(z)$ has a triple pole at 0, and no other poles, it has three zeros in F. Therefore,

$$\operatorname{div}(\ell) = [z_1] + [z_2] + [z_3] - 3[0].$$

By Theorem 9.1(4), $z_1 + z_2 + z_3 \in L$. Therefore,

$$\wp(z_1 + z_2) = \wp(-z_3) = \wp(z_3) = x_3.$$

We obtain

$$\wp(z_1 + z_2) = \frac{1}{4}\left(\frac{\wp'(z_2) - \wp'(z_1)}{\wp(z_2) - \wp(z_1)}\right)^2 - \wp(z_1) - \wp(z_2). \tag{9.7}$$

By continuity, this formula, which we proved with certain values of the z_i excluded, now holds for all z_i for which it is defined.

We now need to consider the y-coordinate. This means that we need to compute $\wp'(z_1 + z_2)$. We sketch the method (the interested and careful reader may check the details). Differentiating (9.7) with respect to z_2 yields an expression for $\wp'(z_1 + z_2)$ in terms of x_1, x_2, y_1, y_2, and $\wp''(z_2)$. We need to express $\wp''$ in terms of $\wp$ and $\wp'$. Differentiating (9.5) yields

$$2\wp''\wp' = (12\wp^2 - g_2)\wp'.$$

Dividing by $\wp'(z)$ (this is all right if $\wp'(z) \neq 0$; the other cases are filled in by continuity) yields

$$2\wp''(z_2) = 12\wp(z_2)^2 - g_2. \tag{9.8}$$

Substituting this into the expression obtained for $\wp'(z_1 + z_2)$ yields an expression for $\wp'(z_1 + z_2)$ in terms of $\wp(z_1), \wp'(z_1), \wp(z_2), \wp'(z_2)$. Some algebraic manipulation shows that this equals the value for $-y_3$ obtained from the addition law for $(x_1, y_1) + (x_2, y_2) = (x_3, -y_3)$. Therefore,

$$(\wp(z_1), \wp'(z_1)) + (\wp(z_2), \wp'(z_2)) = (\wp(z_1 + z_2), \wp'(z_1 + z_2)).$$

This is exactly the statement that

$$\Phi(z_1) + \Phi(z_2) = \Phi(z_1 + z_2). \tag{9.9}$$

It remains to check (9.9) in the cases where (9.7) is not defined. The cases where $\wp(z_i) = \infty$ and where $z_1 \equiv -z_2 \mod L$ are easily checked. The remaining case is when $z_1 = z_2$. Let $z_2 \to z_1$ in (9.7), use l'Hôpital's rule, and use (9.8) to obtain

$$
\begin{aligned}
\wp(2z_1) &= \frac{1}{4}\left(\frac{\wp''(z_1)}{\wp'(z_1)}\right)^2 - 2\wp(z_1) \\
&= \frac{1}{4}\left(\frac{6\wp(z_1)^2 - \frac{1}{2}g_2}{\wp'(z_1)}\right)^2 - 2\wp(z_1) \\
&= \frac{1}{4}\left(\frac{6x_1^2 - \frac{1}{2}g_2}{y_1}\right)^2 - 2x_1.
\end{aligned}
\tag{9.10}
$$

This is the formula for the coordinate x_3 that is obtained from the addition law on E. Differentiating with respect to z_1 yields the correct formula for the y-coordinate, as above. Therefore,

$$\Phi(z_1) + \Phi(z_1) = \Phi(2z_1).$$

This completes the proof of the theorem. ∎

The theorem shows that the natural group law on the torus $\mathbf{C}/L$ matches the group law on the elliptic curve, which perhaps looks a little unnatural. Also, the classical formulas (9.7) and (9.10) for the Weierstrass $\wp$-function, which look rather complicated, are now seen to be expressing the group law for E.

9.3   Elliptic Curves over C

In the preceding section, we showed that a torus yields an elliptic curve. In the present section, we'll show the converse, namely, that every elliptic curve over $\mathbf{C}$ comes from a torus.

Let $L = \mathbf{Z}\omega_1 + \mathbf{Z}\omega_2$ be a lattice and let

$$\tau = \omega_1/\omega_2.$$

Since ω_1 and ω_2 are linearly independent over $\mathbf{R}$, the number τ cannot be real. By switching ω_1 and ω_2 if necessary, we may assume that the imaginary part of τ is positive:

$$\Im(\tau) > 0.$$

In other words, we assume τ lies in the **upper half plane**

$$\mathcal{H} = \{x + iy \in \mathbf{C} \,|\, y > 0\}.$$

The lattice

$$L_\tau = \mathbf{Z}\tau + \mathbf{Z}$$

is **homothetic** to L. This means that there exists a nonzero complex number λ such that $L = \lambda L_\tau$. In our case, $\lambda = \omega_2$.

For integers $k \geq 3$, define

$$G_k(\tau) = G_k(L_\tau) = \sum_{(m,n)\neq(0,0)} \frac{1}{(m\tau + n)^k}. \tag{9.11}$$

We have

$$G_k(\tau) = \omega_2^k G_k(L),$$

where $G_k(L)$ is the Eisenstein series defined for $L = \mathbf{Z}\omega_1 + \mathbf{Z}\omega_2$ by (9.4). Let

$$q = e^{2\pi i\tau}.$$

It will be useful to express certain functions as sums of powers of q. If $\tau = x + iy$ with $y > 0$, then $|q| = e^{-2\pi y} < 1$. This implies that the expressions we obtain will converge.

PROPOSITION 9.11

Let $\zeta(x) = \sum_{n=1}^{\infty} n^{-x}$ and let

$$\sigma_\ell(n) = \sum_{d|n} d^\ell$$

be the sum of the ℓth powers of the positive divisors of n. If $k \geq 1$ is an integer, then

$$G_{2k}(\tau) = 2\zeta(2k) + 2\frac{(2\pi i)^{2k}}{(2k-1)!} \sum_{n=1}^{\infty} \sigma_{2k-1}(n)q^n$$

$$= 2\zeta(2k) + 2\frac{(2\pi i)^{2k}}{(2k-1)!} \sum_{j=1}^{\infty} \frac{j^{2k-1}q^j}{1-q^j}.$$

PROOF   We have

$$\pi \frac{\cos \pi \tau}{\sin \pi \tau} = \pi i \frac{e^{\pi i \tau} + e^{-\pi i \tau}}{e^{\pi i \tau} - e^{-\pi i \tau}}$$

$$= \pi i \frac{q+1}{q-1} = \pi i + \frac{2\pi i}{q-1}$$

$$= \pi i - 2\pi i \sum_{j=0}^{\infty} q^j. \tag{9.12}$$

Recall the product expansion

$$\sin \pi \tau = \pi \tau \prod_{n=1}^{\infty} \left(1 - \frac{\tau}{n}\right)\left(1 + \frac{\tau}{n}\right)$$

(see [2]). Taking the logarithmic derivative yields

$$\pi \frac{\cos \pi \tau}{\sin \pi \tau} = \frac{1}{\tau} + \sum_{n=1}^{\infty} \left(\frac{1}{\tau - n} + \frac{1}{\tau + n}\right). \tag{9.13}$$

Differentiating (9.12) and (9.13) $2k - 1$ times with respect to τ yields

$$-\sum_{j=1}^{\infty}(2\pi i)^{2k} j^{2k-1} q^j = (-1)^{2k-1}(2k-1)! \sum_{n=-\infty}^{\infty} \frac{1}{(\tau + n)^{2k}}.$$

Consider (9.11) with $2k$ in place of k. Since $2k$ is even, the terms for (m, n) and $(-m, -n)$ are equal, so we only need to sum for $m = 0, n > 0$ and for $m > 0, n \in \mathbf{Z}$, then double the answer. We obtain

$$G_{2k}(\tau) = 2\sum_{n=1}^{\infty} \frac{1}{n^{2k}} + 2\sum_{m=1}^{\infty}\sum_{n=-\infty}^{\infty} \frac{1}{(m\tau + n)^{2k}}$$

$$= 2\zeta(2k) + 2\sum_{m=1}^{\infty}\sum_{j=1}^{\infty} \frac{(2\pi i)^k j^{2k-1}}{(2k-1)!} q^{mj}$$

$$= 2\zeta(2k) + 2\frac{(2\pi i)^{2k}}{(2k-1)!} \sum_{m=1}^{\infty}\sum_{j=1}^{\infty} j^{2k-1} q^{mj}.$$

Let $n = mj$ in the last expression. Then, for a given n, the sum over j can be regarded as the sum over the positive divisors of n. This yields the first expression in the statement of the proposition. The expansion $\sum_{m \geq 1} q^{mj} = q^j/(1 - q^j)$ yields the second expression.   ∎

Recall that we defined $g_2 = g_2(L) = 60G_4(L)$ and $g_3 = g_3(L) = 140G_6(L)$ for arbitrary lattices L. Restricting to L_τ, we define

$$g_2(\tau) = g_2(L_\tau), \quad g_3(\tau) = g_3(L_\tau).$$

Using the facts that

$$\zeta(4) = \frac{\pi^4}{90} \quad \text{and} \quad \zeta(6) = \frac{\pi^6}{945},$$

we obtain

$$g_2(\tau) = \frac{4\pi^4}{3}(1 + 240q + \cdots) = \frac{4\pi^4}{3}\left(1 + 240 \sum_{j=1}^{\infty} \frac{j^3 q^j}{1 - q^j}\right)$$

$$g_3(\tau) = \frac{8\pi^6}{27}(1 - 504q + \cdots) = \frac{8\pi^6}{27}\left(1 - 504 \sum_{j=1}^{\infty} \frac{j^5 q^j}{1 - q^j}\right).$$

Since $\Delta = g_2^3 - 27g_3^2$, a straightforward calculation shows that

$$\Delta(\tau) = (2\pi)^{12}(q + \cdots).$$

Define

$$j(\tau) = 1728\frac{g_2^3}{\Delta}.$$

Then $j(\tau) = \frac{1}{q} + \cdots$. Including a few more terms in the above calculations yields

$$j(\tau) = \frac{1}{q} + 744 + 196884q + 21493760q^2 + \cdots.$$

For computational purposes, this series converges slowly since the coefficients are large. It is usually better to use the following.

PROPOSITION 9.12

$$j(\tau) = 1728 \frac{\left(1 + 240 \sum_{j=1}^{\infty} \frac{j^3 q^j}{1-q^j}\right)^3}{\left(1 + 240 \sum_{j=1}^{\infty} \frac{j^3 q^j}{1-q^j}\right)^3 - \left(1 - 504 \sum_{j=1}^{\infty} \frac{j^5 q^j}{1-q^j}\right)^2}.$$

PROOF Substitute the above expressions for g_2, g_3 into the definition of the j-function. The powers of π and other constants cancel to yield the present expression. ∎

It can be shown (see [55, p. 249]) that

$$\Delta = (2\pi)^{12} q \prod_{k=1}^{\infty} (1 - q^k)^{24}.$$

This yields the expression

$$j = \frac{\left(1 + 240\sum_{j=1}^{\infty} \frac{j^3 q^j}{1 - q^j}\right)^3}{q\prod_{k=1}^{\infty}(1 - q^k)^{24}},$$

which also works very well for computing j.

More generally, if L is a lattice, define

$$j(L) = 1728\frac{g_2(L)^3}{g_2(L)^3 - 27g_3(L)^2}.$$

If $\lambda \in \mathbf{C}^{\times}$, then the definitions of G_4 and G_6 easily imply that

$$g_2(\lambda L) = \lambda^{-4}g_2(L) \quad \text{and} \quad g_3(\lambda L) = \lambda^{-6}g_3(L). \tag{9.14}$$

Therefore

$$j(L) = j(\lambda L).$$

Letting $L = \mathbf{Z}\omega_1 + \mathbf{Z}\omega_2$ and $\lambda = \omega_2^{-1}$, we obtain

$$j(\mathbf{Z}\omega_1 + \mathbf{Z}\omega_2) = j(\tau),$$

where $\tau = \omega_1/\omega_2$.

Recall that

$$SL_2(\mathbf{Z}) = \left\{\begin{pmatrix} a & b \\ c & d \end{pmatrix} \middle| a, b, c, d \in \mathbf{Z}, ad - bc = 1\right\}$$

acts on the upper half plane $\mathcal{H}$ by

$$\begin{pmatrix} a & b \\ c & d \end{pmatrix}\tau = \frac{a\tau + b}{c\tau + d}$$

for all $\tau \in \mathcal{H}$.

PROPOSITION 9.13

Let $\tau \in \mathcal{H}$ and let $\begin{pmatrix} a & b \\ c & d \end{pmatrix} \in SL_2(\mathbf{Z})$. Then

$$j\left(\frac{a\tau + b}{c\tau + d}\right) = j(\tau).$$

PROOF   We first compute what happens with G_k:

$$G_k\left(\frac{a\tau + b}{c\tau + d}\right) = \sum_{(m,n)\neq(0,0)} \frac{1}{(m\frac{a\tau+b}{c\tau+d} + n)^k}$$

$$= (c\tau + d)^k \sum_{(m,n)\neq(0,0)} \frac{1}{(m(a\tau + b) + n(c\tau + d))^k}$$

$$= (c\tau + d)^k \sum_{(m,n)\neq(0,0)} \frac{1}{((ma + nc)\tau + (mb + nd))^k}.$$

Since $\begin{pmatrix} a & b \\ c & d \end{pmatrix}$ has determinant 1, we have

$$\begin{pmatrix} a & b \\ c & d \end{pmatrix}^{-1} = \begin{pmatrix} d & -b \\ -c & a \end{pmatrix}.$$

Let

$$(m', n') = (m, n) \begin{pmatrix} a & b \\ c & d \end{pmatrix} = (ma + nc, mb + nd).$$

Then

$$(m, n) = (m', n') \begin{pmatrix} d & -b \\ -c & a \end{pmatrix},$$

so there is a one-to-one correspondence between pairs of integers (m, n) and pairs of integers (m', n'). Therefore,

$$G_k\left(\frac{a\tau + b}{c\tau + d}\right) = (c\tau + d)^k \sum_{(m', n') \neq (0,0)} \frac{1}{(m'\tau + n')^k}$$

$$= (c\tau + d)^k G_k(\tau).$$

Since g_2 and g_3 are multiples of G_4 and G_6, we have

$$g_2\left(\frac{a\tau + b}{c\tau + d}\right) = (c\tau + d)^4 g_2(\tau), \quad g_3\left(\frac{a\tau + b}{c\tau + d}\right) = (c\tau + d)^6 g_3(\tau).$$

Therefore, when we substitute these expressions into the definition of j, all the factors $(c\tau + d)$ cancel. ▮

Let $\mathcal{F}$ be the subset of $z \in \mathcal{H}$ such that

$$|z| \geq 1, \quad -\frac{1}{2} \leq \Re(z) < \frac{1}{2}, \quad z \neq e^{i\theta} \text{ for } \frac{\pi}{3} < \theta < \frac{\pi}{2}.$$

Figure 9.3 is a picture of $\mathcal{F}$. Since we will need to refer to it several times, we let

$$\rho = e^{2\pi i/3}.$$

PROPOSITION 9.14
  Given $\tau \in \mathcal{H}$, there exists

$$\begin{pmatrix} a & b \\ c & d \end{pmatrix} \in SL_2(\mathbf{Z})$$

such that

$$\frac{a\tau + b}{c\tau + d} = z \in \mathcal{F}.$$

Moreover, $z \in \mathcal{F}$ is uniquely determined by τ.

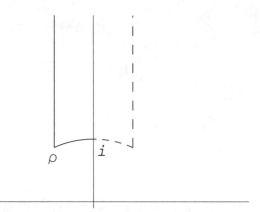

Figure 9.3
The Fundamental Domain for $SL_2(\mathbf{Z})$

The proposition says that $\mathcal{F}$ is a **fundamental domain** for the action of $SL_2(\mathbf{Z})$ on $\mathcal{H}$. For a proof of the proposition, see [86] or [89].

COROLLARY 9.15
Let $L \subset \mathbf{C}$ be a lattice. There exists a basis $\{\omega_1, \omega_2\}$ of L with $\omega_1/\omega_2 \in \mathcal{F}$. In other words,

$$L = (\lambda)(\mathbf{Z}\tau + \mathbf{Z})$$

for some $\lambda \in \mathbf{C}^\times$ and some uniquely determined $\tau \in \mathcal{F}$.

PROOF   Let $\{\alpha, \beta\}$ be a basis for L and let $\tau_0 = \alpha/\beta$. By changing the sign of α if necessary, we may assume that $\tau_0 \in \mathcal{H}$. Let

$$\begin{pmatrix} a & b \\ c & d \end{pmatrix} \in SL_2(\mathbf{Z})$$

be such that

$$\frac{a\tau_0 + b}{c\tau_0 + d} = \tau \in \mathcal{F}.$$

Let

$$\omega_1 = a\alpha + b\beta, \quad \omega_2 = c\alpha + d\beta.$$

Since the matrix is in $SL_2(\mathbf{Z})$,

$$L = \alpha + \mathbf{Z}\beta = \mathbf{Z}\omega_1 + \mathbf{Z}\omega_2 = \omega_2(\mathbf{Z}\tau + \mathbf{Z}).$$

This proves the corollary.   ∎

If $z \in \mathbf{C}$, recall that $\mathrm{ord}_z f$ is the order of f at z. That is,

$$f(\tau) = (\tau - z)^{\mathrm{ord}_z(f)} g(\tau),$$

with $g(z) \neq 0, \infty$. We can also define the order of f at $i\infty$. Suppose

$$f(\tau) = a_n q^n + a_{n+1} q^{n+1} + \cdots, \tag{9.15}$$

with $n \in \mathbf{Z}$ and $a_n \neq 0$, and assume that this series converges for all q close to 0 (with $q \neq 0$ when $n < 0$). Then

$$\mathrm{ord}_{i\infty}(f) = n.$$

Note that $q \to 0$ as $\tau \to i\infty$, so $\mathrm{ord}_{i\infty}(f)$ expresses whether f vanishes $(n > 0)$ or blows up $(n < 0)$ as $\tau \to i\infty$.

PROPOSITION 9.16
Let f be a function meromorphic in $\mathcal{H}$ such that f is not identically zero and such that

$$f\left(\frac{a\tau + b}{c\tau + d}\right) = f(\tau) \quad \text{for all} \quad \begin{pmatrix} a & b \\ c & d \end{pmatrix} \in SL_2(\mathbf{Z}).$$

Then

$$\mathrm{ord}_{i\infty}(f) + \frac{1}{3}\mathrm{ord}_\rho(f) + \frac{1}{2}\mathrm{ord}_i(f) + \sum_{z \neq i, \rho, i\infty} \mathrm{ord}_z(f) = 0.$$

REMARK 9.17   The function f can be regarded as a function on the surface obtained as follows. Identify the left and right sides on $\mathcal{F}$ to get a tube, then fold the part with $|z| = 1$ at i. Then pinch the open end at $i\infty$ to a point. This gives a surface that is topologically a sphere. The proposition expresses the fact that the number of poles of f equals the number of zeros on such a surface, just as occurred for doubly periodic functions in Theorem 9.1. The point i is special since a small neighborhood around i contains only half of a disc inside $\mathcal{F}$. Similarly, a small neighborhood around ρ includes only 1/3 of a disc from $\mathcal{F}$ (namely, 1/6 near ρ and 1/6 near $1 + \rho$, which is folded over to meet ρ). This explains the factors 1/2 and 1/3 in the proposition. For a related phenomenon, see Exercise 9.3.   ∎

PROOF   Let C be the path shown in Figure 9.4. Essentially, C goes around the edge of $\mathcal{F}$. However, it consists of a small circular arc past each of $\rho, 1 + \rho$, and i. If there is a pole or zero of f at a point on the path, we make a small detour around it and a corresponding detour at the corresponding point on the other side of $\mathcal{F}$.

The arcs near $\rho, 1 + \rho$, and i have radius ϵ, where ϵ is chosen small enough that there are no zeros or poles of f inside the circles, except possibly at ρ,

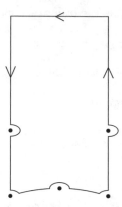

Figure 9.4

$1 + \rho$, or i. Similarly, the top part of C is chosen to have imaginary part N, where N is large enough that $f(z)$ has no zeros or poles with imaginary part greater than N, except perhaps at $i\infty$. This is possible since

$$f(z) = q^n(a_n + a_{n+1}q + \cdots).$$

Since the series $a_n + a_{n+1}q + \cdots$ is assumed to converge for q small, it is finite and is approximately equal to $a_n \neq 0$ for sufficiently small q.

As in the proof of Theorem 9.1, we have

$$\frac{1}{2\pi i} \int_C \frac{f'(z)}{f(z)}\, dz = \sum_{\substack{z \in \mathcal{F} \\ z \neq i, \rho}} \operatorname{ord}_z(f).$$

Since $\begin{pmatrix} 1 & 1 \\ 0 & 1 \end{pmatrix} \in SL_2(\mathbf{Z})$ gives the map $z \mapsto z + 1$, we have

$$f(z) = f(z + 1). \tag{9.16}$$

Therefore, the integrals over the left and right vertical parts of C are the same, except that they are in opposite directions, so they cancel each other.

Now we'll show that the integral over the part of the unit circle to the left of i cancels the part to the right. This is proved by using the fact that $\begin{pmatrix} 0 & -1 \\ 1 & 0 \end{pmatrix} \in SL_2(\mathbf{Z})$ gives the map $z \mapsto -1/z$, which interchanges the left and right arcs of the unit circle. In addition, differentiating the relation $f(-1/z) = f(z)$ yields

$$\frac{f'}{f}\left(\frac{-1}{z}\right) d\left(\frac{-1}{z}\right) = \frac{f'}{f}(z)\, dz.$$

Therefore, the integral over C from ρ to i equals the integral from $-1/\rho = 1+\rho$ to $-1/i = i$, which is the negative of the integral from i to $1 + \rho$. Therefore, the two parts cancel.

All that remains are the parts of C near ρ, $1 + \rho$, i, and $i\infty$. Near i, we have $f(z) = (z - i)^k g(z)$ for some k, with $g(i) \neq 0, \infty$. Therefore,

$$\frac{f'(z)}{f(z)} = \frac{k}{z - i} + \frac{g'(z)}{g(z)}. \tag{9.17}$$

The integral over the small semicircle near i is

$$\frac{1}{2\pi i} \int_\theta \frac{f'(i + \epsilon e^{i\theta})}{f(i + \epsilon e^{i\theta})} \epsilon i e^{i\theta} \, d\theta, \tag{9.18}$$

where θ ranges from slightly more than π to slightly less than 0. (Note that C is traveled clockwise. Because of the curvature of the unit circle, the limits are 0 and π only in the limit as $\epsilon \to 0$.) Substitute (9.17) into (9.18) and let $\epsilon \to 0$. Since g'/g is continuous at i, the integral of g'/g goes to 0. The integral of $k/(z - i)$ yields

$$\frac{1}{2\pi i} \int_{\theta=\pi}^0 ki \, d\theta = -\frac{1}{2}k = -\frac{1}{2}\mathrm{ord}_i(f).$$

Similarly, the contributions from the parts of C near ρ and $1 + \rho$ add up to $-(1/3)\mathrm{ord}_\rho(f)$ (we are using the fact that $f(\rho) = f(\rho + 1)$, by (9.16)).

Finally, the integral along the top part of C is

$$\frac{1}{2\pi i} \int_{t=\frac{1}{2}}^{-\frac{1}{2}} \frac{f'(t + iN)}{f(t + iN)} \, dt.$$

Since $f(\tau) = q^n(a_n + a_{n+1}q + \cdots)$, we have

$$\frac{f'(\tau)}{f(\tau)} = 2\pi i n + \frac{2\pi i a_{n+1}q + \cdots}{a_n + \cdots}.$$

The second term goes to 0 as $q \to 0$, hence as $N \to \infty$. The limit of the integral as $N \to \infty$ is therefore

$$\frac{1}{2\pi i} \int_{t=\frac{1}{2}}^{-\frac{1}{2}} 2\pi i n \, dt = -n = -\mathrm{ord}_{i\infty}(f).$$

Combining all of the above calculations yields the theorem. ∎

COROLLARY 9.18

If $z \in \mathbf{C}$, then there is exactly one $\tau \in \mathcal{F}$ such that $j(\tau) = z$.

PROOF   First, we need to calculate $j(\rho)$ and $j(i)$. Recall that τ corresponds to the lattice $L_\tau = \mathbf{Z}\tau + \mathbf{Z}$. Since $\rho^2 = -1 - \rho$, it follows easily that $\rho L_\rho \subseteq L_\rho$. Therefore,

$$L_\rho = \rho^3 L_\rho \subseteq \rho^2 L_\rho \subseteq \rho L_\rho \subseteq L_\rho,$$

so $\rho L_\rho = L_\rho$. It follows from (9.14) that

$$g_2(L_\rho) = g_2(\rho L_\rho) = \rho^{-4}g_2(L_\rho) = \rho^{-1}g_2(L_\rho).$$

Since $\rho \neq 1$, we have $g_2(\rho) = g_2(L_\rho) = 0$. Therefore,

$$j(\rho) = 1728\frac{g_2(L_\rho)^3}{g_2(L_\rho)^3 - 27g_3(L_\rho)^2} = 0$$

(note that the denominator is nonzero, by Proposition 9.9).

Similarly, $\tau = i$ corresponds to the lattice $L_i = \mathbf{Z}i + \mathbf{Z}$, and $iL_i = L_i$. Therefore,

$$g_3(L_i) = g_3(iL_i) = i^{-6}g_3(L_i) = -g_3(L_i),$$

so $g_3(i) = g_3(L_i) = 0$. Therefore,

$$j(i) = 1728\frac{g_2(L_i)^3}{g_2(L_i)^3 - 27g_3(L_i)^2} = 1728.$$

We now look at the other values of τ. Consider the function $h(\tau) = j(\tau) - z$. Then h has a pole of order 1 at $i\infty$ and no other poles. By Proposition 9.16, we have

$$\frac{1}{3}\mathrm{ord}_\rho(h) + \frac{1}{2}\mathrm{ord}_i(h) + \sum_{z \neq i, \rho, \infty} \mathrm{ord}_z(h) = 1.$$

If $z \neq 0, 1728$, then h has order 0 at ρ and at i. Therefore, h has a unique zero in $\mathcal{F}$, so $j(\tau) = z$ has a unique solution in $\mathcal{F}$. If $z = 1728$, then $(1/2)\mathrm{ord}_i(h) > 0$. Since the order of h at a point is an integer, the order must be 0 when $z \neq i, \rho$; otherwise, the sum would be larger than 1. Also, there is no combination of $m/2 + n/3$ that equals 1 except when either $m = 0$ or $n = 0$. Therefore, $j(\tau) - 1728$ has a double zero at i and no other zero in $\mathcal{F}$. Similarly, $j(\tau)$ has a triple zero at ρ and no other zero in $\mathcal{F}$.   ∎

We can now show that every elliptic curve over $\mathbf{C}$ corresponds to a torus.

THEOREM 9.19
Let $y^2 = 4x^3 - Ax - B$ define an elliptic curve E over $\mathbf{C}$. Then there is a lattice L such that

$$g_2(L) = A \quad \text{and} \quad g_3(L) = B.$$

There is an isomorphism of groups

$$\mathbf{C}/L \simeq E(\mathbf{C}).$$

PROOF   Let

$$j = 1728 \frac{A^3}{A^3 - 27B^2}.$$

By Corollary 9.18, there exists a lattice $L = \mathbf{Z}\tau + \mathbf{Z}$ such that $j(\tau) = j(L) = j$. Assume first that $g_2(L) \neq 0$. Then $j = j(L) \neq 0$, so $A \neq 0$. Choose $\lambda \in \mathbf{C}^\times$ such that

$$g_2(\lambda L) = \lambda^{-4} g_2(L) = A.$$

The equality $j = j(L)$ implies that

$$g_3(\lambda L)^2 = B^2,$$

so $g_3(\lambda L) = \pm B$. If $g_3(\lambda L) = B$, we're done. If $g_3(\lambda L) = -B$, then

$$g_3(i\lambda L) = i^{-6} g_3(\lambda L) = B \quad \text{and} \quad g_2(i\lambda L) = i^4 g_2(\lambda L) = A.$$

Therefore, either λL or $i\lambda L$ the desired lattice.

If $g_2(L) = 0$, then $j = j(L) = 0$, so $A = 0$. Since $A^3 - 27B^2 \neq 0$ by assumption and since $g_2(L)^3 - 27g_3(L)^2 \neq 0$ by Proposition 9.9, we have $B \neq 0$ and $g_3(L) \neq 0$. Choose $\mu \in \mathbf{C}^\times$ such that

$$g_3(\mu L) = \mu^{-6} g_3(L) = B.$$

Then $g_2(\mu L) = \mu^{-2} g_2(L) = 0 = A$, so μL is the desired lattice.

By Theorem 9.10, the map

$$\mathbf{C}/L \longrightarrow E(\mathbf{C}).$$

is an isomorphism.   ∎

The elements of L are called the **periods** of L.

Theorem 9.19 gives us a good way to work with elliptic curves over $\mathbf{C}$. For example, let n be a positive integer and let E be an elliptic curve over $\mathbf{C}$. By Theorem 9.19, there exists a lattice $L = \mathbf{Z}\omega_1 + \mathbf{Z}\omega_2$ such that $\mathbf{C}/L$ is isomorphic to $E(\mathbf{C})$. It is easy to see that the n-torsion on $\mathbf{C}/L$ is given by the points

$$\frac{j}{n}\omega_1 + \frac{k}{n}\omega_2, \quad 0 \leq j, k \leq n - 1.$$

It follows that

$$E[n] \simeq \mathbf{Z}_n \oplus \mathbf{Z}_n.$$

In fact, we can use this observation to give a proof of Theorem 3.2 for all fields of characteristic 0.

COROLLARY 9.20

   Let K be a field of characteristic 0, and let E be an elliptic curve over K. Then

$$E[n] = \{P \in E(\overline{K}) \mid nP = \infty\} \simeq \mathbf{Z}_n \oplus \mathbf{Z}_n.$$

PROOF   Let L be the field generated by $\mathbf{Q}$ and the coefficients of the equation of E. Then L has finite transcendence degree over $\mathbf{Q}$, hence can be embedded into $\mathbf{C}$ (see Appendix C). Therefore, we can regard E as an elliptic curve over $\mathbf{C}$. Therefore, the n-torsion is $\mathbf{Z}_n \oplus \mathbf{Z}_n$.

There is a technical point to worry about. The definition of $E[n]$ that we have used requires the coordinates of the n-torsion to lie in the algebraic closure of the base field. How can we be sure that the field $\overline{K}$ isn't so large that it allows more torsion points than $\mathbf{C}$? Suppose that $E[n] \subset E(\overline{K})$ has order larger than n^2. Then we can choose $n^2 + 1$ of these points and adjoin their coordinates to L. Then L still has finite transcendence degree over $\mathbf{Q}$, hence can be embedded into $\mathbf{C}$. The coordinates of the $n^2 + 1$ points will yield $n^2 + 1$ points in $E(\mathbf{C})$ that are n-torsion points. This is impossible. Therefore, $E[n]$ is no larger than it should be.

There is also the reverse possibility. How do we know that $\overline{K}$ is large enough to account for all the n-torsion points that we found in $E(\mathbf{C})$? We need to show that the n-torsion points in $E(\mathbf{C})$ have coordinates that are algebraic over L (where L is regarded as a subfield of $\mathbf{C}$). Let $P = (x, y)$ be an n-torsion point in $E(\mathbf{C})$, and suppose that x and y are transcendental over L (since x and y satisfy the polynomial defining E, they are both algebraic or both transcendental over K). Let σ be an automorphism of $\mathbf{C}$ such that $\sigma(x) = x + 1$, and such that σ is the identity on L. Such an automorphism exists: take σ to be the desired automorphism of $K(x)$, then use Zorn's Lemma to extend σ to all of $\mathbf{C}$ (see Appendix C). The points $\sigma^m(P)$ for $m = 1, 2, 3, \ldots$, have distinct x-coordinates $x + 1, x + 2, x + 3, \ldots$, hence are distinct points. Each must be an n-torsion point of E in $E(\mathbf{C})$. But there are only n^2 such points, so we have a contradiction. Therefore, the coordinates of the n-torsion points are algebraic over L, hence are algebraic over K, since $L \subseteq K$. Therefore, the passage from K to $\mathbf{C}$ does not affect $E[n]$.   ∎

Suppose we have an elliptic curve E defined over the real numbers $\mathbf{R}$. Usually, it is represented by a graph, as in Chapter 2 (see Figure 2.1 on page 10). It is interesting to see how the torus we obtain relates to this graph. It can be shown (Exercise 9.4) that the lattice L for E has one of two shapes. Suppose first that the lattice is rectangular: $L = \mathbf{Z}\omega_1 + \mathbf{Z}\omega_2$ with $\omega_1 \in i\mathbf{R}$ and $\omega_2 \in \mathbf{R}$. Then

$$(\wp(z),\ \wp'(z)) \in E(\mathbf{R})$$

when

$$(I)\quad z = t\omega_2 \quad \text{with } 0 \le t < 1,$$

and also when

$$(II)\quad z = (1/2)\omega_1 + t\omega_2 \quad \text{with } 0 \le t < 1.$$

The first of these is easy to see: if z is real and the lattice L is preserved by complex conjugation, then conjugating the defining expression for $\wp(z)$ leaves it unchanged, so $\wp$ maps reals to reals. The second is a little more subtle: conjugating $z = (1/2)\omega_1 + t\omega_2$ yields $\bar{z} = -(1/2)\omega_1 + t\omega_2$, which is equivalent to $z \bmod L$. Therefore, the defining expression for $\wp(z)$ is again unchanged by complex conjugation, so $\wp$ maps reals to reals.

Fold the parallelogram into a torus by connecting the right and left sides to form a tube, then connecting the ends. The paths (I) and (II) (see Figure 9.5) yield circles on the torus.

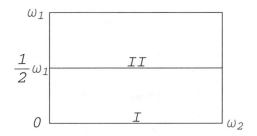

Figure 9.5

The Real Points on C/L

When the ends of path (II) are disconnected at 0, we obtain a slightly deformed version of the graph of Figure 2.1(a) on page 10.

In the case of a skewed parallelogram, that is, when ω_2 is real and the imaginary part of ω_1 is half of ω_2, the real axis is mapped to the reals, but the analogue of path (II) is not mapped to the reals. This corresponds to the situation of Figure 2.1(b) on page 10.

9.4 Computing Periods

Suppose E is an elliptic curve over **C**. From Theorem 9.19, we know that E corresponds to a lattice $L = \mathbf{Z}\omega_1 + \mathbf{Z}\omega_2$ via the doubly periodic functions $\wp$ and $\wp'$, but how do we find the periods ω_1 and ω_2?

For simplicity, let's consider the case where E is defined over **R** and $E[2] \subset E(\mathbf{R})$. Then the equation for E can be put in the form

$$y^2 = 4x^3 - g_2 x - g_3 = 4(x - e_1)(x - e_2)(x - e_3) \quad \text{with } e_1 < e_2 < e_3.$$

We may assume $\omega_2 \in \mathbf{R}$ with $\omega_2 > 0$ and $\omega_1 \in i\mathbf{R}$ with $\Im(\omega_1) > 0$, as in Figure 9.5. The graph of E is as in Figure 2.1(a) on page 10. The Weierstrass

$\wp$-function and its derivative map $\mathbf{C}/L$ to E via

$$(x,\, y) = (\wp(z),\, \wp'(z)).$$

As z goes from 0 to $\omega_2/2$, the function $\wp(z)$ takes on real values, starting with $x = \infty$. The first point of order two is encountered when $z = \omega_2/2$. Which point $(e_i, 0)$ is it? The graph of the real points of E has two components. The one connected to ∞ contains the point $(e_3, 0)$ of order two, so $x = \wp(z)$ must run from ∞ to e_3 as z goes from 0 to $\omega_2/2$. The expansion of $\wp'(z)$ starts with the term $-2/z^3$, from which it follows that $y = \wp'(z) < 0$ near $z = 0$, hence $\wp'(z) < 0$ for $0 < z < \omega_2/2$.

Consider now the integral

$$\int_{e_3}^{\infty} \frac{dx}{\sqrt{4(x - e_1)(x - e_2)(x - e_3)}}.$$

Substitute $x = \wp(z)$. The denominator becomes $\sqrt{\wp'(z)^2} = -\wp'(z)$ (recall that $\wp'(z) < 0$) and the limits of integration are from $z = \omega_2/2$ to 0. Adjusting the direction of integration and the sign yields

$$\int_0^{\omega_2/2} dz = \frac{\omega_2}{2}.$$

Therefore,

$$\omega_2 = \int_{e_3}^{\infty} \frac{dx}{\sqrt{(x - e_1)(x - e_2)(x - e_3)}}.$$

The change of variables

$$x = \frac{\left(e_3 - \sqrt{(e_3 - e_1)(e_3 - e_2)}\right) t + \left(e_3 + \sqrt{(e_3 - e_1)(e_3 - e_2)}\right)}{t + 1}$$

(plus a lot of algebraic manipulation) changes the integral to

$$\omega_2 = \frac{2}{\sqrt{e_3 - e_1} + \sqrt{e_3 - e_2}} \int_{-1}^{1} \frac{dt}{\sqrt{(1 - t^2)(1 - k^2 t^2)}},$$

where

$$k = \frac{\sqrt{e_3 - e_1} - \sqrt{e_3 - e_2}}{\sqrt{e_3 - e_1} + \sqrt{e_3 - e_2}}. \tag{9.19}$$

Since the integrand is an even function, we can take twice the integral over the interval from 0 to 1 and obtain

$$\omega_2 = \frac{4}{\sqrt{e_3 - e_1} + \sqrt{e_3 - e_2}} \int_0^{1} \frac{dt}{\sqrt{(1 - t^2)(1 - k^2 t^2)}}.$$

This integral is called an **elliptic integral** (more precisely, an elliptic integral of the first kind). It is usually denoted by

$$K(k) = \int_0^1 \frac{dt}{\sqrt{(1 - t^2)(1 - k^2 t^2)}}.$$

In the following, we'll see how to compute $K(k)$ numerically very accurately and quickly, but first let's find an expression for ω_1.

When z runs along the vertical line from $\omega_2/2$ to $\omega_2/2 + \omega_1/2$, the function $\wp(z)$ takes on real values (see Exercise 9.5) from e_3 to e_2, and its derivative $\wp'(z)$ takes on purely imaginary values. Reasoning similar to that above (including the same change of variables) yields

$$\omega_1 = \frac{2i}{\sqrt{e_3 - e_1} + \sqrt{e_3 - e_2}} \int_1^{1/k} \frac{dt}{\sqrt{(t^2 - 1)(1 - k^2 t^2)}}.$$

Let $k' = \sqrt{1 - k^2}$ and make the substitution

$$t = (1 - k'^2 u^2)^{-1/2}.$$

The integral becomes

$$\int_0^1 \frac{dt}{\sqrt{(1 - t^2)(1 - k'^2 t^2)}} = K(k') = K(\sqrt{1 - k^2}).$$

Therefore,

$$\omega_1 = \frac{2i}{\sqrt{e_3 - e_1} + \sqrt{e_3 - e_2}} K(\sqrt{1 - k^2}).$$

9.4.1 The Arithmetic-Geometric Mean

In this subsection, we introduce the arithmetic-geometric mean. It yields a very fast and ingenious method, due to Gauss, for computing elliptic integrals.

Start with two positive real numbers a, b. Define a_n and b_n by

$$\begin{aligned} a_0 &= a, \quad b_0 = b \\ a_n &= \frac{1}{2}(a_{n-1} + b_{n-1}) \\ b_n &= \sqrt{a_{n-1} b_{n-1}}. \end{aligned} \tag{9.20}$$

Then a_n is the arithmetic mean (=average) of a_{n-1} and b_{n-1}, and b_n is their geometric mean.

Example 9.1

Let $a = \sqrt{2}$ and $b = 1$. Then

$$a_1 = 1.20710678118654752440084436 2 \ldots$$
$$b_1 = 1.18920711500272106671749997 0 \ldots$$

$$a_2 = 1.19815694809463429555917216 6 \ldots$$
$$b_2 = 1.19812352149312012260658557 1 \ldots$$

$$a_3 = 1.19814023479387720908287886 9 \ldots$$
$$b_3 = 1.19814023467730720579838378 8 \ldots$$

$$a_4 = 1.19814023473559220744063132 8 \ldots$$
$$b_4 = 1.19814023473559220743921365 5 \ldots .$$

The sequences are converging very quickly to the limit

$$a_\infty = b_\infty = 1.19814023473559220743992249 2 \ldots .$$

▯

The rapid convergence is explained by the following.

PROPOSITION 9.21

Suppose $a \geq b > 0$. Then

$$b_{n-1} \leq b_n \leq a_n \leq a_{n-1}$$

and

$$0 \leq a_n - b_n \leq \frac{1}{2}(a_{n-1} - b_{n-1}). \tag{9.21}$$

Therefore

$$M(a, b) = \lim_{n \to \infty} a_n = \lim_{n \to \infty} b_n$$

exists. Moreover, if $b \geq 1$ then

$$a_{n+m} - b_{n+m} \leq 8 \left(\frac{a_n - b_n}{8} \right)^{2^m} \tag{9.22}$$

for all $m, n \geq 0$.

PROOF   The fact that $a_n \geq b_n$ for all n is the arithmetic-geometric mean inequality, or the fact that

$$a_n - b_n = \frac{1}{2}(\sqrt{a_{n-1}} - \sqrt{b_{n-1}})^2 \geq 0.$$

Therefore, since $a_{n-1} \geq b_{n-1}$, it follows immediately from (9.20) that

$$a_n \leq \frac{1}{2}(a_{n-1} + a_{n-1}) = a_{n-1} \quad \text{and} \quad b_n \leq \sqrt{a_{n-1}a_{n-1}} = a_{n-1}.$$

Also,

$$
\begin{aligned}
a_n - b_n &= \frac{1}{2}\left(\sqrt{a_{n-1}} - \sqrt{b_{n-1}}\right)^2 \\
&\leq \frac{1}{2}\left(\sqrt{a_{n-1}} - \sqrt{b_{n-1}}\right)\left(\sqrt{a_{n-1}} + \sqrt{b_{n-1}}\right)
\end{aligned}
$$

$$= \frac{1}{2}(a_{n-1} - b_{n-1}).$$

Since the a_n's are a decreasing sequence bounded below by the increasing sequence of b_n's, it follows immediately that the two sequences converge to the same limit, so $M(a,b)$ exists. If $b_{n-1} \geq 1$, then $\sqrt{a_{n-1}} + \sqrt{b_{n-1}} \geq 2$, so

$$
\begin{aligned}
\frac{a_n - b_n}{8} &= \frac{1}{16}\left(\sqrt{a_{n-1}} - \sqrt{b_{n-1}}\right)^2 \\
&\leq \frac{1}{16}\left(\sqrt{a_{n-1}} - \sqrt{b_{n-1}}\right)^2 \frac{\left(\sqrt{a_{n-1}} + \sqrt{b_{n-1}}\right)^2}{4} \\
&= \left(\frac{a_{n-1} - b_{n-1}}{8}\right)^2.
\end{aligned}
$$

Inequality 9.22 follows easily by induction.   ∎

The limit $M(a,b)$ is called the **arithmetic-geometric mean** of a and b. Since

$$M(ca, cb) = cM(a,b),$$

we can always rescale a and b to make $b \geq 1$. Also, since $M(b,a) = M(a,b)$ (because a_1 and b_1 are symmetric in a, b), we may always arrange that $a \geq b$. By Inequality (9.21), $a_n - b_n < 1$ for sufficiently large n. The numbers a_{n+m} and b_{n+m} give approximations to $M(a,b)$. Inequality (9.22) predicts that the number of decimal places of accuracy doubles with each iteration. This phenomenon occurs in the above example.

The reasons we are interested in the arithmetic-geometric mean are the following two propositions.

PROPOSITION 9.22

Let a, b be positive real numbers. Define

$$I(a,b) = \int_0^{\pi/2} \frac{d\theta}{\sqrt{a^2\cos^2\theta + b^2\sin^2\theta}}.$$

Then

$$I\left(\frac{a+b}{2}, \sqrt{ab}\right) = I(a, b).$$

Moreover,

$$I(a, b) = \frac{\pi/2}{M(a, b)}.$$

PROOF   Let $u = b \tan \theta$. The integral becomes

$$I(a, b) = \int_0^\infty \frac{du}{\sqrt{(u^2 + a^2)(u^2 + b^2)}} = \frac{1}{2} \int_{-\infty}^\infty \frac{du}{\sqrt{(u^2 + a^2)(u^2 + b^2)}}.$$

Therefore,

$$I\left(\frac{a+b}{2}, \sqrt{ab}\right) = \frac{1}{2} \int_{-\infty}^\infty \frac{du}{\sqrt{(u^2 + (\frac{a+b}{2})^2)(u^2 + ab)}}.$$

Let

$$u = \frac{1}{2}\left(v - \frac{ab}{v}\right), \quad 0 < v < \infty.$$

Then $v = u + \sqrt{u^2 + ab}$. Since

$$u^2 + \left(\frac{a+b}{2}\right)^2 = \frac{1}{4v^2}(v^2 + a^2)(v^2 + b^2),$$

it is straightforward to obtain

$$I\left(\frac{a+b}{2}, \sqrt{ab}\right) = \int_0^\infty \frac{dv}{\sqrt{(v^2 + a^2)(v^2 + b^2)}} = I(a, b).$$

By induction, we obtain

$$I(a, b) = I(a_1, b_1) = I(a_2, b_2) = \cdots.$$

Let

$$a_\infty = b_\infty = M(a, b) = \lim_{n \to \infty} a_n = \lim_{n \to \infty} b_n.$$

It is fairly easy to justify taking the limit inside the integral sign to obtain

$$
\begin{aligned}
I(a, b) &= \lim_{n \to \infty} I(a_n, b_n) \\
&= I(a_\infty, b_\infty) \\
&= \int_0^{\pi/2} \frac{d\theta}{\sqrt{a_\infty^2 \cos^2 \theta + b_\infty^2 \sin^2 \theta}} \\
&= \frac{1}{M(a, b)} \int_0^{\pi/2} \frac{d\theta}{\sqrt{\cos^2 \theta + \sin^2 \theta}} \\
&= \frac{\pi/2}{M(a, b)}. \quad \blacksquare
\end{aligned}
$$

PROPOSITION 9.23

If $0 < k < 1$, then

$$K(k) = I\left(1, \sqrt{1 - k^2}\right) = I(1 + k, 1 - k).$$

PROOF

$$K(k) = \int_0^1 \frac{dt}{\sqrt{(1 - t^2)(1 - k^2 t^2)}}$$

$$= \int_0^{\pi/2} \frac{d\theta}{\sqrt{1 - k^2 \sin^2 \theta}} \qquad (\text{let } t = \sin \theta)$$

$$= \int_0^{\pi/2} \frac{d\theta}{\sqrt{\cos^2 \theta + (1 - k^2) \sin^2 \theta}}$$

$$= I(1, \sqrt{1 - k^2})$$

$$= I(1 + k, 1 - k).$$

The last equation follows from Proposition 9.22, with $a = 1 + k$ and $b = 1 - k$.

∎

Putting everything together, we can now express the periods ω_1 and ω_2 in terms of arithmetic-geometric means.

THEOREM 9.24

Suppose E is given by

$$y^2 = 4x^3 - g_2 x - g_3 = 4(x - e_1)(x - e_2)(x - e_3)$$

with real numbers $e_1 < e_2 < e_3$. Then $\mathbf{Z}\omega_1 + \mathbf{Z}\omega_2$ is a lattice for E, where

$$\omega_1 = \frac{\pi i}{M\left(\sqrt{e_3 - e_1}, \sqrt{e_2 - e_1}\right)}$$

$$\omega_2 = \frac{\pi}{M\left(\sqrt{e_3 - e_1}, \sqrt{e_3 - e_2}\right)}.$$

PROOF   We have, with k as in (9.19),

$$\omega_2 = \frac{4}{\sqrt{e_3 - e_1} + \sqrt{e_3 - e_2}} K(k)$$

$$= \frac{4}{\sqrt{e_3 - e_1} + \sqrt{e_3 - e_2}} I(1 + k, 1 - k).$$

Use the definition (9.19) of k and the relation $cI(ca, cb) = I(a, b)$ with

$$c = \frac{\sqrt{e_3 - e_1} + \sqrt{e_3 - e_2}}{2}$$

to obtain

$$\omega_2 = 2I(\sqrt{e_3 - e_1}, \sqrt{e_3 - e_2})$$
$$= \frac{\pi}{M(\sqrt{e_3 - e_1}, \sqrt{e_3 - e_2})}.$$

The proof of the formula for ω_1 uses similar reasoning to obtain

$$\omega_1 = \frac{2i}{\sqrt{e_3 - e_1} + \sqrt{e_3 - e_2}} K(\sqrt{1 - k^2})$$
$$= \frac{2i}{\sqrt{e_3 - e_1} + \sqrt{e_3 - e_2}} I(1, k)$$
$$= 2iI(\sqrt{e_3 - e_1} + \sqrt{e_3 - e_2}, \quad \sqrt{e_3 - e_1} - \sqrt{e_3 - e_2}).$$

If we let

$$a = \sqrt{e_3 - e_1} + \sqrt{e_3 - e_2}, \quad b = \sqrt{e_3 - e_1} - \sqrt{e_3 - e_2},$$

then (9.20) yields

$$a_1 = \sqrt{e_3 - e_1}, \quad b_1 = \sqrt{e_2 - e_1}.$$

Proposition 9.22 therefore implies that

$$\omega_1 = 2iI(\sqrt{e_3 - e_1}, \sqrt{e_2 - e_1})$$
$$= \frac{\pi i}{M(\sqrt{e_3 - e_1}, \sqrt{e_2 - e_1})}. \qquad \blacksquare$$

Example 9.2
Consider the elliptic curve E given by

$$y^2 = 4x^3 - 4x.$$

Then $e_1 = -1$, $e_2 = 0$, $e_3 = 1$, so

$$\omega_1 = \frac{\pi i}{M(\sqrt{2}, 1)} = i2.6220575542921198104648395 9\ldots$$
$$\omega_2 = \frac{\pi}{M(\sqrt{2}, 1)} = 2.6220575542921198104648395 9\ldots.$$

Therefore, the fundamental parallelogram for the lattice is a square. This also follows from the fact that E has complex multiplication by $\mathbf{Z}[i]$. See Chapter 10. The number $2.622\ldots$ can be shown (see Exercise 9.7) to equal

$$\int_{-1}^{1} \frac{dx}{\sqrt{1 - x^4}} = \frac{\Gamma(1/4)\Gamma(1/2)}{2\,\Gamma(3/4)},$$

where Γ is the gamma function (for its definition, see Section 12.2). This is a special case of the Chowla-Selberg formula, which expresses the periods of elliptic curves with complex multiplication in terms of values of the gamma function (see [83]).   □

The formulas in Theorem 9.24 can also be used in the case where $g_2, g_3 \in \mathbf{R}$ but $4x^3 - g^2x - g_3$ has only one real root. In this case, e_1 and e_2 are complex conjugate complex numbers. After one iteration of arithmetic and geometric means, the numbers a_1, b_1 obtained are positive reals, so the algorithm converges. For more on the arithmetic-geometric mean, including how it has been used to compute π very accurately and how it behaves for complex arguments, see [12] and [23].

9.5 Division Polynomials

In this section, we prove Theorem 3.6, which gives a formula for $n(x, y)$, where $n > 1$ is an integer and (x, y) is a point on an elliptic curve. We'll start with the case of an elliptic curve in characteristic zero, then use this to deduce the case of positive characteristic.

Let E be an elliptic curve over a field of characteristic 0, given by an equation $y^2 = x^3 + Ax + B$. All of the equations describing the group law are defined over $\mathbf{Q}(A, B)$. Since $\mathbf{C}$ is algebraically closed and has infinite transcendence degree over $\mathbf{Q}$, it is easy to see that $\mathbf{Q}(A, B)$ may be considered as a subfield of $\mathbf{C}$. Therefore, we regard E as an elliptic curve defined over $\mathbf{C}$. By Theorem 9.19, there is a lattice L corresponding to E. Let $\wp(z)$ be the associated Weierstrass $\wp$-function, which satisfies the relation

$$(\wp')^2 = 4\wp^3 - g_2\wp - g_3,$$

with $g_2 = -4A$, $g_3 = -4B$. We'll derive formulas for $\wp(nz)$ and $\wp'(nz)$, then use $x = \wp(z)$ and $y = \wp'(z)/2$ to obtain the desired formulas for $n(x, y)$.

LEMMA 9.25
There is a doubly periodic function $f_n(z)$ such that

$$f_n(z)^2 = n^2 \prod_{0 \neq u \in (\mathbf{C}/L)[n]} (\wp(z) - \wp(u)).$$

The sign of f_n can be chosen so that

1. *if n is odd, $f_n = P_n(\wp)$, where $P_n(X)$ is a polynomial of degree $(n^2-1)/2$ with leading coefficient n,*

2. *if n is even, $f_n = \wp' P_n(\wp)$, where $P_n(X)$ is a polynomial of degree $(n^2 - 4)/2$ with leading coefficient $n/2$.*

The expansion of f_n at 0 is

$$f_n(z) = \frac{(-1)^{n+1}n}{z^{n^2-1}} + \cdots.$$

The zeros of f_n are at the points $0 \neq u \in (\mathbf{C}/L)[n]$, and these are simple zeros.

PROOF   The product is over the nonzero n-torsion in $\mathbf{C}/L$. Since $\wp(u) = \wp(-u)$, the factors for u and $-u$ are equal. Suppose n is odd. Then u is never congruent to $-u$ mod L, so every factor in the product occurs twice. Therefore, f_n can be taken to be $n \prod(\wp(z) - \wp(u))$, where we use only one member of each pair $(u, -u)$. This is clearly a polynomial in $\wp(z)$ of degree $(n^2 - 1)/2$ and leading coefficient n. When n is even, there are three values of u that are congruent to their negatives mod L, namely, $\omega_j/2$ for $j = 1, 2, 3$. Since

$$(\wp')^2 = 4 \prod_j (\wp - \wp(\omega_j/2)),$$

these factors contribute $\wp'/2$ to f_n. The remaining factors can be paired up, as in the case when n is odd, to obtain a polynomial in $\wp$ of degree $(n^2 - 4)/2$ and leading coefficient n. Therefore, f_n has the desired form.

Since $\wp(z) = z^{-2} + \cdots$ and $\wp'(z) = -2z^{-3} + \cdots$, we immediately obtain the expansion of f_n at 0.

Clearly f_n has a zero at each nonzero $u \in (\mathbf{C}/L)[n]$. There are $n^2 - 1$ such points. Since the only pole mod L of f_n is one of order $n^2 - 1$ at $z = 0$, and since the number of zeros equals the number of poles (counting multiplicities), these zeros must all be simple.   ∎

LEMMA 9.26
 Let $n \geq 2$. Then

$$\wp(nz) = \wp(z) - \frac{f_{n-1}(z)f_{n+1}(z)}{f_n(z)^2}.$$

PROOF   Let $g(z) = \wp(nz) - \wp(z)$. We'll show that g and $f_{n-1}f_{n+1}/f_n^2$ have the same divisors.

The function $g(z)$ has a double pole at each $u \in (\mathbf{C}/L)[n]$ with $u \neq 0$. At $z = 0$, it has the expansion

$$g(z) = \frac{1}{n^2 z^2} - \frac{1}{z^2} + \cdots,$$

so g also has a double pole at 0. Therefore, g has a total of $2n^2$ poles, counting multiplicities.

The function g has a zero at $z = w$ when $nw \equiv \pm w \not\equiv 0 \pmod{L}$. For such w,

$$\frac{d}{dz}g(z)\Bigg|_{z=w} = n\wp'(nw) - \wp'(w) = \pm n\wp'(w) - \wp'(w) = (\pm n - 1)\wp'(w).$$

Since the zeros of $\wp'(z)$ occur when $z = \omega_j/2$, we have $g'(w) \neq 0$ when $w \neq \omega_j/2$, so such w are simple zeros of g. Moreover, when n is odd, $n(\omega_j/2) \equiv \omega_i/2$, so the points $\omega_j/2$ are at least double zeros of g in this case.

If $nw \equiv w \pmod{L}$, then $(n-1)w \equiv 0$. Let $\delta = 0$ if n is even and $\delta = 1$ if n is odd. There are $(n-1)^2 - 1 - 3\delta$ points w with $(n-1)w = 0$ and $w \neq 0, \omega_j/2$. Similarly, there are $(n+1)^2 - 1 - 3\delta$ points w with $(n+1)w = 0$ and $w \neq 0, \omega_j/2$. There are at least 6δ zeros (counting multiplicities) at the points $\omega_j/2$. Therefore, we have accounted for at least

$$(n-1)^2 - 1 - 3\delta + (n+1)^2 - 1 - 3\delta + 6\delta = 2n^2$$

zeros. Since $g(z)$ has exactly $2n^2$ poles, we have found all the zeros and their multiplicities.

The function

$$\frac{f_{n-1}f_{n+1}}{f_n^2}$$

has a double pole at each of the zeros of f_n. If $w \neq 0$ and $(n \pm 1)w \equiv 0$ then $f_{n\pm 1}$ has a simple zero at w. If both $(n+1)w \equiv 0$ and $(n-1)w \equiv 0$, then $2w \equiv 0$, so $w \equiv \omega_j/2$ for some j. Therefore, $f_{n-1}f_{n+1}$ has a simple zero at each w with $(n \pm 1)w \equiv 0$, except for those where $w \equiv \omega_j/2$, at which points it has a double zero. At $z = 0$, the expansions of the functions yield

$$\frac{f_{n-1}f_{n+1}}{f_n^2} =$$

$$\left(\frac{(-1)^n(n-1)}{z^{(n-1)^2-1}} + \cdots\right)\left(\frac{(-1)^{n+2}(n+1)}{z^{(n+1)^2-1}} + \cdots\right)\Bigg/\left(\frac{(-1)^{n+1}n}{z^{n^2-1}} + \cdots\right)$$

$$= -\left(1 - \frac{1}{n^2}\right)z^{-2} + \cdots,$$

so there is a double zero at $z = 0$. Therefore, $-f_{n-1}f_{n+1}/f_n^2$ has the same divisor as $\wp(nz) - \wp(z)$, so the two functions are constant multiples of each other. Since their expansions at 0 have the same leading coefficient, they must be equal. This proves the lemma. ∎

LEMMA 9.27
$f_{2n+1} = f_{n+2}f_n^3 - f_{n+1}^3 f_{n-1}.$

PROOF   As in the proof of Lemma 9.26, we see that

$$\wp((n+1)z) - \wp(nz) = -\frac{f_{2n+1}}{f_{n+1}^2 f_n^2}$$

since the two sides have the same divisors and their expansions at 0 have the same leading coefficient. Since

$$\wp((n+1)z) - \wp(nz) = (\wp((n+1)z) - \wp(z)) - (\wp(nz) - \wp(z))$$
$$= -\frac{f_{n+2}f_n}{f_{n+1}^2} + \frac{f_{n+1}f_{n-1}}{f_n^2},$$

the result follows by equating the two expressions for $\wp((n+1)z) - \wp(nz)$.
∎

LEMMA 9.28
$\wp' f_{2n} = (f_n)(f_{n+2}f_{n-1}^2 - f_{n-2}f_{n+1}^2).$

PROOF   As in the proofs of the previous two lemmas, we have

$$\wp((n+1)z) - \wp((n-1)z) = -\frac{\wp' f_{2n}}{f_{n-1}^2 f_{n+1}^2}.$$

(A little care is needed to handle the points $\omega_j/2$.) Since

$$\wp((n+1)z) - \wp((n-1)z) = (\wp((n+1)z) - \wp(z)) - (\wp((n-1)z) - \wp(z))$$
$$= -\frac{f_{n+2}f_n}{f_{n+1}^2} + \frac{f_n f_{n-2}}{f_{n-1}^2},$$

the result follows.   ∎

LEMMA 9.29
For all $n \geq 1$,

$$f_n(z) = \psi_n\left(\wp(z), \frac{1}{2}\wp'(z)\right)$$

where ψ_n is defined in Section 3.2.

PROOF   Since $\psi_1 = 1$ and $\psi_2 = 2y$, the lemma is easily seen to be true for $n = 1, 2$. From Equations (9.10) and (9.8) in Section 9.2, we have

$$-\frac{f_3}{(\wp')^2} = -\frac{f_3}{f_2^2} = \wp(2z) - \wp(z) \tag{9.23}$$
$$= \frac{1}{4}\left(\frac{\wp''(z)}{\wp'(z)}\right)^2 - 2\wp(z) - \wp(z)$$
$$= -\frac{3\wp^4 - \frac{3}{2}g_2\wp^2 - 3g_3\wp - \frac{1}{16}g_2^2}{(\wp')^2}.$$

Therefore,

$$f_3 = 3\wp^4 - \frac{3}{2}g_2\wp^2 - 3g_3\wp - \frac{1}{16}g_2^2$$
$$= 3\wp^4 + 6A\wp^2 + 12B\wp - A^2 = \psi_3(\wp).$$

This proves the lemma for $n = 3$.

By Equation (9.7) in Section 9.2, we have

$$\wp(2z + z) = \frac{1}{4}\left(\frac{\wp'(2z) - \wp'(z)}{\wp(2z) - \wp(z)}\right)^2 - \wp(2z) - \wp(z)$$

and

$$\wp(2z - z) = \frac{1}{4}\left(\frac{\wp'(2z) + \wp'(z)}{\wp(2z) - \wp(z)}\right)^2 - \wp(2z) - \wp(z).$$

Therefore,

$$-\frac{f_4 f_2}{f_3^2} = \wp(3z) - \wp(z)$$

$$= \frac{1}{4}\left(\frac{\wp'(2z) - \wp'(z)}{\wp(2z) - \wp(z)}\right)^2 - \frac{1}{4}\left(\frac{\wp'(2z) + \wp'(z)}{\wp(2z) - \wp(z)}\right)^2$$

$$= -\frac{\wp'(2z)\wp'(z)}{(\wp(2z) - \wp(z))^2}$$

$$= -\frac{\wp'(2z)\wp'(z)}{(-f_3/\wp'(z)^2)^2} \qquad \text{(by Equation 9.23)}$$

$$= -\frac{\wp'(2z)\wp'(z)^5}{f_3^2}.$$

This yields $f_4 f_2 = \wp'(2z)\wp'(z)^5$. We know that $\frac{1}{2}\wp'(2z)$ is the y-coordinate of $2(\wp(z), \frac{1}{2}\wp'(z))$, which means that $\wp'(2z)$ can be expressed in terms of $\wp(z)$ and $\wp'(z)$, using the formulas for the group law. When this is done, we obtain

$$f_4 = \psi_4\left(\wp, \frac{1}{2}\wp'\right),$$

so the lemma is true for $n = 4$.

Since the f_n's satisfy the same recurrence relations as the ψ_n's (see Lemmas 9.27 and 9.28 and the definition of the ψ_n's), and since the lemma holds for enough small values of n, the lemma now follows for all n. ∎

LEMMA 9.30

$$\wp'(nz) = \frac{f_{2n}}{f_n^4}.$$

PROOF   The function $\wp'(nz)$ has triple poles at all points of $(\mathbf{C}/L)[n]$. Therefore, there are $3n^2$ poles. Since the zeros of $\wp'$ are at the points in $(\mathbf{C}/L)[2]$ other than 0, the zeros of $\wp'(nz)$ are at the points that are in $(\mathbf{C}/L)[2n]$ but not in $(\mathbf{C}/L)[n]$. There are $3n^2$ such points. Since the number of zeros equals the number of poles, all of these zeros are simple. The expansion of $\wp'(nz)$ at $z = 0$ is

$$\wp'(nz) = \frac{-2}{n^3 z^3} + \cdots .$$

The function f_{2n}/f_n^4 is easily seen to have the same divisor as $\wp'(nz)$ and their expansions at $z = 0$ have the same leading coefficients. Therefore, the functions are equal.   ∎

Finally, we can prove the main result of this section.

THEOREM 9.31
Let E be an elliptic curve over a field of characteristic not 2, let n be a positive integer, and let (x, y) be a point on E. Then

$$n(x, y) = \left(\frac{\phi_n}{\psi_n^2} , \frac{\omega_n}{\psi_n^3} \right) ,$$

where ϕ_n, ψ_n, and ω_n are defined in Section 3.2.

PROOF   First, assume E is defined over a field of characteristic 0. As above, we regard E as being defined over $\mathbf{C}$. We have

$$(x, y) = \left(\wp(z), \frac{1}{2}\wp'(z) \right), \quad n(x, y) = \left(\wp(nz), \frac{1}{2}\wp'(nz) \right)$$

for some z. Therefore,

$$\wp(nz) = \wp(z) - \frac{f_{n-1}f_{n+1}}{f_n^2}$$

$$= \frac{\wp f_n^2 - f_{n-1}f_{n+1}}{f_n^2}$$

$$= \frac{x\psi_n^2 - \psi_{n-1}\psi_{n+1}}{\psi_n^2} \quad \text{(by Lemma 9.29)}$$

$$= \frac{\phi_n}{\psi_n^2} .$$

This proves the formula for the x-coordinate.

For the y-coordinate, observe that the definition of ω_n can be rewritten as

$$\omega_n = \frac{1}{2} \frac{\psi_{2n}}{\psi_n} .$$

Therefore, by Lemmas 9.30 and 9.29,

$$\frac{1}{2}\wp'(nz) = \frac{1}{2}\frac{\psi_{2n}}{\psi_n^4} = \frac{\omega_n}{\psi_n^3}.$$

This completes the proof of the theorem when the characteristic of the field is 0.

Suppose now that E is defined over a field K of arbitrary characteristic (not 2) by $y^2 = x^3 + Ax + B$. Let $(x,y) \in E(K)$. Let α, β, and X be three independent transcendental elements of $\mathbf{C}$ and let Y satisfy $Y^2 = X^3 + \alpha X + \beta$. There is a ring homomorphism

$$\rho : \mathbf{Z}[\alpha, \beta, X, Y] \longrightarrow K(x,y)$$

such that

$$g(\alpha, \beta, X, Y) \mapsto g(A, B, x, y)$$

for all polynomials g. Let $R = \mathbf{Z}[\alpha, \beta, X, Y]$ and let $\tilde{E}$ be the elliptic curve over R defined by $y^2 = x^3 + \alpha x + \beta$. We want to say that by Corollary 2.32, ρ induces a homomorphism

$$\rho : \tilde{E}(R) \longrightarrow E(K(x,y)).$$

But we need to have R satisfy Conditions (1) and (2) of Section 2.10. The easiest way to accomplish this is to let $\mathcal{M}$ be the kernel of the map $R \to K(x,y)$. Since $K(x,y)$ is a field, $\mathcal{M}$ is a maximal ideal of R. Let $R_{\mathcal{M}}$ be the localization of R at $\mathcal{M}$ (this means, we invert all elements of R not in $\mathcal{M}$). Then $R \subseteq R_{\mathcal{M}}$ and the map ρ extends to a map

$$\rho : R_{\mathcal{M}} \longrightarrow K(x,y).$$

Since $R_{\mathcal{M}}$ is a local ring, and projective modules over local rings are free, it can be shown that $R_{\mathcal{M}}$ satisfies Condition (2). Since we are assuming that $K(x,y)$ has characteristic not equal to 2, it follows that 2 is not in $\mathcal{M}$, hence is invertible in $R_{\mathcal{M}}$. Therefore, $R_{\mathcal{M}}$ satisfies Condition (1). Now we can apply Corollary 2.32.

The point $n(X,Y)$ in $\tilde{E}(R_{\mathcal{M}})$ is described by the polynomials ψ_j, ϕ_j, and ω_j, which are polynomials in X, Y with coefficients in $\mathbf{Z}[\alpha, \beta]$. Applying ρ shows that these polynomials, regarded as polynomials in x, y with coefficients in K, describe $n(x,y)$ on E. Therefore, the theorem holds for E.   ∎

As an application of the division polynomials, we prove the following result, which will be used in Chapter 11.

PROPOSITION 9.32

Let E be an elliptic curve over a field K. Let $f(x,y)$ be a function from E to $\overline{K} \cup \{\infty\}$ and let $n \geq 1$ be an integer not divisible by the characteristic of K.

Suppose $f(P + T) = f(P)$ for all $P \in E(\overline{K})$ and all $T \in E[n]$. Then there is a function h on E such that $f(P) = h(nP)$ for all P.

PROOF   The case $n = 1$ is trivial, so we assume $n > 1$. Let $T \in E[n]$. There are rational functions $R(x, y), S(x, y)$ depending on T such that

$$(x, y) + T = (R(x, y), S(x, y)).$$

Let $y^2 = x^3 + Ax + B$ be the equation of E and regard $\overline{K}(x, y)$ as the quadratic extension of $\overline{K}(x)$ given by adjoining $\sqrt{x^3 + Ax + B}$. Since (R, S) lies on E, we have $S^2 = R^3 + AR + B$. The map

$$\sigma_T : \overline{K}(x, y) \to \overline{K}(x, y)$$
$$f(x, y) \mapsto f(R, S)$$

is a homomorphism from $\overline{K}(x, y)$ to itself. Since σ_{-T} is the inverse of σ_T, the map σ_T is an automorphism. Because $(x, y) + T \neq (x, y) + T'$ when $T \neq T'$, we have $\sigma_T(x, y) \neq \sigma_{T'}(x, y)$ when $T \neq T'$. Therefore, we have a group of n^2 distinct automorphisms σ_T, where T runs through $E[n]$, acting on $\overline{K}(x, y)$. A basic result in Galois theory says that if G is a group of distinct automorphisms of a field L, then the fixed field F of G satisfies $[L : F] = |G|$. Therefore, the field F of functions f satisfying the conditions of the proposition satisfies

$$[\overline{K}(x, y) : F] = n^2. \tag{9.24}$$

Let $n(x, y) = (g_n(x), y\, h_n(x))$ for rational functions g_n, h_n. Then

$$\overline{K}(g_n(x), y\, h_n(x)) \subseteq F. \tag{9.25}$$

Moreover,

$$[\overline{K}(g_n(x), y\, h_n(x)) : \overline{K}(g_n(x))] \geq 2 \tag{9.26}$$

since clearly $y\, h_n(x) \notin \overline{K}(g_n(x))$. Therefore, by (9.24), (9.25), and (9.26),

$$[\overline{K}(x, y) : \overline{K}(g_n(x))] \geq 2n^2.$$

From Theorem 3.6,

$$g_n(x) = \frac{\phi_n}{\psi_n^2},$$

and ϕ_n and ψ_n^2 are polynomials in x. Therefore, $X = x$ is a root of the polynomial

$$P(X) = \phi_n(X) - g_n(x)\psi_n^2(X) \in \overline{K}[g_n(x)][X].$$

By Lemma 3.5,

$$\phi_n(X) = X^{n^2} + \cdots$$

and $\psi_n^2(X)$ has degree $n^2 - 1$. Therefore,

$$P(X) = X^{n^2} + \cdots,$$

so x is of degree at most n^2 over $\overline{K}(g_n(x))$. Since

$$[\overline{K}(x, y) : \overline{K}(x)] = 2,$$

we obtain

$$[\overline{K}(x, y) : \overline{K}(g_n(x))] \le 2n^2.$$

Combined with the previous inequality from above, we obtain equality, which means that we had equality in all of our calculations. In particular,

$$F = \overline{K}(g_n(x), \, y\, h_n(x)).$$

The functions in F are those that are invariant under translation by elements of $E[n]$. Those on the right are those that are of the form $h(n(x, y))$. Therefore, we have proved the proposition. ∎

Exercises

9.1 (a) Show that $d^3 \equiv d^5 \pmod{12}$ for all integers d.

 (b) Show that

$$5 \sum_{d|n} d^3 + 7 \sum_{d|n} d^5 \equiv 0 \pmod{12}$$

   for all positive integers n.

 (c) Show that

$$g_2 = \frac{(2\pi)^4}{12}(1 + 240X),$$

   where $X = \sum_{n=1}^{\infty} \sigma_3(n)q^n$.

 (d) Show that

$$g_3 = \frac{(2\pi)^6}{216}(1 - 504Y),$$

   where $Y = \sum_{n=1}^{\infty} \sigma_5(n)q^n$.

 (e) Show that

$$1728(2\pi)^{-12}\Delta = (1 + 240X)^3 - (1 - 504Y)^2. \qquad (9.27)$$

 (f) Show that the right side of (9.27) is congruent to $144(5X + 7Y)$ mod 1728.

(g) Conclude that $(2\pi)^{-12}\Delta = \sum_{n=0}^{\infty} d_n q^n$, with $d_n \in \mathbf{Z}$.

(h) Compute enough coefficients to obtain that

$$(2\pi)^{-12}\Delta = q(1 + \sum_{n=1}^{\infty} e_n q^n)$$

with $e_n \in \mathbf{Z}$.

(i) Show that $(2\pi)^{12}/\Delta = q^{-1}\sum_{n=1}^{\infty} f_n q^n$ with $f_n \in \mathbf{Z}$.

(j) Show that

$$j = \frac{1}{q} + \sum_{n=1}^{\infty} c_n q^n$$

with $c_n \in \mathbf{Z}$.

9.2 Let $k \geq 0$ be an integer. Let f be a meromorphic function on the upper half plane such that f has a q-expansion at $i\infty$ (as in Equation (9.15)) and such that

$$f\left(\frac{a\tau + b}{c\tau + d}\right) = (c\tau + d)^k f(\tau)$$

for all $\tau \in \mathcal{H}$ and for all $\begin{pmatrix} a & b \\ c & d \end{pmatrix} \in SL_2(\mathbf{Z})$. Show that

$$\mathrm{ord}_{i\infty}(f) + \frac{1}{3}\mathrm{ord}_{\rho}(f) + \frac{1}{2}\mathrm{ord}_i(f) + \sum_{z \neq i,\rho,\infty} \mathrm{ord}_z(f) = \frac{k}{12}.$$

9.3 The stabilizer in $SL_2(\mathbf{Z})$ of a point $z \in \mathcal{H}$ is the set of matrices $\begin{pmatrix} a & b \\ c & d \end{pmatrix}$ such that $(az + b)/(cz + d) = z$.

(a) Show that the stabilizer of i has order 4.

(b) Show that the stabilizer of ρ has order 6.

(c) Show that the stabilizer of $i\infty$ consists of the matrices of the form $\pm\begin{pmatrix} 1 & b \\ 0 & 1 \end{pmatrix}$ with $b \in \mathbf{Z}$.

(d) Show that the stabilizer of each $z \in \mathcal{H}$ has order at least 2.

It can be shown that the stabilizer of each element in the fundamental domain $\mathcal{F}$ has order 2 except for i and ρ.

9.4 (a) Show that $\overline{j(\tau)} = j(-\overline{\tau})$.

(b) Show that if τ is in the fundamental domain $\mathcal{F}$, then either $-\overline{\tau} \in \mathcal{F}$ or $\Re(\tau) = -1/2$.

(c) Suppose that $\tau \in \mathcal{F}$ and that $j(\tau) \in \mathbf{R}$. Show that $\Re(\tau) = 0$ or $-1/2$. (*Hint:* Use Corollary 9.18.)

(d) Let E be an elliptic curve defined over $\mathbf{R}$ and let L be the lattice associated to E by Theorem 9.19. Assume that $j(E) \neq 0, 1728$ (equivalently, $g_2, g_3 \neq 0$). Show that there exists $\lambda \in \mathbf{C}$ such that $\lambda^2 \in \mathbf{R}$ and $\lambda L = \mathbf{Z}\tau + \mathbf{Z}$, with $\Re(\tau) = 0$ or $-1/2$. (*Hint:* Use Equations (9.14).)

(e) Let $0 \neq y \in \mathbf{R}$. Let M be the lattice $(\frac{1}{2} + iy)\mathbf{Z} + \mathbf{Z}$. Show that iM has $\{y + \frac{1}{2}i, \, 2y\}$ as a basis.

(f) Let E be as in (d). Show that L has a basis $\{\omega_1, \, \omega_2\}$ with $\omega_2 \in \mathbf{R}$ and $\Re(\omega_1) = 0$ or $\frac{1}{2}\omega_2$.

(g) Use the facts that $j(\rho) = 0$ and $j(i) = 1728$ to prove (d) in the cases that $j(E) = 0$ and $j(E) = 1728$. (The condition that $\lambda^2 \in \mathbf{R}$ gets replaced by $\lambda^6 \in \mathbf{R}$ and $\lambda^4 \in \mathbf{R}$, respectively. However, the lattices for $\tau = \rho$ and $\tau = i$ have extra symmetries.)

9.5 Let L be a lattice such that is stable under complex conjugation (that is, if $\omega \in L$ then $\overline{\omega} \in L$). This is the same as requiring that the elliptic curve associated to L is defined over $\mathbf{R}$ (see Exercise 9.4).

(a) Show that $\overline{\wp(z)} = \wp(\overline{z})$.

(b) Show that if $t \in \mathbf{R}$ and $\omega_2 \in \mathbf{R}$ is a real period, then

$$\wp\left(\frac{1}{2}\omega_2 + it\right) \in \mathbf{R}.$$

(*Hint:* Use (a), the periodicity of $\wp$, and the fact that $\wp(-z) = \wp(z)$.)

(c) Differentiate the result of (b) to show that $\wp'(z) \in i\mathbf{R}$ for the points $\frac{1}{2}\omega_2 + it$ in (b). This path, for $0 \leq t \leq \omega_1$, corresponds to x moving along the x-axis between the two parts of the graph in Figure 2.1(a) on page 10. The points don't appear on the graph because y is imaginary. For the curve in Figure 2.1(b) on page 10, x moves to the left along the x-axis, from the point on the x-axis back to the point at infinity, corresponding to the fact that $\omega_1 = \frac{1}{2}\omega_2 + it$ for appropriate t (see Exercise 9.4).

9.6 Define the **elliptic integral of the second kind** to be

$$E(k) = \int_0^1 \frac{\sqrt{1 - k^2 x^2}}{\sqrt{1 - x^2}} \, dx, \quad -1 < k < 1.$$

(a) Show that

$$E(k) = \int_0^{\pi/2} (1 - k^2 \sin^2 \theta)^{1/2} \, d\theta.$$

(b) Show that the arc length of the ellipse

$$\frac{x^2}{a^2} + \frac{y^2}{b^2} = 1$$

with $b \geq a > 0$ equals $4bE(\sqrt{1 - (a/b)^2})$.

This connection with ellipses is the origin of the name "elliptic integral." The relation between elliptic integrals and elliptic curves, as in Section 9.4, is the origin of the name "elliptic curve." For more on elliptic integrals, see [63].

9.7 Let E be the elliptic curve $y^2 = 4x^3 - 4x$. Show that

$$\omega_2 = \int_1^\infty \frac{dx}{\sqrt{x(x^2 - 1)}} = \frac{1}{2} \int_0^1 t^{-3/4}(1 - t)^{-1/2}\, dt = \beta(1/4, 1/2),$$

where $\beta(p, q) = \int_0^1 t^{p-1}(1 - t)^{q-1}\, dt$ is the **beta function**. A classical result says that

$$\beta(p, q) = \frac{\Gamma(p)\Gamma(q)}{\Gamma(p + q)}.$$

Therefore,

$$\omega_2 = \frac{1}{2}\frac{\Gamma(1/4)\Gamma(1/2)}{\Gamma(3/4)}.$$

Chapter 10

Complex Multiplication

The endomorphisms of an elliptic curve E always include multiplication by arbitrary integers. When the endomorphism ring of E is strictly larger than $\mathbf{Z}$, we say that E has **complex multiplication**. As we'll see, elliptic curves over $\mathbf{C}$ with complex multiplication correspond to lattices with extra symmetry. Over finite fields, all elliptic curves have complex multiplication, and often the Frobenius provides one of the additional endomorphisms. In general, elliptic curves with complex multiplication form an interesting and important class of elliptic curves, partly because of their extra structure and partly because of their frequent occurrence.

10.1 Elliptic Curves over C

Consider the elliptic curve E given by $y^2 = 4x^3 - 4x$ over $\mathbf{C}$. As we saw in Section 9.4, E corresponds to the torus $\mathbf{C}/L$, where $L = \mathbf{Z}\omega + \mathbf{Z}i\omega$, for a certain $\omega \in \mathbf{R}$. Since L is a square lattice, it has extra symmetries. For example, rotation by 90° sends L into itself. This can be expressed by saying that $iL = L$. Using the definition of the Weierstrass $\wp$-function, we easily see that

$$
\wp(iz) = \frac{1}{(iz)^2} + \sum_{\omega \neq 0} \left(\frac{1}{(iz - \omega)^2} - \frac{1}{\omega^2} \right)
$$

$$
= \frac{1}{(iz)^2} + \sum_{i\omega \neq 0} \left(\frac{1}{(iz - i\omega)^2} - \frac{1}{(i\omega)^2} \right)
$$

$$
= -\wp(z).
$$

Differentiation yields

$$
\wp'(iz) = i\wp(z).
$$

On the elliptic curve E, we obtain the endomorphism given by

$$
i(x, y) = (-x, iy).
$$

Therefore, the map

$$z \mapsto iz$$

gives a map

$$(x, y) = (\wp(z), \wp'(z)) \mapsto (\wp(iz), \wp'(iz)) = (-x, iy).$$

This is a homomorphism from $E(\mathbf{C})$ to $E(\mathbf{C})$ and it is clearly given by rational functions. Therefore, it is an endomorphism of E, as in Section 2.8. Let

$$\mathbf{Z}[i] = \{a + bi \,|\, a, b \in \mathbf{Z}\}.$$

Then $\mathbf{Z}[i]$ is a ring, and multiplication by elements of $\mathbf{Z}[i]$ sends L into itself. Correspondingly, if $a + bi \in \mathbf{Z}[i]$ and $(x, y) \in E(\mathbf{C})$, then we obtain an endomorphism of E defined by

$$(x, y) \mapsto (a + bi)(x, y) = a(x, y) + b(-x, iy).$$

Since multiplication by a and b can be expressed by rational functions, multiplication of points by $a + bi$ is an endomorphism of E, as in Section 2.8. Therefore,

$$\mathbf{Z}[i] \subseteq \mathrm{End}(E),$$

where $\mathrm{End}(E)$ denotes the ring of endomorphisms of E. (We'll show later that this is an equality.) Therefore, $\mathrm{End}(E)$ is strictly larger than $\mathbf{Z}$, so E has complex multiplication. Just as $\mathbf{Z}[i]$ is the motivating example for a lot of ring theory, so is E the prototypical example for complex multiplication.

We now consider endomorphism rings of arbitrary elliptic curves over $\mathbf{C}$. Let E be an elliptic curve over $\mathbf{C}$, corresponding to the lattice

$$L = \mathbf{Z}\omega_1 + \mathbf{Z}\omega_2.$$

Let α be an endomorphism of E. Recall that this means that α is a homomorphism from $E(\mathbf{C})$ to $E(\mathbf{C})$, and that α is given by rational functions:

$$\alpha(x, y) = (R(x), yS(x))$$

for rational functions R, S. The map

$$\Phi : \mathbf{C}/L \to E(\mathbf{C}), \quad \Phi(z) = (\wp(z), \wp'(z)),$$

(see Theorem 9.10) is an isomorphism of groups. The map

$$\tilde{\alpha}(z) = \Phi^{-1}(\alpha(\Phi(z)))$$

is therefore a homomorphism from $\mathbf{C}/L$ to $\mathbf{C}/L$. If we restrict to a sufficiently small neighborhood U of $z = 0$, we obtain an analytic map from U to $\mathbf{C}$ such that

$$\tilde{\alpha}(z_1 + z_2) \equiv \tilde{\alpha}(z_1) + \tilde{\alpha}(z_2) \pmod{L}$$

for all $z_1, z_2 \in U$. By subtracting an appropriate element of L, we may assume that $\tilde{\alpha}(0) = 0$. By continuity, $\tilde{\alpha}(z)$ is near 0 when z is near 0. If U is sufficiently small, we may therefore assume that

$$\tilde{\alpha}(z_1 + z_2) = \tilde{\alpha}(z_1) + \tilde{\alpha}(z_2)$$

for all $z_1, z_2 \in U$ (since both sides are near 0, they can differ only by the element $0 \in L$). Therefore, for $z \in U$, we have

$$\tilde{\alpha}'(z) = \lim_{h \to 0} \frac{\tilde{\alpha}(z + h) - \tilde{\alpha}(z)}{h}$$
$$= \lim_{h \to 0} \frac{\tilde{\alpha}(z) + \tilde{\alpha}(h) - \tilde{\alpha}(z)}{h}$$
$$= \lim_{h \to 0} \frac{\tilde{\alpha}(z) - \tilde{\alpha}(0)}{h} = \tilde{\alpha}'(0).$$

Let $\beta = \tilde{\alpha}'(0)$. Since $\tilde{\alpha}'(z) = \beta$ for all $z \in U$, we must have

$$\tilde{\alpha}(z) = \beta z$$

for all $z \in U$.

Now let $z \in \mathbf{C}$ be arbitrary. There exists an integer n such that $z/n \in U$. Therefore,

$$\tilde{\alpha}(z) \equiv n\tilde{\alpha}(z/n) = n(\beta z/n) = \beta z \pmod{L},$$

so the endomorphism $\tilde{\alpha}$ is given by multiplication by β. Since $\tilde{\alpha}(L) \subseteq L$, it follows that

$$\beta L \subseteq L.$$

We have proved half of the following.

THEOREM 10.1
Let E be an elliptic curve over $\mathbf{C}$ corresponding to the lattice L. Then

$$\text{End}(E) \simeq \{\beta \in \mathbf{C} \,|\, \beta L \subseteq L\}.$$

PROOF   We have shown that all endomorphisms are given by numbers β. We need to show that all such β's give endomorphisms. Suppose $\beta \in \mathbf{C}$ satisfies $\beta L \subseteq L$. Then multiplication by β gives a homomorphism

$$\beta : \mathbf{C}/L \to \mathbf{C}/L.$$

We need to show that the corresponding map on E is given by rational functions in x, y.

The functions $\wp(\beta z)$ and $\wp'(\beta z)$ are doubly periodic with respect to L, since $\beta L \subseteq L$. By Theorem 9.3, there are rational functions R and S such that

$$\wp(\beta z) = R(\wp(z)), \quad \wp'(\beta z) = \wp'(z) S(\wp(z)).$$

Therefore, multiplication by β on $\mathbf{C}/L$ corresponds to the map

$$(x, y) \mapsto (R(x),\, yS(x))$$

on E. This is precisely the statement that β induces an endomorphism of E.
∎

Theorem 10.1 imposes rather severe restrictions on the endomorphism ring of E. We'll show below that $\mathrm{End}(E)$ is either $\mathbf{Z}$ or an order in an imaginary quadratic field. First, we need to say what this means. We'll omit the proofs of the following facts, which can be found in many books on algebraic number theory. Let $d > 0$ be a squarefree integer and let

$$K = \mathbf{Q}(\sqrt{-d}) = \{a + b\sqrt{-d} \mid a, b \in \mathbf{Q}\}.$$

Then K is called an **imaginary quadratic field**. The largest subring of K that is also a finitely generated abelian group is

$$\mathcal{O}_K = \begin{cases} \mathbf{Z}\left[\frac{1+\sqrt{-d}}{2}\right] & \text{if } d \equiv 3 \pmod 4 \\[2mm] \mathbf{Z}\left[\sqrt{-d}\right] & \text{if } d \equiv 1, 2 \pmod 4, \end{cases}$$

where, in these two cases, $\mathbf{Z}[\delta] = \{a + b\delta \mid a, b \in \mathbf{Z}\}$. An **order** in an imaginary quadratic field is a ring R such that $\mathbf{Z} \subset R \subseteq \mathcal{O}_K$ and $\mathbf{Z} \neq R$. Such an order is a finitely generated abelian group and has the form

$$R = \mathbf{Z} + \mathbf{Z}f\delta,$$

where $f > 0$ and where $\delta = (1 + \sqrt{-d})/2$ or $\sqrt{-d}$, corresponding respectively to the two cases given above. The integer f is called the **conductor** of R and is the index of R in $\mathcal{O}_K$. The **discriminant** of R is

$$D_R = \begin{cases} -f^2 d & \text{if } d \equiv 3 \pmod 4 \\ -4f^2 d & \text{if } d \equiv 1, 2 \pmod 4. \end{cases}$$

It is the discriminant of the quadratic polynomial satisfied by δ.

A complex number β is an **algebraic integer** if it is a root of a monic polynomial with integer coefficients. The only algebraic integers in $\mathbf{Q}$ are the elements of $\mathbf{Z}$. If β is an algebraic integer in a quadratic field, then there are integers b, c such that $\beta^2 + b\beta + c = 0$. The set of algebraic integers in an imaginary quadratic field K is precisely the ring $\mathcal{O}_K$ defined above. An order is therefore a subring (not equal to $\mathbf{Z}$) of the ring of algebraic integers in K. If $\beta \in \mathbf{C}$ is an algebraic number (that is, a root of a polynomial with rational coefficients), then there is an integer $u \neq 0$ such that $u\beta$ is an algebraic integer.

THEOREM 10.2

Let E be an elliptic curve over $\mathbf{C}$. Then $\mathrm{End}(E)$ is isomorphic either to $\mathbf{Z}$ or to an order in an imaginary quadratic field.

PROOF   Let $L = \mathbf{Z}\omega_1 + \mathbf{Z}\omega_2$ be the lattice corresponding to E, and let

$$R = \{\beta \in \mathbf{C} \mid \beta L \subseteq L\}.$$

It is easy to see that $\mathbf{Z} \subseteq R$ and that R is closed under addition, subtraction, and multiplication. Therefore, R is a ring. Suppose $\beta \in R$. There exist integers j, k, m, n such that

$$\beta\omega_1 = j\omega_1 + k\omega_2, \quad \beta\omega_2 = m\omega_1 + n\omega_2.$$

Then

$$\begin{pmatrix} \beta - j & -k \\ -m & \beta - n \end{pmatrix} \begin{pmatrix} \omega_1 \\ \omega_2 \end{pmatrix} = 0,$$

so the determinant of the matrix is 0. This implies that

$$\beta^2 - (j + n)\beta + (jn - km) = 0.$$

Since j, k, m, n are integers, this means that β is an algebraic integer, and that β lies in some quadratic field K.

Suppose $\beta \in \mathbf{R}$. Then $(\beta - j)\omega_1 - k\omega_2 = 0$ gives a dependence relation between ω_1 and ω_2 with real coefficients. Since ω_1 and ω_2 are linearly independent over $\mathbf{R}$, we have $\beta = a \in \mathbf{Z}$. Therefore, $R \cap \mathbf{R} = \mathbf{Z}$.

Suppose now that $R \neq \mathbf{Z}$. Let $\beta \in R$ with $\beta \notin \mathbf{Z}$. Then β is an algebraic integer in a quadratic field K. Since $\beta \notin \mathbf{R}$, the field K must be imaginary quadratic, say $K = \mathbf{Q}(\sqrt{-d})$. Let $\beta' \notin \mathbf{Z}$ be another element of R. Then $\beta' \in K' = \mathbf{Q}(\sqrt{-d'})$ for some d'. Since $\beta + \beta'$ also must lie in a quadratic field, it follows (see Exercise 10.1) that $K = K'$. Therefore, $R \subset K$, and since all elements of R are algebraic integers, we have

$$R \subseteq \mathcal{O}_K.$$

Therefore, if $R \neq \mathbf{Z}$, then R is an order in an imaginary quadratic field.   ∎

Example 10.1
Let E be the elliptic curve $y^2 = 4x^4 - 4x$. We showed in Section 10.1 that $\mathbf{Z}[i] \subseteq \text{End}(E)$. Since $\text{End}(E)$ is an order in $\mathbf{Q}(i)$ and every such order is contained in the ring $\mathbf{Z}[i]$ of algebraic integers in $\mathbf{Q}(i)$, we must have

$$\text{End}(E) = \mathbf{Z}[i].$$

☐

Suppose from now on that E has complex multiplication, which means that $R = \text{End}(E)$ is an order in an imaginary quadratic field K. Rescaling L does not change R, so we may consider

$$\omega_2^{-1} L = \mathbf{Z} + \mathbf{Z}\tau,$$

with $\tau \in \mathcal{H} = \{z \in \mathbf{C} \mid \Im(z) > 0\}$. Let $\beta \in R$ with $\beta \notin \mathbf{Z}$. Since $1 \in \omega_2^{-1}L$, we have $\beta \cdot 1 = m \cdot 1 + n\tau$ with $m, n \in \mathbf{Z}$ and $n \neq 0$. Therefore,

$$\tau = (\beta - m)/n \in K. \tag{10.1}$$

Let u be an integer such that $u\tau \in R$. Such an integer exists since τ multiplied by n is in $\mathcal{O}_K$, and R is of finite index in $\mathcal{O}_K$. Then

$$L' = u\omega_2^{-1}L = \mathbf{Z}u + \mathbf{Z}u\tau \subseteq R.$$

Then L' is a nonempty subset of R that is closed under addition and subtraction, and is closed under multiplication by elements of R (since L' is a rescaling of L). This is exactly what it means for L' to be an ideal of R. We have proved the first half of the following.

PROPOSITION 10.3

Let R be an order in an imaginary quadratic field. Let L be a lattice such that $R = End(\mathbf{C}/L)$. Then there exists $\gamma \in \mathbf{C}^{\times}$ such that γL is an ideal of R. Conversely, if L is a subset of $\mathbf{C}$ and $\gamma \in \mathbf{C}^{\times}$ is such that γL is an ideal of R, then L is a lattice and $R \subseteq End(\mathbf{C}/L)$.

PROOF   By $End(\mathbf{C}/L)$, we mean $End(E)$, where E is the elliptic curve corresponding to L under Theorem 9.10.

We proved the first half of the proposition above. For the converse, assume that γL is an ideal of R. Let $0 \neq x \in \gamma L$. Then

$$Rx \subseteq \gamma L \subseteq R.$$

Since R and therefore also Rx are abelian groups of rank 2 (that is, isomorphic to $\mathbf{Z} \oplus \mathbf{Z}$), the same must be true for γL. This means that there exist $\omega_1', \omega_2' \in L$ such that

$$\gamma L = \gamma \mathbf{Z}\omega_1' + \gamma \mathbf{Z}\omega_2'.$$

Since R contains two elements linearly independent over $\mathbf{R}$, so does Rx, and therefore so does L. It follows that ω_1' and ω_2' are linearly independent over $\mathbf{R}$. Therefore, $L = \mathbf{Z}\omega_1' + \mathbf{Z}\omega_2'$ is a lattice. Since γL is an ideal of R, we have $R\gamma L \subseteq \gamma L$, and therefore $RL \subseteq L$. Therefore $R \subseteq End(\mathbf{C}/L)$.   ∎

Note that sometimes R is not all of $End(\mathbf{C}/L)$. For example, suppose $R = \mathbf{Z}[2i] = \{a + 2bi \mid a, b \in \mathbf{Z}\}$ and let $L = \mathbf{Z}[i]$. Then R is an order in $\mathbf{Q}(i)$ and $RL \subseteq L$, but $End(\mathbf{C}/L) = \mathbf{Z}[i] \neq R$.

We say that two lattices L_1, L_2 are **homothetic** if there exists $\gamma \in \mathbf{C}^{\times}$ such that $\gamma L_1 = L_2$. We say that two ideals I_1, I_2 of R are equivalent if there exists $\lambda \in K^{\times}$ such that $\lambda I_1 = I_2$. Regard I_1 and I_2 as lattices, and suppose I_1 and I_2 are homothetic. Then $\gamma I_1 = I_2$ for some γ. Choose any $x \neq 0$ in I_1.

Then $\gamma x \in I_2 \subset K$, so $\gamma \in K$. It follows that I_1 and I_2 are equivalent ideals. Therefore, we have a bijection

$$\begin{array}{ccc} \text{Homothety classes of lattices } L & & \text{Equivalence classes of} \\ \text{with } RL \subseteq L & \longleftrightarrow & \text{nonzero ideals of } R \end{array}$$

It can be shown that the set of equivalence classes of ideals is finite (when $R = \mathcal{O}_K$, this is just the finiteness of the class number). Therefore, the set of homothety classes is finite. This observation has the following consequence.

PROPOSITION 10.4

Let R be an order in an imaginary quadratic field and let L be a lattice such that $RL \subseteq L$. Then $j(L)$ is algebraic over $\mathbf{Q}$.

PROOF   Let E be the elliptic curve corresponding to L. We may assume that E is given by an equation $y^2 = 4x^3 - g_2 x - g_3$. Let σ be an automorphism of $\mathbf{C}$. Let E^σ be the curve $y^2 = 4x^3 - \sigma(g_2)x - \sigma(g_3)$. If α is an endomorphism of E, then α^σ is an endomorphism of E^σ, where α^σ means applying σ to all of the coefficients of the rational functions describing α. This implies that

$$\text{End}(E) \simeq \text{End}(E^\sigma).$$

Therefore, the lattice corresponding to E^σ belongs to one of the finitely many homothety classes of lattices containing R in their endomorphism rings (there is a technicality here; see Exercise 10.2). Since $\sigma(j(L))$ is the j-invariant of E^σ, we conclude that $j(L)$ has only finitely many possible images under automorphisms of $\mathbf{C}$. This implies (see Appendix C) that $j(L)$ is algebraic over $\mathbf{Q}$.   ∎

In Section 10.3, we'll prove the stronger result that $j(L)$ is an algebraic integer.

COROLLARY 10.5

Let K be an imaginary quadratic field.

1. *Let $\tau \in \mathcal{H}$. Then $\mathbf{C}/(\mathbf{Z}\tau + \mathbf{Z})$ has complex multiplication by some order in K if and only if $\tau \in K$.*

2. *If $\tau \in \mathcal{H}$ is contained in K, then $j(\tau)$ is algebraic.*

PROOF   We have already shown (see (10.1)) that if there is complex multiplication by an order in K then $\tau \in K$. Conversely, suppose $\tau \in K$. Then τ satisfies a relation

$$a\tau^2 + b\tau + c,$$

where a, b, c are integers and $a \neq 0$. It follows that multiplication by $a\tau$ maps the lattice $L_\tau = \mathbf{Z}\tau + \mathbf{Z}$ into itself (for example, $a\tau \cdot \tau = -b\tau - c \in L_\tau$). Therefore, $\mathbf{C}/L_\tau$ has complex multiplication. This proves (1).

Suppose $\tau \in K$. Let R be the endomorphism ring of $\mathbf{C}/L_\tau$. By (1), $R \neq \mathbf{Z}$, so R is an order in K. By Proposition 10.4, $j(\tau)$ is algebraic. This proves (2).
∎

10.2 Elliptic Curves over Finite Fields

An elliptic curve E over a finite field $\mathbf{F}_q$ always has complex multiplication. In most cases, this is easy to see. The Frobenius endomorphism ϕ_q is a root of

$$X^2 - aX + q = 0,$$

where $|a| \leq 2\sqrt{q}$. If $|a| < 2\sqrt{q}$, then this polynomial has only complex roots, so $\phi_q \notin \mathbf{Z}$. Therefore,

$$\mathbf{Z} \neq \mathbf{Z}[\phi_q] \subseteq \text{End}(E).$$

When $a = \pm 2\sqrt{q}$, the ring of endomorphisms is still larger than $\mathbf{Z}$, so there is complex multiplication in this case, too. In fact, as we'll discuss below, the endomorphism ring is an order in a quaternion algebra, hence is larger than an order in a quadratic field.

Recall the **Hamiltonian quaternions**

$$\mathbf{H} = \{a + b\mathbf{i} + c\mathbf{j} + d\mathbf{k} \mid a, b, c, d \in \mathbf{Q}\},$$

where $\mathbf{i}^2 = \mathbf{j}^2 = \mathbf{k}^2 = -1$ and $\mathbf{ij} = \mathbf{k} = -\mathbf{ji}$. This is a noncommutative ring in which every nonzero element has a multiplicative inverse. If we allow the coefficients a, b, c, d to be real numbers or 2-adic numbers, then we still obtain a ring where every nonzero element has an inverse. However, if a, b, c, d are allowed to be p-adic numbers (see Appendix A), where p is an odd prime, then the ring contains nonzero elements whose product is 0 (see Exercise 10.4). Such elements cannot have inverses. Corresponding to whether there are zero divisors or not, we say that $\mathbf{H}$ is **split** at all odd primes and is **ramified** at 2 and ∞ (this use of ∞ is the common way to speak about the real numbers when simultaneously discussing p-adic numbers; see Section 8.8).

In general, a **definite quaternion algebra** is a ring of the form

$$\mathcal{Q} = \{a + b\alpha + c\beta + d\alpha\beta \mid a, b, c, d \in \mathbf{Q}\},$$

where

$$\alpha^2, \beta^2 \in \mathbf{Q}, \quad \alpha^2 < 0, \quad \beta^2 < 0, \quad \beta\alpha = -\alpha\beta$$

("definite" refers to the requirement that $\alpha^2 < 0$ and $\beta^2 < 0$). In such a ring, every nonzero element has a multiplicative inverse (see Exercise 10.5). If this is still the case when we allow p-adic coefficients for some $p \le \infty$, then we say that the quaternion algebra is **ramified** at p. Otherwise, it is **split** at p.

A **maximal order** $\mathcal{O}$ in a quaternion algebra $\mathcal{Q}$ is a subring of $\mathcal{Q}$ that is finitely generated as an additive abelian group, and such that if $\mathcal{R}$ is a ring with $\mathcal{O} \subseteq \mathcal{R} \subseteq \mathcal{Q}$ and such that R is finitely generated as an additive abelian group, then $\mathcal{O} = \mathcal{R}$. For example, consider the Hamiltonian quaternions $\mathbf{H}$. The subring $\mathbf{Z} + \mathbf{Z}\mathbf{i} + \mathbf{Z}\mathbf{j} + \mathbf{Z}\mathbf{k}$ is finitely generated as an additive abelian group, but it is not a maximal order since it is contained in

$$\mathcal{O} = \mathbf{Z} + \mathbf{Z}\mathbf{i} + \mathbf{Z}\mathbf{j} + \mathbf{Z}\frac{1 + \mathbf{i} + \mathbf{j} + \mathbf{k}}{2}. \tag{10.2}$$

It is not hard to show that $\mathcal{O}$ is a ring, and it can be shown that it is a maximal order of $\mathbf{H}$.

The main theorem on endomorphism rings is the following. For a proof, see [25].

THEOREM 10.6

Let E be an elliptic curve over a finite field of characteristic p.

1. *If E is ordinary (that is, $\#E[p] = p$), then $\mathrm{End}(E)$ is an order in an imaginary quadratic field.*

2. *If E is supersingular (that is, $\#E[p] = 1$), then $\mathrm{End}(E)$ is a maximal order in a definite quaternion algebra that is ramified at p and ∞ and is split at the other primes.*

If E is an elliptic curve defined over $\mathbf{Q}$ and p is a prime where E has good reduction, then it can be shown that $\mathrm{End}(E)$ injects into $\mathrm{End}(E \bmod p)$. Therefore, if E has complex multiplication by an order R in an imaginary quadratic field, then the endomorphism ring of E mod p contains R. If E mod p is ordinary, then R is of finite index in the endomorphism ring of E mod p. However, if E mod p is supersingular, then there are many more endomorphisms, since the endomorphism ring is noncommutative in this case. The following result shows how to decide when E mod p is ordinary and when it is supersingular.

THEOREM 10.7

Let E be an elliptic curve defined over $\mathbf{Q}$ with good reduction at p. Suppose E has complex multiplication by an order in $\mathbf{Q}(\sqrt{-D})$. If $-D$ is divisible by p, or if $-D$ is not a square mod p, then E mod p is supersingular. If $-D$ is a nonzero square mod p, then E mod p is ordinary.

For a proof, see [55, p. 182].

Example 10.2

Let E be the elliptic curve $y^2 = x^3 - x$. It has good reduction for all primes $p \neq 2$. The endomorphism ring R of E is $\mathbf{Z}[i]$, where

$$i(x, y) = (-x, iy)$$

(see Section 10.1). This endomorphism ring is contained in $\mathbf{Q}(\sqrt{-4})$, where we use $-D = -4$ since it is the discriminant of R. We know that -4 is a square mod an odd prime p if and only if $p \equiv 1 \pmod 4$. Therefore, E mod p is ordinary if and only if $p \equiv 1 \pmod 4$. This is exactly what we obtained in Proposition 4.35.

When $p \equiv 3 \pmod 4$, it is easy to see that the endomorphism ring of E mod p is noncommutative. Since $i^p = -i$, we have

$$\phi_p(i(x, y)) = \phi_p(-x, iy) = (-x^p, -iy^p),$$

and

$$i(\phi_p(x, y)) = i(x^p, y^p) = (-x^p, iy^p).$$

Therefore,

$$i\phi_p = -\phi_p i,$$

so i and ϕ_p do not commute.   □

The following result, known as **Deuring's Lifting Theorem**, shows that the method given in Theorem 10.7 for obtaining ordinary elliptic curves mod p with complex multiplication is essentially the only way. Namely, it implies that an elliptic curve with complex multiplication over a finite field can be obtained by reducing an elliptic curve with complex multiplication in characteristic zero.

THEOREM 10.8

Let E be an elliptic curve defined over a finite field and let α be an endomorphism of E. Then there exists an elliptic curve $\tilde{E}$ defined over a finite extension K of $\mathbf{Q}$ and an endomorphism $\tilde{\alpha}$ of $\tilde{E}$ such that E is the reduction of $\tilde{E}$ mod some prime ideal of the ring of algebraic integers of K and the reduction of $\tilde{\alpha}$ is α.

For a proof in the ordinary case, see [55, p. 184].

It is not possible to extend the theorem to lifting two arbitrary endomorphisms simultaneously. For example, the endomorphisms i and ϕ_p in the above example cannot be simultaneously lifted to characteristic 0 since they do not commute. All endomorphism rings in characteristic 0 are commutative.

Finally, we give an example of a supersingular curve in characteristic 2. In particular, we'll show how to identify the maximal order of **H** in the endomorphism ring.

Example 10.3
Let E be the elliptic curve defined over $\mathbf{F}_2$ by

$$y^2 + y = x^3.$$

An easy calculation shows that $E(\mathbf{F}_2)$ consists of 3 points, so

$$a = 2 + 1 - \#E(\mathbf{F}_2) = 2 + 1 - 3 = 0.$$

Therefore, E is supersingular and the Frobenius endomorphism ϕ_2 satisfies

$$\phi_2^2 + 2 = 0.$$

If $(x, y) \in E(\overline{\mathbf{F}}_2)$, then

$$2(x, y) = -\phi_2^2(x, y) = -(x^4, y^4) = (x^4, y^4 + 1),$$

since negation on E is given by

$$-(x, y) = (x, y + 1).$$

By Theorem 10.6, the endomorphism ring is a maximal order in a quaternion algebra ramified at only 2 and ∞. We gave such a maximal order in (10.2) above. Let's start by finding endomorphisms corresponding to $\mathbf{i}, \mathbf{j}, \mathbf{k}$. Let $\omega \in \mathbf{F}_4$ satisfy

$$\omega^2 + \omega + 1 = 0.$$

Define endomorphisms $\mathbf{i}, \mathbf{j}, \mathbf{k}$ by

$$\begin{aligned}
\mathbf{i}(x, y) &= (x + 1, \, y + x + \omega) \\
\mathbf{j}(x, y) &= (x + \omega, \, y + \omega^2 x + \omega) \\
\mathbf{k}(x, y) &= (x + \omega^2, \, y + \omega x + \omega).
\end{aligned}$$

An easy calculation shows that

$$\mathbf{i}(\mathbf{j}(x, y)) = \mathbf{k}(x, y), \quad \mathbf{j}(\mathbf{i}(x, y)) = -\mathbf{k}(x, y)$$

and that

$$\mathbf{i}^2 = \mathbf{k}^2 = \mathbf{k}^2 = -1.$$

A straightforward calculation yields

$$(1 + \mathbf{i} + \mathbf{j} + \mathbf{k})(x, y) = (\omega x^4, y^4) = \phi_2^2(\omega x, y) = -2(\omega(x, y)),$$

where ω is used to denote the endomorphism $(x, y) \mapsto (\omega x, y)$. Therefore,

$$\frac{1 + \mathbf{i} + \mathbf{j} + \mathbf{k}}{2} = -\omega \in \text{End}(E).$$

It follows that

$$\mathbf{Z} + \mathbf{Z}\mathbf{i} + \mathbf{Z}\mathbf{j} + \mathbf{Z}\frac{1 + \mathbf{i} + \mathbf{j} + \mathbf{k}}{2} \subseteq \text{End}(E).$$

In fact, by Theorem 10.6, this is the whole endomorphism ring. ⬜

10.3 Integrality of j-invariants

At the end of Section 10.1, we showed that the j-invariant of a lattice, or of a complex elliptic curve, with complex multiplication by an order in an imaginary quadratic field is algebraic over $\mathbf{Q}$. This means that the j-invariant is a root of a polynomial with rational coefficients. In the present section, we show that this j-invariant is an algebraic integer, so it is a root of a monic polynomial with integer coefficients.

THEOREM 10.9

Let R be an order in an imaginary quadratic field and let L be a lattice with $RL \subseteq L$. Then $j(L)$ is an algebraic integer. Equivalently, let E be an elliptic curve over $\mathbf{C}$ with complex multiplication. Then $j(E)$ is an algebraic integer.

The proof of the theorem will occupy the remainder of this section. The theorem has an amusing consequence. The ring $R = \mathbf{Z}\left[\frac{1+\sqrt{-163}}{2}\right]$ is a principal ideal domain (see [11]), so there is only one equivalence class of ideals of R, namely the one represented by R. The proof of Proposition 10.4 shows that all automorphisms of $\mathbf{C}$ must fix $j(R)$, where R is regarded as a lattice. Therefore, $j(R) \in \mathbf{Q}$. The only algebraic integers in $\mathbf{Q}$ are the elements of $\mathbf{Z}$, so $j(R) \in \mathbf{Z}$. Recall that $j(\tau)$ is the j-invariant of the lattice $\mathbf{Z}\tau + \mathbf{Z}$, and that

$$j(\tau) = \frac{1}{q} + 744 + 196884q + 21493760q^2 + \cdots,$$

where $q = e^{2\pi i \tau}$. When $\tau = \frac{1+\sqrt{-163}}{2}$, we have $R = \mathbf{Z}\tau + \mathbf{Z}$ and

$$q = -e^{-\pi\sqrt{163}}.$$

Therefore,

$$-e^{\pi\sqrt{163}} + 744 - 196884e^{-\pi\sqrt{163}} + 21493760e^{-2\pi\sqrt{163}} + \cdots \in \mathbf{Z}.$$

Since

$$196884e^{-\pi\sqrt{163}} - 21493760e^{-2\pi\sqrt{163}} + \cdots < 10^{-12},$$

we find that $e^{\pi\sqrt{163}}$ differs from an integer by less than 10^{-12}. In fact,

$$e^{\pi\sqrt{163}} = 262537412640768743.999999999999250\ldots,$$

as predicted. In the days when high precision calculation was not widely available, it was often claimed as a joke that $e^{\pi\sqrt{163}}$ was an integer. Any calculation with up to 30 places of accuracy seemed to indicate that this was the case. This was in contradiction to the Gelfond-Schneider theorem, which implies that such a number must be transcendental.

We now start the proof of the theorem. If $L = \mathbf{Z}\omega_1 + \mathbf{Z}\omega_2$ is a lattice, we may divide by ω_2 and thus assume that

$$L = \mathbf{Z}\tau + \mathbf{Z},$$

with $\tau \in \mathcal{H}$. If $\beta \in R$, then $\beta L \subseteq L$ implies that there exist integers j, k, m, n with

$$\beta \begin{pmatrix} \tau \\ 1 \end{pmatrix} = \begin{pmatrix} j & k \\ m & n \end{pmatrix} \begin{pmatrix} \tau \\ 1 \end{pmatrix}.$$

Let $N = jn - km$ be the determinant of the matrix. Rather than concentrating only on β, it is convenient to consider all 2×2 matrices with determinant N simultaneously.

LEMMA 10.10

Let N be a positive integer and let S_N be the set of matrices of the form

$$\begin{pmatrix} a & b \\ 0 & d \end{pmatrix}$$

with $a, b, d \in \mathbf{Z}$, $ad = N$, and $0 \le b < d$. If M is a 2×2 matrix with integer entries and determinant N, then there is a unique matrix $S \in S_N$ such that

$$MS^{-1} \in SL_2(\mathbf{Z}).$$

In other words, if we say that two matrices M_1, M_2 are left $SL_2(\mathbf{Z})$-equivalent when there exists a matrix $X \in SL_2(\mathbf{Z})$ with $XM_1 = M_2$, then S_N contains exactly one element in each equivalence class of the set of integer matrices of determinant N.

PROOF   Let $\begin{pmatrix} p & q \\ r & s \end{pmatrix}$ be an integer matrix with determinant N. Write

$$-\frac{p}{r} = \frac{x}{y}$$

with $\gcd(x, y) = 1$. There exist $w, z \in \mathbf{Z}$ such that $xz - wy = 1$. Then

$$\begin{pmatrix} z & w \\ y & x \end{pmatrix} \in SL_2(\mathbf{Z})$$

and

$$\begin{pmatrix} z & w \\ y & x \end{pmatrix} \begin{pmatrix} p & q \\ r & s \end{pmatrix} = \begin{pmatrix} * & * \\ 0 & * \end{pmatrix}.$$

Therefore, we may assume at the start that $r = 0$, and hence $ps = N$. Choose $t \in \mathbf{Z}$ such that

$$0 \le q + ts < s.$$

Then

$$\begin{pmatrix} 1 & t \\ 0 & 1 \end{pmatrix} \begin{pmatrix} p & q \\ 0 & s \end{pmatrix} = \begin{pmatrix} p & q+ts \\ 0 & s \end{pmatrix} \in S_N.$$

Therefore, the elements of S_N represent all $SL_2(\mathbf{Z})$-equivalence classes for matrices of determinant N.

For the uniqueness, suppose that $M_i = \begin{pmatrix} a_i & b_i \\ 0 & d_i \end{pmatrix}$ for $i = 1, 2$ are left $SL_2(\mathbf{Z})$-equivalent. Then,

$$\begin{pmatrix} a_1/a_2 & (b_1 a_2 - a_1 b_2)/N \\ 0 & d_1/d_2 \end{pmatrix} = \begin{pmatrix} a_1 & b_1 \\ 0 & d_1 \end{pmatrix} \begin{pmatrix} a_2 & b_2 \\ 0 & d_2 \end{pmatrix}^{-1} \in SL_2(\mathbf{Z}).$$

Therefore, a_1/a_2 and d_1/d_2 are positive integers with product equal to 1, so they are both equal to 1. Consequently, $a_1 = a_2$ and $d_1 = d_2$. This implies that

$$\frac{b_1 a_2 - a_1 b_2}{N} = \frac{b_1 a_1 - a_1 b_2}{a_1 d_1} = \frac{b_1 - b_2}{d_1}.$$

Since this must be an integer (because the matrix is in $SL_2(\mathbf{Z})$), we have

$$b_1 \equiv b_2 \pmod{d_1}.$$

Since $0 \le b_1, b_2 < d_1 = d_2$, we have $b_1 = b_2$. Therefore, $M_1 = M_2$. This proves the uniqueness. ∎

For $S = \begin{pmatrix} a & b \\ 0 & d \end{pmatrix} \in S_N$, the function

$$(j \circ S)(\tau) = j\left(\frac{a\tau + b}{d}\right)$$

is analytic in $\mathcal{H}$. Define

$$F_N(X, \tau) = \prod_{S \in S_N} (X - j \circ S) = \sum_k a_k(\tau) X^k,$$

so F_N is a polynomial in the variable X with coefficients $a_k(\tau)$ that are analytic functions for $\tau \in \mathcal{H}$.

LEMMA 10.11
$a_k(M\tau) = a_k(\tau)$ for all $M \in SL_2(\mathbf{Z})$.

PROOF   If $S \in S_N$, then SM has determinant N, so there exists $A_S \in SL_2(\mathbf{Z})$ and a uniquely determined $M_S \in S_N$ such that $A_S M_S = SM$. If $S_1, S_2 \in S_N$ and $M_{S_1} = M_{S_2}$, then

$$A_{S_1}^{-1} S_1 M = M_{S_1} = M_{S_2} = A_{S_2}^{-1} S_2 M,$$

which implies that $A_{S_2} A_{S_1}^{-1} S_1 = S_2$. By the uniqueness part of Lemma 10.10, $S_1 = S_2$. Therefore, the map $S \mapsto M_S$ is an injection on the finite set S_N, hence is a permutation of the set. Since $j \circ A = j$ for $A \in SL_2(\mathbf{Z})$, we have

$$
\begin{aligned}
F_N(X, M\tau) &= \prod_{S \in S_N} (X - j(SM\tau)) \\
&= \prod_{S \in S_N} (X - j(A_S M_S \tau)) \\
&= \prod_{S \in S_N} (X - j(M_S \tau)) \\
&= \prod_{S \in S_N} (X - j(S\tau)) \\
&= F_N(X, \tau).
\end{aligned}
$$

The next to last equality expresses the fact that $S \mapsto M_S$ is a permutation of S_N, hence does not change the product over all of S_N.

Since F_N is invariant under $\tau \mapsto M\tau$, the same must hold for its coefficients $a_k(\tau)$. ∎

LEMMA 10.12
For each k, there exists an integer n such that

$$a_k(\tau) \in q^{-n} \mathbf{Z}[[q]],$$

where $\mathbf{Z}[[q]]$ denotes power series in q with integer coefficients. In other words, $a_k(\tau)$ can be expressed as a Laurent series with only finitely many negative terms, and the coefficients are integers.

PROOF   The j-function has the expansion

$$j(\tau) = \frac{1}{q} + 744 + 196884q + \cdots = \sum_{k=-1}^{\infty} c_k q^k = P(q),$$

where the coefficients c_k are integers (see Exercise 9.1). Therefore,

$$j((a\tau + b)/d) = \sum_{k=-1}^{\infty} c_k (\zeta^b e^{2\pi i a\tau/d})^k = P(\zeta^b e^{2\pi i a\tau/d}),$$

where $\zeta = e^{2\pi i/d}$. Fix a and d with $ad = N$.

CLAIM 10.13
$$\prod_{b=0}^{d-1} (X - P(\zeta^b e^{2\pi i a\tau/d})) = \sum_{\ell=0}^{d} p_\ell(e^{2\pi i a\tau/d}) X^\ell$$

is a polynomial in X whose coefficients p_k are Laurent series in $e^{2\pi i a \tau/d}$ with integer coefficients.

In the statement of the claim and in the following, a Laurent series will always be one with only finitely many negative terms (in other words, a power series plus finitely many terms with negative exponents). Everything in the claim is obvious except the fact that the coefficients of the Laurent series p_k are integers. One proof of this is as follows. The coefficients of each p_k lie in $\mathbf{Z}[\zeta]$. The Galois group of $\mathbf{Q}(\zeta)/\mathbf{Q}$ permutes the factors of the product, hence leaves the coefficients of p_k unchanged. Therefore, they are in $\mathbf{Q}$. But the elements of $\mathbf{Z}[\zeta] \cap \mathbf{Q}$ are algebraic integers in $\mathbf{Q}$, hence are in $\mathbf{Z}$. This proves the claim.

For a proof of the claim that does not use Galois theory, consider the matrix

$$
Z = \begin{pmatrix} 0 & 1 & 0 & \cdots & 0 \\ 0 & 0 & 1 & \cdots & 0 \\ \vdots & \vdots & \vdots & \ddots & \vdots \\ 1 & 0 & 0 & \cdots & 0 \end{pmatrix}.
$$

Let $0 \le b < d$ and let

$$
v_b = \begin{pmatrix} 1 \\ \zeta^b \\ \zeta^{2b} \\ \vdots \\ \zeta^{b(d-1)} \end{pmatrix}.
$$

Then $Z v_b = \zeta^b v_b$. It follows that

$$
P(e^{2\pi i a \tau/d} Z) v_b = P(\zeta^b e^{2\pi i a \tau/d}) v_b.
$$

Therefore, the numbers $P(\zeta^b e^{2\pi i a \tau/d})$, for $0 \le b < d$, are a complete set of eigenvalues for the $d \times d$ matrix $P(e^{2\pi i a \tau/d} Z)$, so the characteristic polynomial is

$$
\prod_{b=0}^{d-1} (X - P(\zeta^b e^{2\pi i a \tau/d})).
$$

But the entries of the matrix $P(e^{2\pi i a \tau/d} Z)$ are Laurent series in $e^{2\pi i a \tau/d}$ with integer coefficients. Therefore, the coefficients of the characteristic polynomial are power series in $e^{2\pi i a \tau/d}$ with integer coefficients. This proves the claim.

Since $ad = N$ for each matrix in S_N,

$$
e^{2\pi i a \tau/d} = e^{2\pi i a^2 \tau/N}.
$$

Therefore, the $p_k(\tau)$ in the claim can be regarded as a Laurent series in $e^{2\pi i \tau/N}$. The claim implies that the coefficients $a_k(\tau)$ of $F_N(X, \tau)$ are Laurent series in $e^{2\pi i \tau / N}$ with integer coefficients. To prove the lemma, we need to remove the N. The matrix

$$\begin{pmatrix} 1 & 1 \\ 0 & 1 \end{pmatrix} \in SL_2(\mathbf{Z})$$

acts on $\mathcal{H}$ by $\tau \mapsto \tau + 1$. Lemma 10.11 implies that $a_k(\tau)$ is invariant under $\tau \mapsto \tau + 1$. Since $(e^{2\pi i \tau / N})^\ell$ is invariant under $\tau \mapsto \tau + 1$ only when $N | \ell$, the Laurent series for a_k must be a Laurent series in $(e^{2\pi i \tau / N})^N = e^{2\pi i \tau}$. This proves Lemma 10.12. ∎

PROPOSITION 10.14
Let $f(\tau)$ be analytic for $\tau \in \mathcal{H}$, and suppose

$$f\left(\frac{a\tau + b}{c\tau + d}\right) = f(\tau)$$

for all $\begin{pmatrix} a & b \\ c & d \end{pmatrix} \in SL_2(\mathbf{Z})$ and all $\tau \in \mathcal{H}$. Also, assume

$$f(\tau) \in q^{-n}\mathbf{Z}[[q]]$$

for some integer n. Then $f(\tau)$ is a polynomial in j with integer coefficients:

$$f(\tau) \in \mathbf{Z}[j].$$

PROOF   Recall that
$$j(\tau) - \frac{1}{q} \in \mathbf{Z}[[q]].$$

Write
$$f(\tau) = \frac{b_n}{q^n} + \cdots,$$

with $b_n \in \mathbf{Z}$. Then
$$f(\tau) - b_n j^n = \frac{b_{n-1}}{q^{n-1}} + \cdots,$$

with $b_{n-1} \in \mathbf{Z}$. Therefore,
$$f(\tau) - b_n j^n - b_{n-1} j^{n-1} = \frac{b_{n-2}}{q^{n-2}} + \cdots.$$

Continuing in this way, we obtain
$$g(\tau) = f(\tau) - b_n j^n - \cdots b_0 \in q\mathbf{Z}[[q]]$$

for integers $b_n, \ldots, b_0$. The function $g(\tau)$ is analytic in $\mathcal{H}$ and vanishes at $i\infty$. Also, $g(\tau)$ is invariant under the action of $SL_2(\mathbf{Z})$. Proposition 9.16 says that if g is not identically zero then a sum of the orders of g at various points is 0. But these orders are all non-negative since g is analytic. Moreover, the order of g at $i\infty$ is positive. Therefore the sum of the orders must be positive, hence cannot be zero. The only possibility is that g is identically zero. This means that

$$g(\tau) = f(\tau) - b_n j^n - \cdots b_0 = 0,$$

so $f(\tau) \in \mathbf{Z}[j]$.   ∎

Combining Lemma 10.12 and Proposition 10.14, we obtain the first part of the following.

THEOREM 10.15

Let N be a positive integer.

1. *There is a polynomial with integer coefficients*

$$\Phi_N(X, Y) \in \mathbf{Z}[X, Y]$$

such that the coefficient of the highest power of X is 1 and such that

$$F_N(X, \tau) = \Phi_N(X, j(\tau)).$$

2. *If N is not a perfect square, then*

$$H_N(X) = \Phi_N(X, X) \in \mathbf{Z}[X]$$

is nonconstant and the coefficient of its highest power of X is ± 1.

PROOF   We have already proved the first part. For the second part, we know that

$$H_N(j) = \Phi_N(j, j) = F_N(j, \tau) = \prod_{S \in S_N} (j - j \circ S)$$

is a polynomial in j with integer coefficients. We need to look at the coefficient of the highest power of j. Let $S = \begin{pmatrix} a & b \\ 0 & d \end{pmatrix} \in S_N$. If we expand the factor $j - j \circ S$ as a Laurent series in $e^{2\pi i \tau / N}$, the first term for j is

$$e^{-2\pi i \tau} = (e^{-2\pi i \tau / N})^N$$

and the first term for $j \circ S$ is

$$\zeta^{-b} e^{-2\pi i a \tau / d} = \zeta^{-b} (e^{-2\pi i \tau / N})^{a^2}.$$

Since N is not a perfect square, $N \neq a^2$. Therefore, these terms represent different powers of $e^{2\pi i \tau / N}$, so they cannot cancel each other. One of them must be the first term of the expansion of $j - j \circ S$, which therefore has coefficient 1 or $-\zeta^b$. In particular, for each factor $j - j \circ S$, the coefficient of the first term of the expansion is a root of unity. The coefficient of the first term of the expansion of $H_N(j)$ is the product of these roots of unity, hence a root of unity. Also, since the terms don't cancel each other, the first term of each factor contains a negative power of $e^{2\pi i \tau / N}$. Therefore, the first term of the expansion $H_N(j)$ is a negative power of q, so $H_N(X)$ is nonconstant.

Suppose $H_N(X) = uX^\ell +$ lower terms. We know that $u \in \mathbf{Z}$. Since the Laurent series for j starts with $1/q$,

$$H_N(j) = uq^{-\ell} + \text{higher terms.}$$

We have shown that u is a root of unity. Since it is an integer, $u = \pm 1$. This completes the proof of (2). ∎

The **modular polynomial** $\Phi_N(X, Y)$ has rather large coefficients. For example,

$$\Phi_2(X, Y) = -X^2 Y^2 + X^3 + Y^3 + 2^4 \cdot 3 \cdot 31\, XY(X + Y)$$
$$+3^4 \cdot 5^3 \cdot 4027\, XY - 2^4 \cdot 3^4 \cdot 5^3 (X^2 + Y^2)$$
$$+2^8 \cdot 3^7 \cdot 5^6 (X + Y) - 2^{12} \cdot 3^9 \cdot 5^9.$$

For Φ_N for higher N, see [37], [40], [41].

We can now prove Theorem 10.9. Let R be an order in an imaginary quadratic field and let L be a lattice with $RL \subseteq L$. By multiplying L by a suitable factor, we may assume that

$$L = \mathbf{Z} + \mathbf{Z}\tau$$

with $\tau \in \mathcal{H}$. The order R is of finite index in $\mathcal{O}_K$ for some imaginary quadratic field $K = \mathbf{Q}(\sqrt{-d})$. Since $\sqrt{-d} \in \mathcal{O}_K$, there is a nonzero integer n such that $n\sqrt{-d} \in R$. Therefore, $n\sqrt{-d}L \subseteq L$, so

$$n\sqrt{-d} \cdot \tau = t\tau + u, \quad n\sqrt{-d} \cdot 1 = v\tau + w \tag{10.3}$$

for some integers t, u, v, w. Dividing the two equations yields

$$\tau = \frac{t\tau + u}{v\tau + w}.$$

As in the proof of Theorem 10.2, the two equations in (10.3) yield

$$(n\sqrt{-d})^2 - (t + w)(n\sqrt{-d}) + (tw - uv) = 0.$$

Therefore, $n\sqrt{-d}$ is a root of $X^2 - (t + w)X + (tw - uv)$ and is also a root of $X^2 + n^2 d$. If these are not the same polynomial, we can subtract them and find that $n\sqrt{-d}$ is a root of a polynomial of degree at most 1 with integer coefficients, which is impossible. Therefore the two polynomials are the same, so

$$\det \begin{pmatrix} t & u \\ v & w \end{pmatrix} = tw - uv = n^2 d.$$

By Lemma 10.10, there exist $M \in SL_2(\mathbf{Z})$ and $S_1 \in S_{n^2 d}$ such that

$$\begin{pmatrix} t & u \\ v & w \end{pmatrix} = M S_1.$$

Then

$$j(\tau) = j\left(\frac{t\tau + u}{v\tau + w}\right) = j(M S_1 \tau) = j(S_1 \tau),$$

since $j \circ M = j$. Therefore,

$$H_{n^2 d}(j(\tau)) = \prod_{S \in S_{n^2 d}} (j(\tau) - j(S\tau)) = 0,$$

since $j(\tau) - j(S_1 \tau) = 0$ is one of the factors.

Assume now that $d \neq 1$. Since $n^2 d$ is not a square, Theorem 10.15 implies that the highest coefficient of $H_{n^2 d}(X)$ is ± 1. Changing the sign of H_N if necessary, we find that $j(\tau)$ is a root of a monic polynomial with integer coefficients. This means that $j(L) = j(\tau)$ is an algebraic integer.

If $d = 1$, then $K = \mathbf{Q}(i)$. Replace $\sqrt{-d}$ in the above argument with $1 + i$. The argument works with a minor modification; namely, $n(1 + i)$ is a root of $X^2 - 2nX + 2n^2$. This yields $tw - uv = 2n^2$, which is not a square. Therefore, we can apply Theorem 10.15 to conclude that $j(\tau)$ is an algebraic integer. This completes the proof of Theorem 10.9. ∎

10.4 Numerical Examples

Suppose we want to evaluate

$$x = j\left(\frac{1 + \sqrt{-171}}{2}\right).$$

This is the j-invariant of an elliptic curve that has complex multiplication by $\mathbf{Z}\left[\frac{1 + \sqrt{-171}}{2}\right]$. The others are $j(\tau_2), j(\tau_3), j(\tau_4)$, which are given below, along with $j\left(\frac{1 + \sqrt{-19}}{2}\right)$, which corresponds to an elliptic curve with a larger endomorphism ring. We can evaluate x numerically using Proposition 9.12. This yields

$$j\left(\frac{1 + \sqrt{-171}}{2}\right) =$$

$$-694282057876536664.0122886867083074260443674536\underline{4}124466\ldots.$$

This number is an algebraic integer by Theorem 10.9. Suppose we want a polynomial that has x as its root. One way to do this is to find the Galois conjugates of x, namely, the other roots of a polynomial satisfied by x. We'll show how to proceed for this particular x, then describe the general method.

Let $\tau_0 = (1 + \sqrt{-171})/2$. Then

$$K = \mathbf{Q}(\tau_0) = \mathbf{Q}(\sqrt{-171}) = \mathbf{Q}(\sqrt{-19}).$$

Let

$$R = \mathbf{Z}\left[\frac{1 + \sqrt{-171}}{2}\right] \subset \mathbf{Z}\left[\frac{1 + \sqrt{-19}}{2}\right] = \mathcal{O}_K.$$

The endomorphism ring of the lattice $R \subset \mathbf{C}$ is R. As we showed in the proof of Proposition 10.4, the Galois conjugates of $j(R)$ are j-invariants of lattices with the same endomorphism ring, namely R. These have the form $j(I)$, where I is an ideal of R. However, I cannot be an ideal for any order larger than R since then I has an endomorphism ring larger than R.

If I is an ideal of R, it has the form

$$I = \gamma(\mathbf{Z}\tau + \mathbf{Z})$$

for some $\gamma \in \mathbf{C}^\times$ and some $\tau \in \mathcal{H}$. By an appropriate change of basis, we can assume $\tau \in \mathcal{F}$, the fundamental domain for $SL_2(\mathbf{Z})$ acting on the upper half plane. See Proposition 9.15. As we saw in Equation 10.1, $\tau \in K$. Let

$$a\tau^2 + b\tau + c = 0,$$

with $a, b, c \in \mathbf{Z}$. We may assume that $\gcd(a, b, c) = 1$ and that $a > 0$. The fact that I is an ideal for R but not for any larger order can be shown to imply that the discriminant is exactly -171:

$$b^2 - 4ac = -171.$$

(On the other hand, the polynomial $X^2 + X + 5$ has a root $\tau = (1 + \sqrt{-19})/2$, which corresponds to the ideal $3\mathcal{O}_K \subset R$. This is an ideal not only of R, but also of $\mathcal{O}_K$.) The fact that $\tau \in \mathcal{F}$ means that

1. $-a < b \le a$

2. $a \le c$,

3. if $a = c$ then $b \ge 0$.

The first of these expresses the condition that $-1/2 \le \Re(\tau) < 1/2$, while the second says that $|\tau| \ge 1$. The case where $a = c$ corresponds to τ lying on the unit circle, and $b > 0$ says that it lies on the left half. It can be shown (see [11]) that there is a one-to-one correspondence between the ideals I that we are considering (endomorphism ring exactly R) and those triples satisfying $a > 0$, $\gcd(a, b, c) = 1$, $b^2 - 4ac = -171$, and conditions (1), (2), and (3).

Let's count these triples. The strategy is to consider $(b^2 + 171)/4$ and try to factor it as ac with a, b, c satisfying (1), (2), and (3):

b	$(b^2 + 171)/4$	a	c
1	43	1	43
± 3	45	5	9
5	49	7	7

The triple $(a, b, c) = (3, 3, 15)$, which arose in the above calculations, is not listed since $\gcd(a, b, c) \neq 1$ (and it corresponds to the ideal $3\mathcal{O}_K$, which is an ideal for the larger ring $\mathcal{O}_K$, as mentioned above). There are no values for a, c when $b = \pm 7$. When $|b| \geq 9$, the condition $|b| \leq a \leq c$ can no longer be satisfied. We have therefore found all triples. They correspond to values of τ, call them $\tau_1, \tau_2, \tau_3, \tau_4$:

$$(a, b, c) = (1, 1, 43) \longleftrightarrow \tau_1 = \frac{-1 + \sqrt{-171}}{2}$$

$$(a, b, c) = (5, 3, 9) \longleftrightarrow \tau_2 = \frac{-3 + \sqrt{-171}}{10}$$

$$(a, b, c) = (5, -3, 9) \longleftrightarrow \tau_3 = \frac{3 + \sqrt{-171}}{10}$$

$$(a, b, c) = (7, 5, 7) \longleftrightarrow \tau_4 = \frac{-5 + \sqrt{-171}}{14}.$$

Note that $j(\tau_0) = j(\tau_1)$ since $\tau_0 = \tau_1 + 1$. Compute the values

$$j(\tau_2) = -417.33569403605596400916623167906655644314607149466\ldots$$
$$+ i3470.1008337250975780924637689706441852341849935 50\ldots$$

$$j(\tau_3) = -417.33569403605596400916623167906655644314607149466\ldots$$
$$- i3470.1008337250975780924637689706441852341849935 50\ldots$$

$$j(\tau_4) = 154.68367675882023544437683081177435754892199372890 6\ldots.$$

We can now form the polynomial

$$(X - j(\tau_1))(X - j(\tau_2))(X - j(\tau_3))(X - j(\tau_4))$$
$$= X^4 + 69428205787653734 4\,X^3 + 472103267541360574464\,X^2$$
$$+ 8391550371275812148084736\,X - 131190152177915577372141158 4.$$

Since we are working with decimals, the numerical coefficients we obtain are not exact integers. But, since the roots $j(\tau_k)$ are a complete set of Galois conjugate algebraic integers, it follows that the coefficients are true integers. Therefore, if the computations are done with enough accuracy, we can round off to obtain the above polynomial.

We now describe the general situation. If we start with $\tau_0 = \frac{x + y\sqrt{-d}}{z}$, then we can use a matrix in $SL_2(\mathbf{Z})$ to move τ_0 to $\tau_1 \in \mathcal{F}$, and we have

$j(\tau_0) = j(\tau_1)$. Therefore, let's assume $\tau_0 \in \mathcal{F}$. Find integers a, b, c such that

$$a\tau_0^2 + b\tau_0 + c = 0$$

and $a > 0$, $\gcd(a, b, c) = 1$. Let $b^2 - 4ac = -D$. Now repeat the procedure used above, with D in place of 171, and obtain values $\tau_1, \ldots, \tau_h$. The polynomial satisfied by $j(\tau_0) = j(\tau_1)$ is

$$\prod_{k=1}^{r} (X - j(\tau_k)) \in \mathbf{Z}[X].$$

The above techniques can be used to find elliptic curves over finite fields with given orders. For example, suppose we want an elliptic curve E over $\mathbf{F}_p$, for some prime p, such that

$$N = \#E(\mathbf{F}_p) = 54323$$

(N is a prime). Because of Hasse's theorem, we must have p fairly close to N. The strategy is to choose a prime p, then let $a_p = p + 1 - N$ and $-D = a_p^2 - 4p$. We then find the polynomial $P(X)$ whose roots are the j-invariants of elliptic curves with complex multiplication by the order R of discriminant $-D$. Find a root of $P(X) \bmod p$. Such a root will be the j-invariant of an elliptic curve $E \bmod p$ that has complex multiplication by R.

The roots of

$$X^2 - a_p X + p = 0$$

lie in R (since $a_p^2 - 4p = -D$) and therefore correspond to endomorphisms of E. It can be shown that one of these endomorphisms is the Frobenius map (up to sign; see below). Therefore, we have found the characteristic polynomial of the Frobenius map. It follows that

$$\#E(\mathbf{F}_p) = p + 1 - a_p = N,$$

as desired. There is a slight complication caused by the fact that we might end up with $-a_p$ in place of a_p. We'll discuss this below.

In order to keep the number of τ_k's small, we want D, in the above notation, to be small. This means that we should have a_p near $\pm 2\sqrt{p}$. A choice that works well for us is

$$p = 54787, \quad a_p = 465, \quad D = 2923.$$

There are six values τ_k, corresponding to the polynomials $aX^2 + bX + c$ with

$$(a, b, c) = (1, 1, 731), (17, \pm 1, 43), (11, \pm 5, 67), (29, 21, 29).$$

We obtain a polynomial $P(X)$ of degree 6 with integer coefficients, as above. One of the roots of $P(X) \bmod p$ is $j = 46514$. Recall (see Section 2.6) that

$$y^2 = x^3 + \frac{3j}{1728 - j}x + \frac{2j}{1728 - j} \tag{10.4}$$

is an elliptic curve E_1 with j-invariant equal to j. In our case, we obtain

$$y^2 = x^3 + 10784x + 43714 \pmod{54787}.$$

The point $Q = (1, 36185)$ lies on E_1. However, we find that

$$54323Q \neq \infty, \quad 55253Q = \infty.$$

Since

$$55253 = p + 1 + 465,$$

we discover that we have obtained a curve E_1 with $a = -465$ instead of $a = 465$. This curve has complex multiplication by the order R of discriminant $-D$ (note that $-D = a^2 - 4p$, so the sign of a is irrelevant for D), so it is natural for it to appear. To obtain the desired curve, we twist by a quadratic nonresidue mod p (see Exercise 4.10). A quick computation shows that 2 is not a square mod p, so we look at the curve E defined by

$$y^2 = x^3 + 4 \cdot 10784x + 8 \cdot 43714 \pmod{54787}.$$

This has N points mod p. Just to be sure, we can compute

$$54323 \, (3, 38039) = \infty.$$

Since 54323 is prime, we find that 54323 divides the number of points in $E(\mathbf{F}_p)$. But

$$2 \cdot 54323 > p + 1 + 2\sqrt{p},$$

so Hasse's theorem implies that

$$\#E(\mathbf{F}_p) = 54323.$$

The above technique can be used to produce an elliptic curve E and a prime p such that $E(\mathbf{F}_p)$ is a desired group (when such a curve exists). For example, suppose we want

$$E(\mathbf{F}_p) \simeq \mathbf{Z}_2 \oplus \mathbf{Z}_2 \oplus \mathbf{Z}_{63}.$$

We take

$$N = 252, \quad p = 271, \quad a = 20,$$

so $N = p + 1 - a$. We choose

$$\tau = \frac{-1 + \sqrt{-171}}{2}.$$

As we'll see below, this choice imposes certain congruence conditions on the Frobenius map that force $E(\mathbf{F}_p)$ to have the desired form. We computed the polynomial satisfied by $j(\tau)$ above. This polynomial has the root 5 mod 271. Putting this value into the formula (10.4) yields the elliptic curve E given by

$$y^2 = x^3 + 70x + 137 \pmod{271}.$$

It has 252 points and has complex multiplication by the order

$$R = \mathbf{Z}\left[\frac{1 + \sqrt{-171}}{2}\right]$$

of discriminant $-171 = a^2 - 4p$. The characteristic polynomial of the Frobenius endomorphism ϕ_p is

$$X^2 - 20X + 271,$$

so ϕ_p corresponds to a root $10 \pm \sqrt{-171}$. The choice of sign is irrelevant for our purposes (it corresponds to how we choose to identify R with the endomorphism ring), so we assume

$$\phi_p = 10 + \sqrt{-171}.$$

Therefore,

$$\phi_p = 1 + 2\left(4 + \frac{1 + \sqrt{-171}}{2}\right) \equiv 1 \pmod{2R}.$$

It follows that ϕ_p acts as the identity on points of order 2, so $E(\mathbf{F}_p)$ has a subgroup isomorphic to $\mathbf{Z}_2 \oplus \mathbf{Z}_2$. In fact,

$$E[2] = \{\infty, (40,0), (56,0), (175,0)\} \subset E(\mathbf{F}_p).$$

Since $252 = 4 \times 63$,

$$E(\mathbf{F}_p) \simeq \mathbf{Z}_2 \oplus \mathbf{Z}_2 \oplus \mathbf{Z}_{63}.$$

If we instead want the group to be cyclic of order 252, we could use $R' = \mathbf{Z}[\sqrt{-171}]$ so that ϕ_p would not be congruent to 1 mod 2 or mod 3. We would then find a new set of τ_k corresponding to the discriminant $-4 \cdot 171$, a new j-invariant mod p, and a new E.

If we had used $R'' = \mathbf{Z}\left[\frac{1+\sqrt{-19}}{2}\right]$, then we would have obtained an elliptic curve with group isomorphic to $\mathbf{Z}_6 \oplus \mathbf{Z}_{42}$, since $\phi_p \equiv 1 \pmod{6R''}$ in this case.

This technique has many uses. For example, in [82], the curve E defined by

$$y^2 = x^3 + 3x - 31846 \pmod{158209}$$

was dedicated to Arjen Lenstra on the occasion of his thesis defense on May 16, 1984. The curve satisfies

$$E(\mathbf{F}_{158209}) \simeq \mathbf{Z}_5 \oplus \mathbf{Z}_{16} \oplus \mathbf{Z}_{1984}.$$

(If the defense had been one month later, such a dedication would have been impossible.) Finding elliptic curves with groups that are cyclic of large prime order is very useful in cryptography (see Chapter 6). Finding elliptic curves of a given order is also useful in primality proving (see Section 7.2). A detailed discussion of the problem, with improvements on the method presented here, is given in [58]. See also [4], [5].

10.5 Kronecker's Jugendtraum

The Kronecker-Weber theorem says that if $K/\mathbf{Q}$ is a finite Galois extension with abelian Galois group, then

$$K \subseteq \mathbf{Q}(e^{2\pi i/n})$$

for some integer n. This can be viewed as saying that the abelian extensions of $\mathbf{Q}$ are generated by the values of an analytic function, namely $e^{2\pi iz}$, at rational numbers. Kronecker's **Jugendtraum** (youthful dream) is that the abelian extensions of an arbitrary number field might similarly be generated by special values of a naturally occurring function. This has been accomplished for imaginary quadratic fields. Some progress has also been made for certain other fields by Shimura using complex multiplication of abelian varieties (higher dimensional analogues of elliptic curves).

If E is an elliptic curve given by $y^2 = x^3 + Ax + B$, then its j-invariant is given by $j = 6912A^3/(4A^3 + 27B^2)$. Therefore, if E is defined over a field L, then the j-invariant of E is contained in L. Conversely, if $j \neq 0$, 1728 lies in some field L, then the elliptic curve

$$y^2 = x^3 + \frac{3j}{1728 - j}x + \frac{2j}{1728 - j}$$

is defined over L and has j-invariant equal to $j \in L$. Therefore, for any j there is an elliptic curve with j-invariant equal to j defined over the field generated by j.

THEOREM 10.16
Let $K = \mathbf{Q}(\sqrt{-D})$ be an imaginary quadratic field, let $\mathcal{O}_K$ be the ring of algebraic integers in K, and let $j = j(\mathcal{O}_K)$, where $\mathcal{O}_K$ is regarded as a lattice in $\mathbf{C}$. Let E be an elliptic curve defined over $K(j)$ with j-invariant equal to j.

1. *Assume $K \neq \mathbf{Q}(i), \mathbf{Q}(e^{2\pi i/3})$. Let F be the field generated over $K(j)$ by the x-coordinates of the torsion points in $E(\overline{\mathbf{Q}})$. Then F/K has abelian Galois group, and every extension of K with abelian Galois group is contained in F.*

2. *If $K = \mathbf{Q}(i)$, the result of (1) holds when F is the extension generated by the squares of the x-coordinates of the torsion points.*

3. *If $K = \mathbf{Q}(e^{2\pi i/3})$, the result of (1) holds when F is the extension generated by the cubes of the x-coordinates of the torsion points.*

For a proof, see, for example, [92, p. 135] or [85]. Note that $j(\mathcal{O}_K)$ is algebraic, by Proposition 10.4. The j-invariant determines the lattice for the elliptic curve up to homothety (Exercise 10.6), so an elliptic curve with invariant $j(\mathcal{O}_K)$ automatically has complex multiplication by $\mathcal{O}_K$.

The x-coordinates of the torsion points are of the form

$$\wp(r_1\omega_1 + r_2\omega_2), \quad r_1, r_2 \in \mathbf{Q},$$

where $\wp$ is the Weierstrass $\wp$-function for the lattice for E. Therefore, the abelian extensions of K are generated by $j(\mathcal{O}_K)$ and special values of the function $\wp$. This is very much the analogue of the Kronecker-Weber theorem.

There is much more that can be said on this subject. See, for example, [92] and [55].

Exercises

10.1 Let $K = \mathbf{Q}(\sqrt{d})$ and $K' = \mathbf{Q}(\sqrt{d'})$ be quadratic fields. Let $\beta \in K$ and $\beta' \in K'$ and assume $\beta, \beta' \notin \mathbf{Q}$. Suppose that $\beta + \beta'$ lies in a quadratic field. Show that $K = K'$. (*Hint:* It suffices to consider the case $\beta = a\sqrt{d}$ and $\beta' = b\sqrt{d'}$. Let $\alpha = \beta + \beta'$. Show that if α is a root of a quadratic polynomial with coefficients in $\mathbf{Q}$, then we can solve for $\sqrt{d}$, say, in terms of $\sqrt{d'}$ and obtain $\sqrt{d} \in K'$.)

10.2 Let R be an order in an imaginary quadratic field. Regard R as a subset of $\mathbf{C}$. Show that if $r \in R$, then its complex conjugate $\bar{r}$ is also in R.

This means that if L is a lattice with complex multiplication by R, then there are two ways to embed R into the endomorphisms of L, namely via the assumed inclusion of R in $\mathbf{C}$ and also via the complex conjugate embedding (that is, if $r \in R$ and $\ell \in L$, define $r * \ell = \bar{r}\ell$). This means that when we say that R is contained in the endomorphism ring of a lattice or of an elliptic curve, we should specify which embedding we are using. For elliptic curves over $\mathbf{C}$, this is not a problem, since we can implicitly regard R as a subset of $\mathbf{C}$ and take the action of R on L as being the usual multiplication. But for elliptic curves over fields of positive characteristic, we cannot use this complex embedding.

10.3 Use the fact that $\mathbf{Z}\left[\frac{1+\sqrt{-43}}{2}\right]$ is a principal ideal domain to show that $e^{\pi\sqrt{43}}$ is very close to an integer.

10.4 Let $x = a + b\mathbf{i} + c\mathbf{j} + d\mathbf{k}$ lie in the Hamiltonian quaternions.

(a) Show that

$$(a + b\mathbf{i} + c\mathbf{j} + d\mathbf{k})(a + b\mathbf{i} + c\mathbf{j} + d\mathbf{k}) = a^2 + b^2 + c^2 + d^2.$$

(b) Show that if $x \neq 0$, then there exists a quaternion y such that $xy = 1$.

(c) Show that if we allow $a, b, c, d \in \mathbf{Q}_2$ (= the 2-adics), then $a^2 + b^2 + c^2 + d^2 = 0$ if and only if $a = b = c = d = 0$. (*Hint:* Clearing denominators reduces this to showing that $a^2 + b^2 + c^2 + d^2 \equiv 0$ (mod 8) implies that $a, b, c, \equiv 0$ (mod 8).)

(d) Show that if x, y are nonzero Hamiltonian quaternions with 2-adic coefficients, then $xy \neq 0$.

(e) Let p be an odd prime. Show that the number of squares a^2 mod p, including 0, is $(p+1)/2$ and that the number of elements of $\mathbf{F}_p$ of the form $1 - b^2$ (mod p) is also $(p+1)/2$.

(f) Show that if p is a prime, then $a^2 + b^2 + 1 \equiv 0$ (mod p) has a solution a, b.

(g) Use Hensel's lemma (see Appendix A) to show that if p is an odd prime, then there exist $a, b \in \mathbf{Q}_p$ such that $a^2 + b^2 + 1 = 0$. (The hypotheses of Hensel's lemma are not satisfied when $p = 2$.)

(h) Let p be an odd prime. Show that there are nonzero Hamiltonian quaternions x, y with p-adic coefficients such that $xy = 0$.

10.5 Show that a nonzero element in a definite quaternion algebra has a multiplicative inverse. (*Hint:* Use the ideas of parts (1) and (2) of Exercise 10.4.)

10.6 Show that if L_1 and L_2 are two lattices such that $j(L_1) = j(L_2)$, then L_1 and L_2 are homothetic. (*Hint:* Use Corollaries 9.15 and 9.18.)

Chapter 11

Divisors

11.1 Definitions and Examples

Let E be an elliptic curve defined over a field K. For each point $P \in E(\overline{K})$, define a formal symbol $[P]$. A **divisor** D on E is a finite linear combination of such symbols with integer coefficients:

$$D = \sum_j a_j [P_j], \quad a_j \in \mathbf{Z}.$$

A divisor is therefore an element of the free abelian group generated by the symbols $[P]$. The group of divisors is denoted $\text{Div}(E)$. Define the **degree** and **sum** of a divisor by

$$\deg\left(\sum_j a_j [P_j]\right) - \sum_j a_j \subset \mathbf{Z}$$

$$\text{sum}\left(\sum_j a_j [P_j]\right) = \sum_j a_j P_j \in E(\overline{K}).$$

The sum function simply uses the group law on E to add up the points that are inside the symbols. The divisors of degree 0 form an important subgroup of $\text{Div}(E)$, denoted $\text{Div}^0(E)$. The sum function gives a surjective homomorphism

$$\text{sum} : \text{Div}^0(E) \to E(\overline{K}).$$

The surjectivity is because

$$\text{sum}([P] - [\infty]) = P.$$

The kernel consists of divisors of functions (see Theorem 11.2 below), which we'll now describe.

Assume E is given by $y^2 = x^3 + Ax + B$. A **function** on E is a rational function

$$f(x, y) \in \overline{K}(x, y)$$

that is defined for at least one point in $E(\overline{K})$ (so, for example, the rational function $1/(y^2 - x^3 - Ax - B)$ is not allowed). The function takes values in $\overline{K} \cup \{\infty\}$.

There is a technicality that is probably best described by an example. Suppose $y^2 = x^3 - x$ is the equation of the elliptic curve. The function

$$f(x, y) = \frac{x}{y}$$

is not defined at $(0, 0)$. However, on E,

$$\frac{x}{y} = \frac{y}{x^2 - 1},$$

which is defined and takes on the value 0 at $(0, 0)$. Similarly, the function y/x can be changed to $(x^2 - 1)/y$, which takes on the value ∞ at $(0, 0)$. It can be shown that a function can always be transformed in this manner so as to obtain an expression that is not $0/0$ and hence gives a uniquely determined value in $\overline{K} \cup \{\infty\}$.

A function is said to have a **zero** at a point P if it takes the value 0 at P, and it has a **pole** at P if it takes the value ∞ at P. However, we need more refined information, namely the order of the zero or pole. Let P be a point. It can be shown that there is a function u_P, called a **uniformizer** at P, with $u(P) = 0$ and such that every function $f(x, y)$ can be written in the form

$$f = u_P^r\, g, \quad \text{with } r \in \mathbf{Z} \text{ and } g(P) \neq 0, \infty.$$

Define the **order** of f at P by

$$\mathrm{ord}_P(f) = r.$$

Example 11.1
On $y^2 = x^3 - x$, it can be shown that the function y is a uniformizer at $(0, 0)$. We have

$$x = y^2 \frac{1}{x^2 - 1},$$

and $1/(x^2 - 1)$ is nonzero and finite at $(0, 0)$. Therefore,

$$\mathrm{ord}_{(0,0)}(x) = 2, \quad \text{and} \quad \mathrm{ord}_{(0,0)}(x/y) = 1.$$

This latter fact agrees with the above computation that showed that x/y vanishes at $(0, 0)$. ☐

Example 11.2
At any finite point $P = (x_0, y_0)$ on an elliptic curve, the uniformizer u_P can be taken from the equation of a line that passes through P but is not tangent

to E. A natural choice is $u_P = x - x_0 = 0$ when $y_0 \neq 0$ and $u_P = y$ when $y_0 = 0$. For example, let $P = (-2, 8)$ on the curve $y^2 = x^3 + 72$. The line

$$x + 2 = 0$$

passes through P, so we take $u_P(x, y) = x + 2$. The function

$$f(x, y) = x + y - 6$$

vanishes at P. Let's find its order of vanishing at P. The equation for the curve can be rewritten as

$$(y + 8)(y - 8) = (x + 2)^3 - 6(x + 2)^2 + 12(x + 2).$$

Therefore,

$$f(x, y) = (x + 2) + (y - 8) = (x + 2)\left(1 + \frac{(x + 2)^2 - 6(x + 2) + 12}{y + 8}\right).$$

The function in parentheses is finite and does not vanish at P, so $\operatorname{ord}_P(f) = 1$. The function

$$t(x, y) = \frac{3}{4}(x + 2) - y + 8$$

comes from the tangent line to E at P. We have

$$
\begin{aligned}
t(x, y) &= (x + 2)\left(\frac{3}{4} - \frac{(x + 2)^2 - 6(x + 2) + 12}{y + 8}\right) \\
&= \frac{(x + 2)}{4(y + 8)}\left(-4(x + 2)^2 + 24(x + 2) + 3(y - 8)\right) \\
&= \frac{(x + 2)^2}{4(y + 8)}\left(-4(x + 2) + 24 + 3\frac{(x + 2)^2 - 6(x + 2) + 12}{y + 8}\right).
\end{aligned}
$$

The expression in parentheses is finite and does not vanish at P, so $\operatorname{ord}_P(t) = 2$. In general, the equation of a tangent line will yield a function that vanishes to order at least 2 (equal to 2 unless $3P = \infty$ in the group law of E, in which case the order is 3).   □

Example 11.3

The point at infinity is a little harder to deal with. If the elliptic curve E is given by

$$y^2 = x^3 + Ax + B,$$

a uniformizer at ∞ is $u_\infty = x/y$. This choice is motivated by the complex situation: The Weierstrass function $\wp$ gives the x-coordinate and $\frac{1}{2}\wp'$ gives the y-coordinate. Recall that the point $0 \in \mathbf{C}/L$ corresponds to ∞ on E. Since $\wp$ has a double pole at 0 and $\wp'$ has a triple pole at 0, the quotient $\wp/\wp'$ has a simple zero at 0, hence can be used as a uniformizer at 0 in $\mathbf{C}/L$.

Let's compute the order of x and y. Rewriting the equation for E as

$$\left(\frac{x}{y}\right)^2 = x^{-1}\left(1 + \frac{A}{x^2} + \frac{B}{x^3}\right)^{-1}$$

shows that x/y vanishes at ∞ and that

$$\mathrm{ord}_\infty(x) = -2$$

(given that x/y is a uniformizer). Since $y = x \cdot (y/x)$, we have

$$\mathrm{ord}_\infty(y) = -3.$$

Note that the orders of x and y at ∞ agree with what we expect from looking at the Weierstrass $\wp$-function. $\square$

If f is a function on E that is not identically 0, define the **divisor** of f to be

$$\mathrm{div}(f) = \sum_{P \in E(\overline{K})} \mathrm{ord}_P(f)[P] \in \mathrm{Div}(E).$$

This is a finite sum, hence a divisor, by the following.

PROPOSITION 11.1
Let E be an elliptic curve and let f be a function on E that is not identically 0.

1. *f has only finitely many zeros and poles*

2. *$\deg(\mathrm{div}(f)) = 0$*

3. *If f has no zeros or poles (so $\mathrm{div}(f) = 0$), then f is a constant.*

For a proof, see [31, Ch.8, Prop. 1] or [36, II, Cor. 6.10]. The complex analytic analogue of the proposition is Theorem 9.1. Note that it is important to look at points with coordinates in $\overline{K}$. It is easy to construct nonconstant functions with no zeros or poles at the points in $E(K)$, and it is easy to construct functions that have zeros but no poles in $E(K)$ (see Exercise 11.1).

The divisor of a function is said to be a **principal divisor**.

Suppose P_1, P_2, P_3 are three points on E that lie on the line $ax + by + c = 0$. Then the function

$$f(x, y) = ax + by + c$$

has zeros at P_1, P_2, P_3. If $b \neq 0$ then f has a triple pole at ∞. Therefore,

$$\mathrm{div}(ax + by + c) = [P_1] + [P_2] + [P_3] - 3[\infty].$$

The line through $P_3 = (x_3, y_3)$ and $-P_3$ is $x - x_3 = 0$. The divisor of the function $x - x_3$ is

$$\text{div}(x - x_3) = [P_3] + [-P_3] - 2[\infty]. \tag{11.1}$$

Therefore,

$$\text{div}\left(\frac{ax + by + c}{x - x_3}\right) = \text{div}(ax + by + c) - \text{div}(x - x_3) = [P_1] + [P_2] - [-P_3] - [\infty].$$

Since $P_1 + P_2 = -P_3$ on E, this may be rewritten as

$$[P_1] + [P_2] = [P_1 + P_2] + [\infty] + \text{div}\left(\frac{ax + by + c}{x - x_0}\right).$$

The following important result is the analogue of Theorem 9.6.

THEOREM 11.2
 Let E be an elliptic curve. Let D be a divisor on E with $\deg(D) = 0$. Then there is a function f on E with

$$\text{div}(f) = D$$

if and only if

$$\text{sum}(D) = \infty.$$

PROOF   We have just shown that a sum $[P_1] + [P_2]$ can be replaced by $[P_1 + P_2] + [\infty]$ plus the divisor of a function, call it g. Note also that

$$\text{sum}(\text{div}(g)) = P_1 + P_2 - (P_1 + P_2) - \infty = \infty.$$

Equation (11.1) shows that $[P_1] + [P_2]$ equals $2[\infty]$ plus the divisor of a function when $P_1 + P_2 = \infty$. Therefore, the sum of all the terms in D with positive coefficients equals a single symbol $[P]$ plus a multiple of $[\infty]$ plus the divisor of a function. A similar result holds for the sum of the terms with negative coefficients. Therefore, there are points P and Q on E, a function g_1, and an integer n such that

$$D = [P] - [Q] + n[\infty] + \text{div}(g_1).$$

Also, since g_1 is the quotient of products of functions g with $\text{sum}(\text{div}(g)) = \infty$, we have

$$\text{sum}(\text{div}(g_1)) = \infty.$$

Since $\deg(\text{div}(g_1)) = 0$ by Proposition 11.1, we have

$$0 = \deg(D) = 1 - 1 + n + 0 = n.$$

Therefore,
$$D = [P] - [Q] + \operatorname{div}(g_1).$$

Also,
$$\operatorname{sum}(D) = P - Q + \operatorname{sum}(\operatorname{div}(g_1)) = P - Q.$$

Suppose $\operatorname{sum}(D) = \infty$. Then $P - Q = \infty$, so $P = Q$ and $D = \operatorname{div}(g_1)$. Conversely, suppose $D = \operatorname{div}(f)$ for some function f. Then

$$[P] - [Q] = \operatorname{div}(f/g_1).$$

The following lemma implies that $P = Q$, and hence $\operatorname{sum}(D) = \infty$. This completes the proof of Theorem 11.2.

LEMMA 11.3
Let $P, Q \in E(\overline{K})$ and suppose there exists a function h on E with

$$\operatorname{div}(h) = [P] - [Q].$$

Then $P = Q$.

Since the proof is slightly long, we postpone it until the end of this section.

COROLLARY 11.4
The map
$$\operatorname{sum} : \ \operatorname{Div}^0(E)/ \ (\textit{principal divisors}) \longrightarrow E(\overline{K})$$
is an isomorphism of groups.

PROOF   Since $\operatorname{sum}([P]-[\infty]) = P$, the sum map from $\operatorname{Div}^0(E)$ to $E(\overline{K})$ is surjective. The theorem says that the kernel is exactly the principal divisors.
∎

Corollary 11.4 shows that the group law on $E(\overline{K})$ corresponds to the very natural group law on $\operatorname{Div}^0(E)$ mod principal divisors.

Example 11.4
The proof of the theorem gives an algorithm for finding a function with a given divisor (of degree 0 and sum equal to ∞). Consider the elliptic curve E over $\mathbf{F}_{11}$ given by
$$y^2 = x^3 + 4x.$$
Let
$$D = [(0,0)] + [(2,4)] + [(4,5)] + [(6,3)] - 4[\infty].$$

Then D has degree 0 and an easy calculation shows that $\text{sum}(D) = \infty$. There-fore, D is the divisor of a function. Let's find the function. The line through $(0,0)$ and $(2,4)$ is $y - 2x = 0$. It is tangent to E at $(2,4)$, so

$$\text{div}(y - 2x) = [(0,0)] + 2[(2,4)] - 3[\infty].$$

The vertical line through $(2,4)$ is $x - 2 = 0$, and

$$\text{div}(x - 2) = [(2,4)] + [(2,-4)] - 2[\infty].$$

Therefore,

$$D = [(2,-4)] + \text{div}\left(\frac{y-2x}{x-2}\right) + [(4,5)] + [(6,3)] - 3[\infty].$$

Similarly, we have

$$[(4,5)] + [(6,3)] = [(2,4)] + [\infty] + \text{div}\left(\frac{y+x+2}{x-2}\right),$$

which yields

$$D = [(2,-4)] + \text{div}\left(\frac{y-2x}{x-2}\right) + [(2,4)] + \text{div}\left(\frac{y+x+2}{x-2}\right) - 2[\infty].$$

Since we have already calculated $\text{div}(x-2)$, we use this to conclude that

$$D = \text{div}(x-2) + \text{div}\left(\frac{y-2x}{x-2}\right) + \text{div}\left(\frac{y+x+2}{x-2}\right)$$

$$= \text{div}\left(\frac{(y-2x)(y+x+2)}{x-2}\right).$$

This function can be simplified. The numerator is

$$(y-2x)(y+x+2) = y^2 - xy - 2x^2 + 2y - 4x$$
$$= x^3 - xy - 2x^2 + 2y \quad (\text{since } y^2 = x^3 + 4x)$$
$$= (x-2)(x^2 - y).$$

Therefore,

$$D = \text{div}(x^2 - y).$$

☐

Proof of Lemma 11.3:
Suppose $P \neq Q$ and $\text{div}(h) = [P] - [Q]$. Then, for any constant c, the function $h - c$ has a simple pole at Q and therefore, by Proposition 11.1, it

has exactly one zero, which must be simple. Let f be any function on E. If f does not have a zero or pole at Q, then

$$g(x,y) = \prod_{R \in E(\overline{K})} (h(x,y) - h(R))^{\operatorname{ord}_R(f)}$$

has the same divisor as f (since we are assuming that $\operatorname{ord}_Q(f) = 0$, the factor for $R = Q$ is defined to be 1). The only thing to check is that the poles of $h(x,y)$ at Q cancel out. Each factor has a pole at $(x,y) = Q$ of order $\operatorname{ord}_R(f)$ (or a zero if $\operatorname{ord}_R(f) < 0$). Since $\sum_R \operatorname{ord}_R(f) = 0$, these cancel.

Since f and g have the same divisor, the quotient f/g has no zeros or poles, and is therefore constant. It follows that f is a rational function of h.

If f has a zero or pole at Q, the factor for $R = Q$ in the above product is not defined. However, $f \cdot h^{\operatorname{ord}_R(f)}$ has no zero or pole at Q. The above reasoning shows that it is therefore a rational function of h, so the same holds for f.

We have shown that every function on $E(\overline{K})$ is a rational function of h. In particular, x and y are rational functions of h. The following result shows that this is impossible. This contradiction means that we must have $P = Q$.

LEMMA 11.5
Let E be an elliptic curve over K (of characteristic not 2) given by

$$y^2 = x^3 + Ax + B.$$

Let t be an indeterminate. There are no nonconstant rational functions $X(t)$ and $Y(t)$ in $\overline{K}(t)$ such that

$$Y(t)^2 = X(t)^3 + AX(t) + B.$$

PROOF   Factor the cubic polynomial as

$$x^3 + Ax + B = (x - e_1)(x - e_2)(x - e_3),$$

where $e_1, e_2, e_3 \in \overline{K}$ are distinct. Suppose $X(t), Y(t)$ exist. Write

$$X(t) = \frac{P_1(t)}{P_2(t)}, \qquad Y(t) = \frac{Q_1(t)}{Q_2(t)},$$

where P_1, P_2, Q_1, Q_2 are polynomials in t. We may assume that $P_1(t)$, $P_2(t)$ have no common roots, and that $Q_1(t)$, $Q_2(t)$ have no common roots. Substituting into the equation for E yields

$$Q_1(t)^2 P_2(t)^3 = Q_2(t)^2 \left(P_1(t)^3 + AP_1(t)P_2(t)^2 + BP_2(t)^3 \right).$$

Since the right side is a multiple of $Q_2(t)^2$, so is the left side. Since Q_1, Q_2 have no common roots, P_2^3 must be a multiple of Q_2^2. A common root of

P_2 and $P_1^3 + AP_1P_2^2 + BP_2^3$ would be a root of P_1^3. Since P_1 and P_2 have no roots in common, this is impossible. Therefore, Q_2^2 must be a multiple of P_2^3. Therefore, P_2^3 and Q_2^2 are multiples of each other, hence are constant multiples of each other. By adjusting P_1 and Q_1 if necessary, we may assume that

$$P_2^3 = Q_2^2.$$

Canceling these from the equation yields

$$Q_1^2 = P_1^3 + AP_1P_2^2 + BP_2^3 = (P_1 - e_1P_2)(P_1 - e_2P_2)(P_3 - e_3P_2).$$

Suppose $i \neq j$ and $P_1 - e_iP_2$ and $P_1 - e_jP_2$ have a common root r. Then r is a root of

$$e_j(P_1 - e_iP_2) - e_i(P_1 - e_jP_2) = (e_j - e_i)P_1 \qquad (11.2)$$

and of

$$(P_1 - e_iP_2) - (P_1 - e_jP_2) = (e_j - e_i)P_2. \qquad (11.3)$$

Since $e_j - e_i \neq 0$, this means that r is a common root of P_1 and P_2, which is a contradiction. Therefore $P_1 - e_iP_2$ and $P_1 - e_jP_2$ have no common roots when $i \neq j$. Since the product

$$(P_1 - e_1P_2)(P_1 - e_2P_2)(P_1 - e_3P_2)$$

is the square of a polynomial, each factor must be a square of a polynomial in $\overline{K}[t]$ (it might seem that each factor is a constant times a square, but all constants are squares in the algebraically closed field $\overline{K}$, hence can be absorbed into the squares of polynomials).

Since $P_2^3 = Q_2^2$, we find that P_2 must also be a square of a polynomial.

LEMMA 11.6

Let P_1 and P_2 be polynomials in $\overline{K}[t]$ with no common roots. Suppose there are four pairs (a_i, b_i), $1 \leq i \leq 4$, with $a_i, b_i \in \overline{K}$ satisfying

1. for each i, at least one of a_i, b_i is nonzero

2. if $i \neq j$, then there does not exist $c \in \overline{K}^\times$ with $(a_i, b_i) = (ca_j, cb_j)$

3. $a_iP_1 + b_iP_2$ is a square of a polynomial for $1 \leq i \leq 4$.

Then P_1, P_2 are constant polynomials.

PROOF   The assumptions imply that any two of the vectors (a_i, b_i) are linearly independent over $\overline{K}$ and therefore span $\overline{K}^2$. Suppose that at least one of P_1, P_2 is nonconstant. We may assume that P_1, P_2 are chosen so that

$$\text{Max}(\deg(P_1), \deg(P_2)) > 0$$

is as small as possible. Since P_1, P_2 have no common roots, it is easy to see that they must be linearly independent over $\overline{K}$. Let

$$a_i P_1 + b_i P_2 = R_i^2, \quad 1 \le i \le 4. \tag{11.4}$$

Note that when $i \ne j$, the polynomial R_i^2 cannot be a constant multiple of R_j^2, since otherwise the linear independence of P_1, P_2 would imply that (a_i, b_i) is a constant multiple of (a_j, b_j).

Since the vectors (a_3, b_3) and (a_4, b_4) are linear combinations of (a_1, b_1) and (a_2, b_2), there are constants $c_1, c_2, d_1, d_2 \in \overline{K}$ such that

$$R_3^2 = c_1 R_1^2 - d_1 R_2^2, \quad R_4^2 = c_2 R_1^2 - d_2 R_2^2.$$

If (c_1, d_1) is proportional to (c_2, d_2), then R_3^2 is a constant times R_4^2, which is not possible. Therefore, (c_1, d_1) and (c_2, d_2) are not proportional. Moreover, since (a_1, b_1) and (a_2, b_2) are linearly independent, Equation (11.4) for $i = 1, 2$ can be solved for P_1 and P_2, showing that P_1 and P_2 are linear combinations of R_1^2 and R_2^2. Therefore, a common root of R_1 and R_2 is a common root of P_1 and P_2, which doesn't exist. It follows that R_1 and R_2 have no common roots. It follows easily (by using equations similar to (11.2) and (11.3)) that

$$\sqrt{c_1} R_1 + \sqrt{d_1} R_2 \quad \text{and} \quad \sqrt{c_1} R_1 - \sqrt{d_1} R_2$$

have no common roots. Since their product is square, namely R_3^2, each factor must be a square. Similarly, both $\sqrt{c_2} R_1 + \sqrt{d_2} R_2$ and $\sqrt{c_2} R_1 - \sqrt{d_2} R_2$ must be squares. Therefore, R_1, R_2 are polynomials satisfying the conditions of the lemma for the pairs

$$(\sqrt{c_1}, \sqrt{d_1}), \quad (\sqrt{c_1}, -\sqrt{d_1}), \quad (\sqrt{c_2}, \sqrt{d_2}), \quad (\sqrt{c_2}, -\sqrt{d_2}).$$

Since (c_1, d_1) and (c_2, d_2) are not proportional, neither of the first two pairs is proportional to either of the last two pairs. If $(\sqrt{c_1}, \sqrt{d_1})$ is proportional to $(\sqrt{c_1}, -\sqrt{d_1})$, then either c_1 or d_1 is zero, which means that R_3^2 is a constant multiple of either R_1^2 or R_2^2. This cannot be the case, as pointed out above. Similarly, the last two pairs are not proportional.

Equation (11.4) implies that

$$\text{Max}(\deg(P_1), \deg(P_2)) \ge 2\text{Max}(\deg(R_1), \deg(R_2)).$$

Since R_1 and R_2 are clearly nonconstant, this contradicts the minimality of $\text{Max}(\deg(P_1), \deg(P_2))$. Therefore, all polynomials P_1, P_2 satisfying the conditions of the lemma must be constant. This proves Lemma 11.6. ∎

Returning to the proof of Lemma 11.5, we have polynomials P_1, P_2 and pairs

$$(1, -e_1), \quad (1, -e_2), \quad (1, -e_3), \quad (0, 1)$$

satisfying the conditions of Lemma 11.6. Therefore, P_1 and P_2 must be constant. But $X(t) = P_1/P_2$ is nonconstant, so we have a contradiction. This completes the proof of Lemma 11.5.   ∎

As pointed out above, Lemma 11.5 completes the proof of Lemma 11.3.

11.2 The Weil Pairing

The goal of this section is to construct the Weil pairing and prove its basic properties that were stated in Section 3.3. Recall that n is an integer not divisible by the characteristic of the field K, and that E is an elliptic curve such that

$$E[n] \subseteq E(K).$$

We want to construct a pairing

$$e_n : E[n] \times E[n] \longrightarrow \mu_n,$$

where μ_n is the set of nth roots of unity in $\overline{K}$ (as we showed in Section 3.3, the assumption $E[n] \subseteq E(K)$ forces $\mu_n \subset K$).

Let $T \in E[n]$. By Theorem 11.2, there exists a function f such that

$$\mathrm{div}(f) = n[T] - n[\infty]. \tag{11.5}$$

Choose $T' \in E[n^2]$ such that $nT' = T$. We'll use Theorem 11.2 to show that there exists a function g such that

$$\mathrm{div}(g) = \sum_{R \in E[n]} ([T' + R] - [R]).$$

We need to verify that the sum of the points in the divisor is ∞. This follows from the fact that there are n^2 points R in $E[n]$. The points R in $\sum[T' + R]$ and $\sum[R]$ cancel, so the sum is $n^2 T' = nT = \infty$. Note that g does not depend on the choice of T' since any two choices for T' differ by an element $R \in E[n]$. Therefore, we could have written

$$\mathrm{div}(g) = \sum_{nT''=T} [T''] - \sum_{nR=\infty} [R].$$

Let $f \circ n$ denote the function that starts with a point, multiplies it by n, then applies f. The points $P = T' + R$ with $R \in E[n]$ are those points P with $nP = T$. It follows from (11.5) that

$$\mathrm{div}(f \circ n) = n\left(\sum_R [T' + R]\right) - n\left(\sum_R [R]\right) = \mathrm{div}(g^n).$$

Therefore, $f \circ n$ is a constant multiple of g^n. By multiplying f by a suitable constant, we may assume that

$$f \circ n = g^n.$$

Let $S \in E[n]$ and let $P \in E(\overline{K})$. Then

$$g(P + S)^n = f(n(P + S)) = f(nP) = g(P)^n.$$

Therefore, $g(P + S)/g(P) \in \mu_n$. In fact, $g(P + S)/g(P)$ is independent of P. The proof of this is slightly technical: In the Zariski topology, $g(P + S)/g(P)$ is a continuous function of P and E is connected. Therefore, the map to the finite discrete set μ_n must be constant.

Define the **Weil pairing** by

$$e_n(S, T) = \frac{g(P + S)}{g(P)}. \tag{11.6}$$

Since g is determined up to a scalar multiple by its divisor, this definition is independent of the choice of g. Note that (11.6) is independent of the choice of the auxiliary point P. The main properties of e_n are given in the following theorem, which was stated in Section 3.3.

THEOREM 11.7

Let E be an elliptic curve defined over a field K and let n be a positive integer. Assume that the characteristic of K does not divide n. Then the Weil pairing

$$e_n : E[n] \times E[n] \to \mu_n$$

satisfies the following properties:

1. *e_n is bilinear in each variable. This means that*

$$e_n(S_1 + S_2, T) = e_n(S_1, T)e_n(S_2, T)$$

   and

$$e_n(S, T_1 + T_2) = e_n(S, T_1)e_n(S, T_2)$$

   for all $S, S_1, S_2, T, T_1, T_2 \in E[n]$.

2. *e_n is nondegenerate in each variable. This means that if $e_n(S, T) = 1$ for all $T \in E[n]$ then $S = \infty$ and also that if $e_n(S, T) = 1$ for all $S \in E[n]$ then $T = \infty$.*

3. *$e_n(T, T) = 1$ for all $T \in E[n]$.*

4. *$e_n(T, S) = e_n(S, T)^{-1}$ for all $S, T \in E[n]$.*

5. $e_n(\sigma S, \sigma T) = \sigma(e_n(S, T))$ *for all automorphisms σ of $\overline{K}$ such that σ is the identity map on the coefficients of E (if E is in Weierstrass form, this means that $\sigma(A) = A$ and $\sigma(B) = B$).*

6. $e_n(\alpha(S), \alpha(T)) = e_n(S, T)^{\deg(\alpha)}$ *for all separable endomorphisms α of E. If the coefficients of E lie in a finite field $\mathbf{F}_q$, then the statement also holds when α is the Frobenius endomorphism ϕ_q. (Actually, the statement holds for all endomorphisms α, separable or not. See [28].)*

PROOF   (1) Since e_n is independent of the choice of P, we use (11.6) with P and with $P + S_1$ to obtain

$$
\begin{aligned}
e_n(S_1, T)e_n(S_2, T) &= \frac{g(P + S_1)}{g(P)}\frac{g(P + S_1 + S_2)}{g(P + S_1)} \\
&= \frac{g(P + S_1 + S_2)}{g(P)} \\
&= e_n(S_1 + S_2, T).
\end{aligned}
$$

This proves linearity in the first variable.

Suppose $T_1, T_2, T_3 \in E[n]$ with $T_1 + T_2 = T_3$. For $1 \le i \le 3$, let f_i, g_i be the functions used above to define $e_n(S, T_i)$. By Theorem 11.2, there exists a function h such that

$$
\operatorname{div}(h) = [T_3] - [T_1] - [T_2] + [\infty].
$$

Equation (11.5) yields

$$
\operatorname{div}\left(\frac{f_3}{f_1 f_2}\right) = n \operatorname{div}(h) = \operatorname{div}(h^n).
$$

Therefore, there exists a constant $c \in \overline{K}^{\times}$ such that

$$
f_3 = c f_1 f_2 h^n.
$$

This implies that

$$
g_3 = c^{1/n}(g_1)(g_2)(h \circ n).
$$

The definition of e_n yields

$$
\begin{aligned}
e_n(S, T_1 + T_2) &= \frac{g_3(P + S)}{g_3(P)} = \frac{g_1(P + S)}{g_1(P)}\frac{g_2(P + S)}{g_2(P)}\frac{h(n(P + S))}{h(nP)} \\
&= e_n(S, T_1)\, e_n(S, T_2),
\end{aligned}
$$

since $nS = \infty$, so $h(n(P + S)) = h(nP)$. This proves linearity in the second variable.

(2) Suppose $T \in E[n]$ is such that $e_n(S, T) = 1$ for all $S \in E[n]$. This means that $g(P + S) = g(P)$ for all P and for all $S \in E[n]$. By Proposition 9.32, there is a function h such that $g = h \circ n$. Then

$$(h \circ n)^n = g^n = f \circ n.$$

Since multiplication by n is surjective on $E(\overline{K})$, we have $h^n = f$. Therefore,

$$n \operatorname{div}(h) = \operatorname{div}(f) = n[T] - n[\infty],$$

so $\operatorname{div}(h) = [T] - [\infty]$. By Theorem 11.2, we have $T = \infty$. This proves half of (2). The nondegeneracy in S follows immediately from (4) plus the nondegeneracy in T.

(3) Let τ_{jT} represent adding jT, so $f \circ \tau_{jT}$ denotes the function $P \mapsto f(P + jT)$. The divisor of $f \circ \tau_{jT}$ is $n[T - jT] - n[-jT]$. Therefore,

$$\operatorname{div}\left(\prod_{j=0}^{n-1} f \circ \tau_{jT}\right) = \sum_{j=0}^{n-1} (n[(1-j)T] - n[-jT]) = 0.$$

This means that $\prod_{j=0}^{n-1} f \circ \tau_{jT}$ is constant. The nth power of the function $\prod_{j=0}^{n-1} g \circ \tau_{jT'}$ is the above product of f's composed with multiplication by n, hence is constant. Since

$$\left(\prod_{j=0}^{n-1} g \circ \tau_{jT'}\right)^n = \prod_{j=0}^{n-1} f \circ n \circ \tau_{jT'}$$

$$= \prod_{j=0}^{n-1} f \circ \tau_{jT} \qquad (\text{since } nT' = T).$$

Since we have proved that this last product is constant, it follows that $\prod_{j=0}^{n-1} g \circ \tau_{jT'}$ is constant (we are again using the connectedness of E in the Zariski topology). Therefore, it has the same value at P and $P + T'$, so

$$\prod_{j=0}^{n-1} g(P + T' + jT') = \prod_{j=0}^{n-1} g(P + jT').$$

Canceling the common terms (we assume P is chosen so that all terms are finite and nonzero) yields

$$g(P + nT') = g(P).$$

Since $nT' = T$, this means that

$$e_n(T, T) = \frac{g(P + T)}{g(P)} = 1.$$

(4) By (1) and (3),

$$1 = e_n(S + T, S + T) = e_n(S, S) \, e_n(S, T) \, e_n(T, S) \, e_n(T, T)$$
$$= e_n(S, T) \, e_n(T, S).$$

Therefore $e_n(T, S) = e_n(S, T)^{-1}$.

(5) Let σ be an automorphism of $\overline{K}$ such that σ is the identity on the coefficients of E. Apply σ to everything in the construction of e_n. Then

$$\operatorname{div}(f^\sigma) - n[\sigma T] - n[\infty]$$

and similarly for g^σ, where f^σ and g^σ denote the functions obtained by applying σ to the coefficients of the rational functions defining f and g (cf. Section 8.9). Therefore,

$$\sigma(e_n(S, T)) = \sigma \left(\frac{g(P + S)}{g(P)} \right) = \frac{g^\sigma(\sigma P + \sigma S)}{g^\sigma(\sigma P)} = e_n(\sigma S, \sigma T).$$

(6) Let $\{Q_1, \ldots, Q_k\} = \operatorname{Ker}(\alpha)$. Since α is a separable morphism, $k = \deg(\alpha)$ by Proposition 2.20. Let

$$\operatorname{div}(f_T) = n[T] - n[\infty], \quad \operatorname{div}(f_{\alpha(T)}) = n[\alpha(T)] - n[\infty]$$

and

$$g_T^n = f_T \circ n, \quad g_{\alpha(T)}^n = f_{\alpha(T)} \circ n.$$

As in (3), let τ_Q denote adding Q. We have

$$\operatorname{div}(f_T \circ \tau_{-Q_i}) = n[T + Q_i] - n[Q_i].$$

Therefore,

$$\operatorname{div}(f_{\alpha(T)} \circ \alpha) = n \sum_{\alpha(T'') = \alpha(T)} [T''] - n \sum_{\alpha(Q) = \infty} [Q]$$
$$= n \sum_i ([T + Q_i] - [Q_i])$$
$$= \operatorname{div}(\prod_i (f_T \circ \tau_{-Q_i})).$$

For each i, choose Q_i' with $nQ_i' = Q_i$. Then

$$g_T(P - Q_i')^n = f_T(nP - Q_i).$$

Consequently,

$$\operatorname{div}\left(\prod_i (g_T \circ \tau_{-Q_i'})^n \right) = \operatorname{div}(\prod_i f_T \circ \tau_{-Q_i} \circ n)$$
$$= \operatorname{div}(f_{\alpha(T)} \circ \alpha \circ n)$$
$$= \operatorname{div}(f_{\alpha(T)} \circ n \circ \alpha)$$
$$= \operatorname{div}(g_{\alpha(T)} \circ \alpha)^n.$$

Therefore, $\prod_i g_T \circ \tau_{-Q_i'}$ and $g_{\alpha(T)} \circ \alpha$ have the same divisor and hence differ by a constant C.

The definition of e_n yields

$$
\begin{aligned}
e_n(\alpha(S), \alpha(T)) &= \frac{g_{\alpha(T)}(\alpha(P+S))}{g_{\alpha(T)}(\alpha(P))} \\
&= \prod_i \frac{g_T(P+S-Q_i')}{g_T(P-Q_i')} \quad \text{(the constant C cancels out)} \\
&= \prod_i e_n(S, T) \\
&\quad \text{(since both P and $P - Q_i'$ give the same value of e_n)} \\
&= e_n(S, T)^k = e_n(S, T)^{\deg(\alpha)}.
\end{aligned}
$$

When $\alpha = \phi_q$ is the Frobenius endomorphism, then (6) implies that

$$
e_n(\phi_q(S), \phi_q(T)) = \phi_q(e_n(S, T)) = e_n(S, T)^q,
$$

since ϕ_q is the qth power map on elements of $\overline{\mathbf{F}}_q$. From Lemma 2.19, we have that $q = \deg(\phi_q)$, which proves (6) when $\alpha = \phi_q$. This completes the proof of Theorem 11.7. ∎

11.3 The Tate-Lichtenbaum Pairing

In this section, we give the definition of the Tate-Lichtenbaum pairing and the modified Tate-Lichtenbaum pairing, which were introduced in Chapter 5.

THEOREM 11.8

Let E be an elliptic curve over $\mathbf{F}_q$. Let n be an integer such that $n|q-1$. Let $E(\mathbf{F}_q)[n]$ denote the elements of $E(\mathbf{F}_q)$ of order dividing n, and let $\mu_n = \{x \in \mathbf{F}_q \,|\, x^n = 1\}$. Assume $E(\mathbf{F}_q)$ contains an element of order n. Then there are nondegenerate bilinear pairings

$$
\langle \cdot, \cdot \rangle_n : E(\mathbf{F}_q)[n] \times E(\mathbf{F}_q)/nE(\mathbf{F}_q) \to \mathbf{F}_q^\times / (\mathbf{F}_q^\times)^n
$$

and

$$
\tau_n : E(\mathbf{F}_q)[n] \times E(\mathbf{F}_q)/nE(\mathbf{F}_q) \to \mu_n.
$$

The first pairing of the theorem is called the **Tate-Lichtenbaum pairing**. We'll refer to τ_n as the **modified Tate-Lichtenbaum pairing**. The pairing τ_n is better suited for computations since it gives a definite answer, rather than a coset in $\mathbf{F}_q^\times$ mod nth powers. As pointed out in Chapter 5, we should write

$\langle P, Q + nE(\mathbf{F}_q)\rangle_n$, and similarly for τ_n, since an element of $E(\mathbf{F}_q)/nE(\mathbf{F}_q)$ has the form $Q + nE(\mathbf{F}_q)$. However, we'll simply write $\langle P, Q\rangle_n$ and $\tau_n(P, Q)$.

PROOF The definition of the first pairing is as follows. Let $P \in E(\mathbf{F}_q)[n]$. Let D_P be a divisor of degree 0 such that $\text{sum}(D_P) = P$. This means that $D_P - [P] + [\infty]$ has degree 0 and sum equal to ∞, hence is the divisor of a function, by Theorem 11.2. Therefore, it would be equivalent to require D_P to be equivalent to $[P] - [\infty]$ mod principal divisors. By Theorem 11.2, there exists a function f such that

$$\text{div}(f) = nD_P.$$

Now let $D_Q = \sum_i a_i[Q_i]$ be a divisor of degree 0 such that $\text{sum}(D_Q) = Q$ and such that D_P and D_Q have no points in common. Define

$$\langle P, Q\rangle_n = f(D_Q) \quad (\text{mod } (\mathbf{F}_q^\times)^n),$$

where, for any function f whose divisor has no points in common with D_Q, we define

$$f(D_Q) = \prod_i f(Q_i)^{a_i}.$$

Note that once we have chosen D_P, the function f is determined up to a constant multiple. Since $0 = \deg(D_Q) = \sum_i a_i$, any such constant cancels out in the definition of the pairing.

Technical point: The function f is defined over $\overline{\mathbf{F}}_q$. However, the value of the pairing is claimed to be in $\mathbf{F}_q$. This can be proved as follows. Let $\sigma \in G = \text{Gal}(\overline{\mathbf{F}}_q/\mathbf{F}_q)$. We can take $D_P = [P] - [\infty]$, so $\sigma(D_P) = D_P$ in this case. Therefore, f^σ has the same divisor as f, so f^σ/f is a constant $c_\sigma \in \overline{\mathbf{F}}_q^\times$. The map $\sigma \mapsto c_\sigma$ represents a cocycle in the Galois cohomology group $H^1(G, \overline{\mathbf{F}}_q^\times)$. Hilbert's Theorem 90 says that this cohomology group is trivial, so there exists $c_1 \in \overline{\mathbf{F}}_q^\times$ such that $c_\sigma = c_1^\sigma/c_1$ for all $\sigma \in G$. This means that f/c_1 is fixed by all σ, hence is defined over $\mathbf{F}_q$. Since f/c_1 has the same divisor as f, we can use it in place of f, and hence assume f is defined over $\mathbf{F}_q$. Now let R be any point in $E(\mathbf{F}_q)$ and let $D_Q = [Q + R] - [R]$. Then $Q + R$ and R have coordinates in $\mathbf{F}_q$ and f is defined over $\mathbf{F}_q$, so

$$f(D_Q) = f(Q + R)/f(R) \in \mathbf{F}_q^\times.$$

There is one remaining technicality. What happens if $\#E(\mathbf{F}_q)$ is so small that we cannot choose the points $\infty, P, R, Q + R$ to be distinct? This situation can be treated by choosing appropriate points defined over extension fields and then using a little Galois theory. We omit the details.

We need to see what happens when we change the choice of D_P or D_Q. Suppose D_P' and D_Q' are divisors of degree 0 with sums P and Q. Then

$$D_P' = D_P + \text{div}(g), \quad D_Q' = D_Q + \text{div}(h),$$

for some functions g and h. For simplicity, we assume that g, h are defined over $\mathbf{F}_q$ and that all points in the divisors D_P, D'_Q are in $E(\mathbf{F}_q)$. (This is to obtain $h(D_P)^n, g(D'_Q)^n \in (\mathbf{F}_q^\times)^n$ below. The general case requires some Galois theory.) We have $\operatorname{div}(f') = nD'_P$ for some function f'. Since

$$\operatorname{div}(f') = \operatorname{div}(fg^n),$$

$f' = cfg^n$ for some constant c. Let's use f' and D'_Q to define a pairing, and denote it by $\langle\, \cdot\, ,\, \cdot\, \rangle'_n$. We obtain

$$\langle P, Q \rangle'_n = f'(D'_Q) = f(D'_Q)\, g(D'_Q)^n = f(D_Q)\, f(\operatorname{div}(h))\, g(D'_Q)^n.$$

Note that the constant c canceled out since $\deg(D'_Q) = 0$. We now need the following result, which is usually called **Weil reciprocity**.

LEMMA 11.9
Let f and h be two functions on E and suppose that $\operatorname{div}(f)$ and $\operatorname{div}(h)$ have no points in common. Then

$$f(\operatorname{div}(h)) = h(\operatorname{div}(f)).$$

For a proof, see [46, p. 427] or [90].

In our situation, Weil reciprocity yields

$$\begin{aligned}
\langle P, Q \rangle'_n &= f(D_Q)\, h(\operatorname{div}(f))\, g(D'_Q)^n \\
&= f(D_Q)\, h(D_P)^n\, g(D'_Q)^n \\
&\equiv \langle P, Q \rangle_n \quad (\operatorname{mod}\, (\mathbf{F}_q^\times)^n).
\end{aligned}$$

Therefore, the pairing is independent mod nth powers of the choice of D_P and D_Q.

If Q_1 and Q_2 are two points and D_{Q_1} and D_{Q_2} are corresponding divisors, then

$$D_{Q_1} + D_{Q_2} \sim [Q_1] - [\infty] + [Q_2] - [\infty] \sim [Q_1 + Q_2] - [\infty]$$

where $\sim$ denotes equivalence of divisors mod principal divisors. The last equivalence is the fact that the sum function in Corollary 11.4 is an homomorphism of groups. Consequently,

$$\langle P, Q_1 + Q_2 \rangle_n = f(D_{Q_1}) f(D_{Q_2}) = \langle P,\, Q_1 \rangle_n \langle P,\, Q_2 \rangle_n.$$

Therefore, the pairing is linear in the second variable.

If $P_1, P_2 \in E(\mathbf{F}_q)[n]$, and D_{P_1}, D_{P_2} are corresponding divisors and f_1, f_2 are the corresponding functions, then

$$D_{P_1} + D_{P_2} \sim [P_1] - [\infty] + [P_2] - [\infty] \sim [P_1 + P_2] - [\infty].$$

Therefore, we can let $D_{P_1+P_2} = D_{P_1} + D_{P_2}$. We have

$$\text{div}(f_1 f_2) = nD_{P_1} + nD_{P_2} - nD_{P_1+P_2},$$

so $f_1 f_2$ can be used to compute the pairing. Therefore,

$$\langle P_1 + P_2, Q \rangle_n = f_1(D_Q) f_2(D_Q) = \langle P_1, Q \rangle_n \langle P_2, Q \rangle_n.$$

Consequently, the pairing is linear in the first variable.

The nondegeneracy is much more difficult to prove. See [30] for details.

Since $\mathbf{F}_q^{\times}$ is cyclic of order $q - 1$, the $(q - 1)/n$th power map gives an isomorphism

$$\mathbf{F}_q^{\times}/(\mathbf{F}_q^{\times})^n \longrightarrow \mu_n.$$

Therefore, define

$$\tau_n(P, Q) = \langle P, Q \rangle_n^{(q-1)/n}.$$

The desired properties of the modified Tate-Lichtenbaum pairing τ_n follow immediately from those of the Tate-Lichtenbaum pairing. ∎

11.4 Computation of the Pairings

In Section 11.1, we showed how to express a divisor of degree 0 and sum ∞ as a divisor of a function. This method suffices to compute the Weil and Tate-Lichtenbaum pairings for small examples. However, for larger examples, a little care is needed to avoid massive calculations. Also, the definition given for the Weil pairing involves a function g whose divisor includes contributions from all of the n^2 points in $E[n]$. When n is large, this can cause computational difficulties. The following result gives an alternate definition of the Weil pairing e_n.

THEOREM 11.10

Let $S, T \in E[n]$. Let D_S and D_T be divisors of degree 0 such that

$$\text{sum}(D_S) = S \quad and \quad \text{sum}(D_T) = T$$

and such that D_S and D_T have no points in common. Let f_S and f_T be functions such that

$$\text{div}(f_S) = nD_S \quad and \quad \text{div}(f_T) = nD_T.$$

Then the Weil pairing is given by

$$e_n(S, T) = \frac{f_T(D_S)}{f_S(D_T)}.$$

(Recall that $f(\sum a_i[P_i]) = \prod_i f(P_i)^{a_i}$.)

For a proof, see [54, Section 6.4] or [38].
A natural choice of divisors is

$$D_S = [S] - [\infty], \quad D_T = [T + R] - [R]$$

for some randomly chosen point R. Then we have

$$e_n(S, T) = \frac{f_S(R)f_T(S)}{f_S(T + R)f_T(\infty)}.$$

Example 11.5
Let E be the elliptic curve over $\mathbf{F}_7$ defined by

$$y^2 = x^3 + 2.$$

Then

$$E(\mathbf{F}_7)[3] \simeq \mathbf{Z}_3 \oplus \mathbf{Z}_3.$$

In fact, this is all of $E(\mathbf{F}_7)$. Let's compute

$$e_3((0, 3), (5, 1)).$$

Let

$$D_{(0,3)} = [(0, 3)] - [\infty], \quad D_{(5,1)} = [(3, 6)] - [(6, 1)].$$

The second divisor was obtained by adding $R = (6, 1)$ to $(5, 1)$ to obtain
$(3, 6) = (5, 1) + (6, 1)$. A calculation (see Section 11.1) shows that

$$\operatorname{div}(y - 3) = 3D_{(0,3)}, \quad \operatorname{div}\left(\frac{4x - y + 1}{5x - y - 1}\right) = 3D_{(5,1)}.$$

Therefore, we take

$$f_{(0,3)} = y - 3, \quad f_{(5,1)} = \frac{4x - y + 1}{5x - y - 1}.$$

We have

$$f_{(0,3)}\left(D_{(5,1)}\right) = \frac{f_{(0,3)}(3, 6)}{f_{(0,3)}(6, 1)} = \frac{6 - 3}{1 - 3} \equiv 2 \quad (\bmod\ 7).$$

Similarly,

$$f_{(5,1)}(D_{(0,3)}) = 4$$

(to evaluate $f_{(5,1)}(\infty)$, see below). Therefore,

$$e_3((0, 3), (5, 1)) = \frac{4}{2} \equiv 2 \quad (\bmod\ 7).$$

The number 2 is a cube root of unity, since $2^3 \equiv 1 \pmod 7$.

There are several ways to evaluate $f_{(5,1)}(\infty)$. The intuitive way is to observe that y has a pole of order 3 at ∞ while x has a pole of order 2. Therefore, the terms $-y$ in the numerator and denominator dominate as $(x, y) \to \infty$, so the ratio goes to 1. Another way is to change to homogeneous form and use projective coordinates:

$$f_{(5,1)}(x : y : z) = \frac{4x - y + z}{5x - y + z}.$$

Then

$$f_{(5,1)}(\infty) = f_{(5,1)}(0 : 1 : 0) = 1.$$

$\square$

The Tate-Lichtenbaum pairing can be calculated as

$$\langle P, Q \rangle_n = \frac{f(Q + R)}{f(R)}$$

for appropriate f (depending on P) and R. We can express the Weil pairing in terms of this pairing:

$$e_n(S, T) = \frac{\langle T, S \rangle_n}{\langle S, T \rangle_n},$$

ignoring the ambiguities (i.e., up to nth powers) in the definition of the terms on the right side, since they cancel out.

Therefore, we see that computing the Weil pairing and computing the Tate-Lichtenbaum pairing both reduce to finding a function f with

$$\mathrm{div}(f) = n[P + R] - n[R]$$

for points $P \in E[n]$ and $R \in E$ and evaluating $f(Q_1)/f(Q_2)$ for points Q_1, Q_2. The following algorithm due to Victor Miller shows how to do this efficiently.

The idea is to use successive doubling (see page 18) to get to n. But the divisors $j[P + R] - j[R]$ for $j < n$ are not divisors of functions, so we introduce the divisors

$$D_j = j[P + R] - j[R] - [jP] + [\infty]. \tag{11.7}$$

Then D_j is the divisor of a function, by Theorem 11.2:

$$\mathrm{div}(f_j) = D_j. \tag{11.8}$$

Suppose we have computed $f_j(Q_1)/f_j(Q_2)$ and $f_k(Q_1)/f_k(Q_2)$. We show how to compute $f_{j+k}(Q_1)/f_{j+k}(Q_2)$. Let

$$ax + by + c = 0$$

be the line through jP and kP (the tangent line if $jP = kP$), and let $x+d = 0$ be the vertical line through $(j + k)P$. Then (see the proof of Theorem 11.2),

$$\operatorname{div}\left(\frac{ax + by + c}{x + d}\right) = [jP] + [kP] - [(j + k)P] - [\infty].$$

Therefore,

$$\operatorname{div}(f_{j+k}) = D_{j+k} = D_j + D_k + \operatorname{div}\left(\frac{ax + by + c}{x + d}\right)$$
$$= \operatorname{div}\left(f_j f_k \frac{ax + by + c}{x + d}\right).$$

This means that there exists a constant γ such that

$$f_{j+k} = \gamma f_j f_k \frac{ax + by + c}{x + d}.$$

Therefore,

$$\frac{f_{j+k}(Q_1)}{f_{j+k}(Q_2)} = \frac{f_j(Q_1)}{f_j(Q_2)} \frac{f_k(Q_1)}{f_k(Q_2)} \frac{(ax + by + c)/(x + d)|_{(x,y)=Q_1}}{(ax + by + c)/(x + d)|_{(x,y)=Q_2}}. \qquad (11.9)$$

We conclude that passing from f_j and f_k to f_{j+k} can be done quite quickly.

For example, this means that if we know $f_j(Q_1)/f_j(Q_2)$ for $j = 2^i$, we can quickly calculate the same expression for $j = 2^{i+1}$. Also, once we have computed some of these, we can combine them to obtain the values when j is a sum of powers of 2. This is what happens when we do successive doubling to reach n. Therefore, we can compute $f_n(Q_1)/f_n(Q_2)$ quickly. But

$$\operatorname{div}(f_n) = n[P + R] - n[R] - [nP] + [\infty] = n[P + R] - n[R],$$

since $nP = \infty$. Therefore, f_n is the function f whose values we are trying to compute, so we have completed the calculation.

The above method can be described in algorithmic form as follows. Let $P \in E[n]$ and let R, Q_1, Q_2 be points on E. Let f_j be as in (11.8). Define

$$v_j = f_j(Q_1)/f_j(Q_2)$$

to be the value of f_j at the divisor $[Q_1] - [Q_2]$.

1. Start with $i = n$, $j = 0$, $k = 1$. Let $f_0 = 1$ and compute f_1 with divisor $[P + R] - [P] - [R] + [\infty]$.

2. If i is even, let $i = i/2$ and compute $v_{2k} = f_{2k}(Q_1)/f_{2k}(Q_2)$ in terms of v_k, using (11.9). Then change k to $2k$, but do not change j. Save the pair (v_j, v_k) for the new value of k.

3. If i is odd, let $i = i - 1$, and compute $v_{j+k} = f_{j+k}(Q_1)/f_{j+k}(Q_2)$ in terms of v_j and v_k, using (11.9). Then change j to $j + k$, but do not change k. Save the pair (v_j, v_k) for the new value of j.

4. If $i \neq 0$, go to step 2.

5. Output v_j.

The output will be $v_n = f_n(Q_1)/f_n(Q_2)$ (this can be proved by induction).

Example 11.6

Suppose we want to calculate v_{13}. At the end of each computation, we have the following values of $i, j, k, (v_j, v_k)$:

1. $i = 13$, $j = 0$, $k = 1$, (v_0, v_1)

2. $i = 12$, $j = 1$, $k = 1$, (v_1, v_1)

3. $i = 6$, $j = 1$, $k = 2$, (v_1, v_2)

4. $i = 3$, $j = 1$, $k = 4$, (v_1, v_4)

5. $i = 2$, $j = 5$, $k = 4$, (v_5, v_4)

6. $i = 1$, $j = 5$, $k = 8$, (v_5, v_8)

7. $i = 0$, $j = 13$, $k = 8$, (v_{13}, v_8) ◻

Example 11.7

Let E be the elliptic curve

$$y^2 = x^3 - x + 1$$

over $\mathbf{F}_{11}$, and let $n = 5$. There are 10 points in $E(\mathbf{F}_{11})$. The point $P = (3, 6)$ has order 5. Let's compute $\langle P, P \rangle_5$. Therefore, in the definition of the Tate-Lichtenbaum pairing, we have $P = Q = (3, 6)$. Let

$$D_P = [(3, 6)] - [\infty], \quad D_Q = [(1, 1)] - [(0, 1)] = [Q_1] - [Q_2].$$

The divisor D_Q was constructed by adding $(0, 1)$ to Q to obtain $(1, 1)$. This was done so that D_P and D_Q have no points in common. We now use the algorithm to compute $f_P(D_Q)$, where

$$\text{div}(f_P) = 5D_P.$$

In Equation (11.7), we have $R = \infty$, so $D_0 = D_1 = 0$. Therefore, we take $f_0 = f_1 = 1$. The algorithm proceeds as follows.

1. Start with $i = 5, j = 0, k = 1, v_0 = 1, v_1 = 1$.

2. Since $i = 5$ is odd, compute $v_{j+k} = v_1$, which is already known to be 1. Update the values of i, j, k to obtain $i = 4, j = 1, k = 1, v_1 = 1, v_1 = 1$.

3. Since $i = 4$ is even, compute the line tangent to E at $kP = P$. This is $4x - y + 5 = 0$. The vertical line through $2kP = 2P$ is $x + 1 = 0$. Therefore, Equation (11.9) becomes

$$v_2 = v_1^2 \cdot \left.\frac{4x - y + 5}{x + 1}\right|_{D_Q} = 1 \cdot 1 = 1.$$

Here we performed the calculation

$$(4x - y + 5)|_{D_Q} = \frac{(4x - y + 5)|_{(1,1)}}{(4x - y + 5)|_{(0,1)}} = \frac{8}{4} = 2$$

and similarly $(x + 1)|_{D_Q} = 2$. Update to obtain $i = 2, j = 1, k = 2, v_1 = 1, v_2 = 1$.

4. Since $i = 2$ is even, use the computation of $4P = 2P + 2P$ to obtain

$$v_4 = v_2 \cdot v_2 \cdot \left.\frac{x + y + 2}{x - 3}\right|_{D_Q} = 1 \cdot 1 \cdot 2 = 2.$$

Update to obtain $i = 1, j = 1, k = 4, v_1 = 1, v_4 = 2$.

5. Since $i = 1$ is odd, use the computation of $5P = P + 4P$ to obtain

$$v_5 = v_1 \cdot v_4 \cdot (x - 3)|_{D_Q} = 1 \cdot 2 \cdot (2/3) \equiv 5 \pmod{11}.$$

Therefore, the Tate-Lichtenbaum pairing of P with P is

$$\langle P, P \rangle_5 = v_5 = 5 \pmod{(\mathbf{F}_{11}^\times)^5},$$

and the modified Tate-Lichtenbaum pairing is

$$\tau_5(P, P) = \langle P, P \rangle_5^{(11-1)/5} \equiv 3 \pmod{11}.$$

Note that, in contrast to the Weil pairing, the Tate-Lichtenbaum pairing of a point with itself can be nontrivial. ☐

11.5 Genus One Curves and Elliptic Curves

Let C be a nonsingular algebraic curve defined over a field K. The curve C is given as the roots in $\mathbf{P}_K^2$ of a polynomial, or as the intersection of surfaces in $\mathbf{P}^3_{\overline{K}}$, for example, and is assumed not to be the union of two smaller such curves. We can define divisors and divisors of functions on C in the same way as we did for elliptic curves. Let

$$D_1 = \sum a_i[P_i], \quad D_2 = \sum b_i[P_i]$$

be divisors on C. We say that

$$D_1 \geq D_2 \Longleftrightarrow a_i \geq b_i \quad \text{for all } i.$$

We say that

$$D_1 \sim D_2 \Longleftrightarrow D_1 - D_2 = \text{div}(f) \quad \text{for some function } f.$$

For a divisor D, define

$$\mathcal{L}(D) = \{\text{functions } f \mid \text{div}(f) + D \geq 0\} \cup \{0\}.$$

Then $\mathcal{L}(D)$ is a vector space over $\overline{K}$. Define

$$\ell(D) = \dim \mathcal{L}(D).$$

For example, let $D = 3[P] - 2[Q]$. A function f in the linear space $\mathcal{L}(D)$ has at most a triple pole at P and at least a double zero at Q. Also, f cannot have any poles other than at P, but it can have zeros other than at Q.

PROPOSITION 11.11

Let C be a nonsingular algebraic curve defined over a field K, and let D, D_1, and D_2 be divisors on C.

1. *If $\deg D < 0$, then $\mathcal{L}(D) = 0$.*

2. *If $D_1 \sim D_2$ then $\mathcal{L}(D_1) \simeq \mathcal{L}(D_2)$.*

3. *$\mathcal{L}(0) = \overline{K}$.*

4. *$\ell(D) < \infty$.*

5. *If $\deg(D) = 0$ then $\ell(D) = 0$ or 1.*

PROOF   Proposition 11.1 holds for all curves (see [28]), not just elliptic curves, and we'll use it in this more general context throughout the present proof. For example, we need that $\deg(\text{div}(f)) = 0$ for functions f that are not identically 0.

If $\mathcal{L} \neq 0$, then there exists $f \neq 0$ with

$$\text{div}(f) + D \geq 0,$$

which implies that

$$\deg(D) = \deg(\operatorname{div}(f) + D) \geq 0.$$

This proves (1).

If $D_1 \sim D_2$, then $D_1 = D_2 + \operatorname{div}(g)$ for some g. The map

$$\mathcal{L}(D_1) \to \mathcal{L}(D_2)$$
$$f \mapsto fg$$

is easily seen to be an isomorphism. This proves (2).

If $0 \neq f \in \mathcal{L}(0)$, then $\operatorname{div}(f) \geq 0$. Since $\deg(\operatorname{div}(f)) = 0$, we must have $\operatorname{div}(f) = 0$, which means that f has no zeros or poles. The analogue of Proposition 11.1 says that f must be a constant. Therefore,

$$\mathcal{L}(0) = \overline{K}$$

and $\ell(0) = 1$. This proves (3) and also proves (4) for $D = 0$.

We can get from 0 to an arbitrary divisor by adding or subtracting one point at a time. We'll show that each such modification changes the dimension by at most one. Therefore, the end result will be a finite dimensional vector space.

Suppose that D_1, D_2 are two divisors with $D_2 = D_1 + [P]$ for some point P. Then

$$\mathcal{L}(D_1) \subseteq \mathcal{L}(D_2).$$

Suppose there exist $g, h \in \mathcal{L}(D_2)$ with $g, h \notin \mathcal{L}(D_1)$. Let $-n$ be the coefficient of $[P]$ in D_2. Then both g and h must have order n at P. (The order of g must be at least n. If it is larger, then $g \in \mathcal{L}(D_1)$. Similarly for h.) Let u be a uniformizer at P. Write

$$g = u^n g_1, \quad h = u^n h_1$$

with $g_1(P) = c \neq 0, \infty$ and $h_1(P) = d \neq 0, \infty$. Then

$$dg - ch = u^n(dg_1 - ch_1),$$

and $(dg_1 - ch_1)(P) = 0$. Therefore, $dg - ch$ has order greater than n at P, so

$$dg - ch \in \mathcal{L}(D_1).$$

Therefore any two such elements $g, h \in \mathcal{L}(D_2)$ are linearly dependent mod $\mathcal{L}(D_1)$. It follows that

$$\ell(D_1) \leq \ell(D_2) \leq \ell(D_1) + 1.$$

As pointed out above, this implies (4).

To prove (5), assume $\deg(D) = 0$. If $\mathcal{L}(D) = 0$, we're done. Otherwise, there exists $0 \neq f \in \mathcal{L}(D)$. Then

$$\operatorname{div}(f) + D \geq 0 \quad \text{and} \quad \deg(\operatorname{div}(f) + D) = 0 + 0 = 0.$$

Therefore,

$$\operatorname{div}(f) + D = 0.$$

Since $D \sim \operatorname{div}(f) + D = 0$, we have

$$\mathcal{L}(D) \simeq \mathcal{L}(0) = \overline{K},$$

by (2) and (3). Therefore, $\ell(D) = 1$. This proves (5). ∎

A very fundamental result concerning divisors is the following.

THEOREM 11.12 (Riemann-Roch)
*Given an algebraic curve C, there exists an integer g (called the **genus** of C) and a divisor $\mathcal{K}$ (called a **canonical divisor**) such that*

$$\ell(D) - \ell(\mathcal{K} - D) = \deg(D) - g + 1$$

for all divisors D.

For a proof, see [31] or [36]. The divisor $\mathcal{K}$ is the divisor of a differential on C.

COROLLARY 11.13

$$\deg(\mathcal{K}) = 2g - 2.$$

PROOF   Letting $D = 0$ and $D = \mathcal{K}$ in the Riemann-Roch theorem, then using (3) in Proposition 11.11, yields

$$\ell(\mathcal{K}) = g, \quad \text{and} \quad \ell(\mathcal{K}) = \deg(\mathcal{K}) - g + 2.$$

Therefore,

$$\deg(\mathcal{K}) = 2g - 2,$$

as desired. ∎

COROLLARY 11.14
If $\deg(D) > 2g - 2$, then $\ell(D) = \deg(D) - g + 1$.

PROOF   Since $\deg(\mathcal{K} - D) < 0$, Proposition 11.11 (1) says that $\ell(\mathcal{K} - D) = 0$. The Riemann-Roch theorem therefore yields the result. ∎

COROLLARY 11.15

Let P, Q be points on C. If $g \geq 1$ and $[P] - [Q] \sim 0$, then $P = Q$.

PROOF   By assumption, $[P] - [Q] = \mathrm{div}(f)$ for some f. Assume $[P] \neq [Q]$. Since f^n has a pole of order n at Q, and since functions with different orders of poles at Q are linearly independent, the set

$$\{1, f, f^2, \ldots, f^{2g-1}\}$$

spans a subspace of $\mathcal{L}((2g-1)[Q])$ of dimension $2g$. Therefore,

$$2g \leq \ell((2g-1)[Q]) = (2g-1) - g + 1 = g,$$

by Corollary 11.14. Since $g \geq 1$, this is a contradiction. Therefore, $P = Q$.
∎

Our goal is to show that a curve C of genus one is isomorphic over $\overline{K}$ to an elliptic curve given by a generalized Weierstrass equation. The following will be used to construct the functions needed to map from C to the elliptic curve.

COROLLARY 11.16

If C has genus $g = 1$ and $\deg(D) > 0$, then

$$\ell(D) = \deg(D).$$

PROOF   This is simply a restatement of Corollary 11.14 in the case $g = 1$.
∎

Choose a point $P \in C(\overline{K})$. If $P \in C(K)$, then it is possible to perform the following construction using only numbers from K rather than from $\overline{K}$. This corresponds to the situation in Chapter 2, where we used rational points to put certain curves into Weierstrass form. However, we'll content ourselves with working over $\overline{K}$.

Corollary 11.16 says that

$$\ell(n[P]) = n \quad \text{for all } n \geq 1.$$

Since $\overline{K} \subseteq \mathcal{L}([P])$, which has dimension 1, we have

$$\mathcal{L}([P]) = \overline{K}.$$

Since $\ell(2[P]) = 2 > \ell([P])$, there exists a function $f \in \mathcal{L}(2[P])$ having a double pole at P and no other poles. Since $\ell(3[P]) = 3 > \ell(2[P])$, there exists a function $g \in \mathcal{L}(3[P])$ with a triple pole at P and no other poles. Since functions with different order poles at P are linearly independent, we can use f and g to give bases for several of the spaces $\mathcal{L}(n[P])$:

$$\mathcal{L}([P]) = \mathrm{span}(1)$$
$$\mathcal{L}(2[P]) = \mathrm{span}(1, f)$$
$$\mathcal{L}(3[P]) = \mathrm{span}(1, f, g)$$
$$\mathcal{L}(4[P]) = \mathrm{span}(1, f, g, f^2)$$
$$\mathcal{L}(5[P]) - \mathrm{span}(1, f, g, f^2, fg).$$

We can write down 7 functions in the 6-dimensional space $\mathcal{L}(6[P])$, namely

$$1, f, g, f^2, fg, f^3, g^2.$$

These must be linearly dependent, so there exist $a_0, a_1, a_2, a_3, a_4, a_6 \in \overline{K}$ with

$$g^2 + a_1 fg + a_3 g = a_0 f^3 + a_2 f^2 + a_4 f + a_6. \tag{11.10}$$

Note that the coefficient of g^2 must be nonzero, hence can be assumed to be 1, since the remaining functions have distinct orders of poles at P and are therefore linearly independent. Similarly, $a_0 \neq 0$. By multiplying f by a suitable constant, we may assume that

$$a_0 = 1.$$

Let E be the elliptic curve defined by

$$y^2 + a_1 xy + a_3 y = x^3 + a_2 x^2 + a_4 x + a_6.$$

We have a map

$$\psi : C(\overline{K}) \to E(\overline{K})$$
$$Q \mapsto (f(Q), g(Q))$$
$$P \mapsto \infty.$$

PROPOSITION 11.17
ψ is a bijection.

PROOF   Suppose $Q_1 \neq Q_2$ are such that $\psi(Q_1) = \psi(Q_2)$, hence

$$f(Q_1) = f(Q_2) = a \quad \text{and} \quad g(Q_1) = g(Q_2) = b$$

for some a, b. Since $f - a$ has a double pole at P and $g - b$ has a triple pole at P,

$$\mathrm{div}(f - a) = [Q_1] + [Q_2] - 2[P]$$
$$\mathrm{div}(g - b) = [Q_1] + [Q_2] + [R] - 3[P]$$

for some R. Subtracting yields

$$[R] - [P] = \operatorname{div}((g - b)/(f - a)) \sim 0.$$

By Corollary 11.15, this means that $R = P$. Therefore,

$$\operatorname{div}(g - b) = [Q_1] + [Q_2] - 2[P],$$

so g has only a double pole at P. This contradiction proves that ψ is an injection.

To prove surjectivity, let $(a, b) \in E(\overline{K})$. We want to find P with $\psi(P) = (a, b)$. Since $f - a$ has a double pole at P and since the divisor of a function has degree 0, there are (not necessarily distinct) points $Q_1, Q_2 \in C(\overline{K})$ such that

$$\operatorname{div}(f - a) = [Q_1] + [Q_2] - 2[P].$$

For a given x-coordinate a, there are two possible y-coordinates b and b' for points on E. If $g(Q_i) = b$ for some $i = 1, 2$, we have $\psi(Q_i) = (a, b)$ and we're done. Therefore, suppose

$$g(Q_1) = g(Q_2) = b'.$$

Then $\psi(Q_1) = \psi(Q_2) = (a, b')$. Since ψ is injective, $Q_1 = Q_2$, so

$$\operatorname{div}(f - a) = 2[Q_1] - 2[P].$$

Let u be a uniformizing parameter at Q_1. Then

$$f - a = u^2 f_1, \quad g - b' = u g_1$$

with $f_1(Q_1) \neq 0, \infty$ and $g_1(Q_1) \neq \infty$ (possibly $g_1(Q_1) = 0$). Substituting into (11.10) and using the fact that $(a, b') \in E$ yields

$$(u g_1)(2b' + a_1 a + a_3) = u^2 h$$

for some function h. Dividing by u and evaluating at Q_1 shows that

$$g_1(Q_1) = 0 \quad \text{or} \quad 2b' + a_1 a + a_3 = 0.$$

If $g_1(Q_1) = 0$, then $g - b'$ has at least a double root at Q_1, so

$$\operatorname{div}(g - b') = 2[Q_1] + [R] - 3[P]$$

for some R. Therefore,

$$\operatorname{div}((g - b')/(f - a)) = [R] - [P].$$

By Corollary 11.15, $R = P$. This means that $g - b'$ has only a double pole at P, which is a contradiction. Therefore, $g_1(Q_1) \neq 0$, so

$$0 = 2b' + a_1 a + a_3 = \frac{\partial}{\partial y}(y^2 + a_1 a y + a_3 y - a^3 - a_2 a^2 - a_4 a - a_6)\Big|_{y=b'}.$$

This means that b' is a double root, so $b = b'$. Therefore, $\psi(Q_1) = (a, b') = (a, b)$. Therefore, ψ is surjective. ■

It is possible to show that not only ψ, but also ψ^{-1}, is given by rational functions. See [90, p. 64]. Since C is assumed to be nonsingular, this implies that the equation for E is nonsingular, so E is actually an elliptic curve.

It is also possible to show that elliptic curves always have genus one. Therefore, over algebraically closed fields, genus one curves, with a base point P specified, are the same as elliptic curves, with P being the origin for the group law. Over non-algebraically closed fields, the situation is more complicated. A genus one curve C such that $C(K)$ is non-empty is an elliptic curve, but there are genus one curves C such that $C(K)$ is empty (see Section 8.8). These curves are not elliptic curves over K, but become elliptic curves over certain extensions of K.

Exercises

11.1 Let E be the elliptic curve $y^2 = x^3 - x$ over Q.

(a) Show that $f(x, y) = (y^4 + 1)/(x^2 + 1)^3$ has no zeros or poles in $E(\mathbf{Q})$.

(b) Show that $g(x, y) = y^4/(x^2 + 1)^3$ has no poles in $E(\mathbf{Q})$ but does have zeros in $E(\mathbf{Q})$.

(c) Find the divisors of f and g (over $\overline{\mathbf{Q}}$).

11.2 Let E be an elliptic curve over a field K and let m, n be positive integers that are not divisible by the characteristic of K. Let $S \in E[mn]$ and $T \in E[n]$. Show that

$$e_{mn}(S, T) = e_n(mS, T).$$

11.3 Suppose f is a function on an algebraic curve C such that $\operatorname{div}(f) = [P] - [Q]$ for points P and Q. Show that f gives a bijection of C with $\mathbf{P}^1$.

11.4 Show that part (3) of Proposition 11.1 follows from part (2). (*Hint:* Let P_0 be any point and look at the function $f - f(P_0)$.)

Chapter 12

Zeta Functions

12.1 Elliptic Curves over Finite Fields

Let E be an elliptic curve over a finite field $\mathbf{F}_q$. Let

$$N_n = \#E(\mathbf{F}_{q^n})$$

be the number of points on E over the field $\mathbf{F}_{q^n}$. The Z-function of E is defined to be

$$Z_E(T) = \exp\left(\sum_{n=1}^{\infty} \frac{N_n}{n} T^n\right).$$

Here $\exp(t) = \sum t^n/n!$ is the usual exponential function. The Z-function encodes certain arithmetic information about E as the coefficients of a generating function. The presence of the exponential function is justified by the simple form for $Z_E(T)$ in the following result.

PROPOSITION 12.1

Let E be an elliptic curve defined over $\mathbf{F}_q$, and let $\#E(\mathbf{F}_q) = q+1-a$. Then

$$Z_E(T) = \frac{qT^2 - aT + 1}{(1-T)(1-qT)}.$$

PROOF   Factor $X^2 - aX + q = (X - \alpha)(X - \beta)$. Theorem 4.12 says that

$$N_n = q^n + 1 - \alpha^n - \beta^n.$$

Therefore, using the expansion $-\log(1-t) = \sum t^n/n$, we have

$$
\begin{aligned}
Z_E(T) &= \exp\left(\sum_{n=1}^{\infty} \frac{N_n}{n} T^n\right) \\
&= \exp\left(\sum_{n=1}^{\infty} (q^n + 1 - \alpha^n - \beta^n)\frac{T^n}{n}\right) \\
&= \exp\left(-\log(1-qT) - \log(1-T) + \log(1-\alpha T) + \log(1-\beta T)\right) \\
&= \frac{(1-\alpha T)(1-\beta T)}{(1-T)(1-qT)} \\
&= \frac{qT^2 - aT + 1}{(1-T)(1-qT)}.
\end{aligned}
$$

∎

Note that the numerator of $Z_E(T)$ is the characteristic polynomial of the Frobenius endomorphism, as in Chapter 4, with the coefficients in reverse order.

A function $Z_C(T)$ can be defined in a similar way for any curve C over a finite field, and, more generally, for any variety over a finite field. It is always a rational function (proved by E. Artin and F. K. Schmidt for curves and by Dwork for varieties).

The **zeta function** of E is defined to be

$$\zeta_E(s) = Z_E(q^{-s}),$$

where s is a complex variable. As we'll see below, $\zeta_E(s)$ can be regarded as an analogue of the classical Riemann zeta function

$$\zeta(s) = \sum_{n=1}^{\infty} \frac{1}{n^s}.$$

One of the important properties of the Riemann zeta function is that it satisfies a functional equation relating the values at s and $1 - s$:

$$\pi^{-s/2}\Gamma(s/2)\zeta(s) = \pi^{-(1-s)/2}\Gamma((1-s)/2)\zeta(1-s).$$

A famous conjecture for $\zeta(s)$ is the Riemann Hypothesis, which predicts that if $\zeta(s) = 0$ with $0 \le \Re(s) \le 1$ then $\Re(s) = 1/2$ (there are also the "trivial" zeros at the negative even integers). The elliptic curve zeta function $\zeta_E(s)$ also satisfies a functional equation, and the analogue of the Riemann Hypothesis holds.

THEOREM 12.2
Let E be an elliptic curve defined over a finite field.

1. $\zeta_E(s) = \zeta_E(1 - s)$

2. *If $\zeta_E(s) = 0$, then $\Re(s) = 1/2$.*

PROOF   The proof of the first statement follows easily from Proposition 12.1:

$$\zeta_E(s) = \frac{q^{1-2s} - aq^{-s} + 1}{(1 - q^{-s})(1 - q^{1-s})}$$

$$= \frac{1 - aq^{s-1} + q^{-1+2s}}{(q^s - 1)(q^{s-1} - 1)}$$

$$= \zeta_E(1 - s).$$

Since the numerator of $Z_E(T)$ is $(1 - \alpha T)(1 - \beta T)$, we have

$$\zeta_E(s) = 0 \Longleftrightarrow q^s = \alpha \text{ or } \beta.$$

By the quadratic formula,

$$\alpha, \beta = \frac{a \pm \sqrt{a^2 - 4q}}{2}.$$

Hasse's theorem (Theorem 4.2) says that

$$|a| \leq 2\sqrt{q},$$

hence $a^2 - 4q \leq 0$. Therefore, α and β are complex conjugates of each other, and

$$|\alpha| = |\beta| = \sqrt{q}.$$

If $q^s = \alpha$ or β, then

$$q^{\Re(s)} = |q^s| = \sqrt{q}.$$

Therefore, $\Re(s) = 1/2$.   ∎

There are infinitely many solutions to $q^s = \alpha$. However, if s_0 is one such solution, all others are of the form $s_0 + 2\pi i n / \log q$ with $n \in \mathbf{Z}$. A similar situation holds for β.

If C is a curve, or a variety, over a finite field, then an analogue of Theorem 12.2 holds. For curves, the functional equation was proved by E. Artin and F. K. Schmidt, and the Riemann Hypothesis was proved by Weil in the 1940s. In 1949, Weil announced what became known as the Weil conjectures, which predicted that analogues of Proposition 12.1 and Theorem 12.2 hold for varieties over finite fields. The functional equation was proved in the 1960s by M. Artin, Grothendieck, and Verdier, and the analogue of the Riemann Hypothesis was proved by Deligne in 1973. Much of Grothendieck's algebraic geometry was developed for the purpose of proving these conjectures.

Finally, we show how $\zeta_E(s)$ can be defined in a way similar to the Riemann zeta function. Recall that the Riemann zeta function has the Euler product expansion

$$\zeta(s) = \prod_p \left(1 - \frac{1}{p^s}\right)$$

when $\Re(s) > 1$. The product is over the prime numbers. We obtain $\zeta_E(s)$ if we replace the primes p by points on E. Consider a point $P \in E(\overline{\mathbf{F}}_q)$. Define $\deg(P)$ to be the smallest n such that $P \in E(\mathbf{F}_{q^n})$. The Frobenius map ϕ_q acts on P, and it is not difficult to show that the set

$$S_P = \{P, \phi_q(P), \phi_q^2(P), \ldots, \phi_q^{n-1}(P)\}$$

has exactly $n = \deg(P)$ elements and that $\phi_q^n(P) = P$. Each of the points in S_P also has degree n.

PROPOSITION 12.3

Let E be an elliptic curve over $\mathbf{F}_q$. Then

$$\zeta_E(s) = \prod_{S_P} \left(1 - \frac{1}{q^{s\deg(P)}}\right)^{-1},$$

where the product is over the points $P \in E(\overline{\mathbf{F}}_q)$, but we take only one point from each set S_P.

PROOF   If $\deg(P) = m$, then P and all the other points in S_P have coordinates in $\mathbf{F}_{q^m}$. Since $\mathbf{F}_{q^m} \subseteq \mathbf{F}_{q^n}$ if and only if $m|n$, we see that S_P contributes m points to $N_n = \#E(\mathbf{F}_{q^n})$ if and only if $m|n$, and otherwise it contributes no points to N_n. Therefore,

$$N_n = \sum_{m|n} \sum_{\substack{S_P \\ \deg(P)=m}} m.$$

Substituting this into the definition of $Z(T)$, we obtain

$$\log Z(T) = \sum_{n=1}^{\infty} \frac{N_n}{n} T^n$$

$$= \sum_{n=1}^{\infty} \frac{1}{n} T^n \sum_{m|n} \sum_{\substack{S_P \\ \deg(P)=m}} m$$

$$= \sum_{j=1}^{\infty} \sum_{m=1}^{\infty} \frac{1}{mj} \sum_{\substack{S_P \\ \deg(P)=m}} m T^{mj} \quad \text{(where } mj = n\text{)}$$

$$= \sum_{j=1}^{\infty} \sum_{S_P} \frac{1}{j} T^j \deg(P)$$

$$= -\sum_{S_P} \log(1 - T^{\deg(P)}).$$

Let $T = q^{-s}$ and exponentiate to obtain the result.   ∎

12.2   Elliptic Curves over Q

Let E be an elliptic curve defined over $\mathbf{Q}$. By changing variables if necessary, we may assume that E is defined by $y^2 = x^3 + Ax + B$ with $A, B \in \mathbf{Z}$. For a prime p, we can reduce the equation $y^2 = x^3 + Ax + B \bmod p$. If $E \bmod p$ is an elliptic curve, then we say that E has **good reduction** mod p. This happens for all but finitely many primes. For each such p, we have

$$\#E(\mathbf{F}_p) = p + 1 - a_p,$$

as in Section 12.1. The **L-function** of E is defined to be approximately the Euler product

$$\prod_{\text{good } p} \left(1 - a_p p^{-s} + p^{1-2s}\right)^{-1}.$$

This definition is good enough for many purposes. However, for completeness, we say a few words below about what happens at the primes of bad reduction. The factor $1 - a_p p^{-s} + p^{1-2s}$ perhaps seems to be rather artificially constructed. However, it is just the numerator of the zeta function for $E \bmod p$, as in Section 12.1. It might seem more natural to use the whole mod p zeta function, but the factors arising from the denominator yield the Riemann zeta function (with a few factors removed) evaluated at s and at $s + 1$. Since the presence of the zeta function would complicate matters, the denominators are omitted in the definition of $L_E(s)$.

For the primes where there is bad reduction, the cubic $x^3 + Ax + B$ has multiple roots mod p. If it has a triple root, we say that E has **additive reduction** mod p. If it has a double root mod p, it has **multiplicative reduction**. Moreover, if the slopes of the tangent lines at the singular point (see Theorem 2.30) are in $\mathbf{F}_p$, we say that E has **split multiplicative reduction** mod p. Otherwise, it has **non-split multiplicative reduction**.

To treat the primes $p = 2$ and $p = 3$, we need to use the general Weierstrass form for E. For simplicity, we have ignored these primes in the preceding discussion. However, in the example below, we'll include them.

There are many possible equations for E with $A, B \in \mathbf{Z}$. We assume that A, B are chosen so that the reduction properties of E are as good as possible. In other words, we assume that A and B are chosen so that the cubic has the largest obtainable number of distinct roots mod p, and the power of p is the discriminant $4A^3 + 27B^2$ is as small as possible, for each p. It can be shown that there is such a choice of A, B. Such an equation is called a **minimal Weierstrass equation** for E.

Example 12.1

Suppose we start with E given by the equation

$$y^2 = x^3 - 270000x + 128250000.$$

The discriminant of the cubic is $-2^8 3^{12} 5^{12} 11$, so E has good reduction except possibly at $2, 3, 5, 11$. The change of variables

$$x = 25x_1, \quad y = 125y_1$$

transforms the equation into

$$y_1^2 = x_1^3 - 432x_1 + 8208.$$

The discriminant of the cubic is $-2^8 3^{12} 11$, so E also has good reduction at 5. This is as far as we can go with the standard Weierstrass model. To treat 2 and 3 we need to allow generalized Weierstrass equations. The change of variables

$$x_1 = 9x_2 - 12, \quad y_1 = 27y_2$$

changes the equation to

$$y_2^2 = x_2^3 - 4x_2^2 + 16.$$

The discriminant of the cubic is $-2^8 11$, so E has good reduction at 3. Since any change of variables can be shown to change the discriminant by a square, this is the best we can do, except possibly at the prime 2. The change of variables

$$x_2 = 4x_3, \quad y_2 = 8y_3 + 4$$

changes the equation of E to

$$y_3^2 + y_3 = x_3^3 - x_3^2.$$

This is nonsingular at 2 (since the partial derivative with respect to y is $2y+1 \equiv 1 \neq 0 \pmod 2$). Therefore, E has good reduction at 2. We conclude that E has good reduction at all primes except $p = 11$, where it has bad reduction. The equation $y_3^2 + y^3 = x_3^3 - x_3^2$ is the minimal Weierstrass equation for E.

Let's analyze the situation at 11 more closely. The polynomial in x_2 factors as

$$x_2^3 - 4x_2^2 + 16 = (x_2 + 1)^2(x_2 + 5).$$

Therefore, E has multiplicative reduction at 11. The method of Section 2.9 shows that the slopes of the tangent lines at the singular point $(x_2, y_2) = (-1, 0)$ are ± 2, which lie in $\mathbf{F}_{11}$. Therefore, E has split multiplicative reduction at 11. ∎

We now give the full definition of the L-series of E. For a prime p of bad reduction, define

$$a_p = \begin{cases} 0 & \text{if } E \text{ has additive reduction at } p \\ 1 & \text{if } E \text{ has split multiplicative reduction at } p \\ -1 & \text{if } E \text{ has non-split multiplicative reduction at } p. \end{cases}$$

The numbers a_p for primes of good reduction are those given above: $a_p = p + 1 - \#E(\mathbf{F}_p)$. Then the **L-function of** E is the Euler product

$$L_E(s) = \prod_{\text{bad } p} \left(1 - a_p p^{-s}\right)^{-1} \prod_{\text{good } p} \left(1 - a_p p^{-s} + p^{1-2s}\right)^{-1}.$$

The estimate $|a_p| < 2\sqrt{p}$ easily implies that the product converges for $\Re(s) > 3/2$ (see Exercise 12.3).

Each good factor can be expanded in the form

$$(1 - a_p p^{-s} + p^{1-2s})^{-1} = 1 + a_p p^{-s} + a_{p^2} p^{-2s} + \cdots,$$

where the a_p on the left equals the a_p on the right (so this is not bad notation) and

$$a_{p^2} = a_p^2 - p. \tag{12.1}$$

The product over all p yields an expression

$$L_E(s) = \sum_{n=1}^{\infty} a_n n^{-s}.$$

If $n = \prod_j p_j^{e_j}$, then

$$a_n = \prod_j a_{p_j^{e_j}}. \tag{12.2}$$

This series for $L_E(s)$ converges for $\Re(s) > 3/2$ (see Exercise 12.3). It is natural to ask whether $L_E(s)$ has an analytic continuation to all of $\mathbf{C}$ and a functional equation, as is the case with the Riemann zeta function. As we'll discuss below, the answer to these questions is yes. However, the proof is much too deep to be included in this book (but see Chapter 13 for a discussion of the proof).

To study the analytic properties of $L_E(s)$, we introduce a new function. Let $\tau \in \mathcal{H}$, the upper half of the complex plane, as in Chapter 9, and let $q = e^{2\pi i \tau}$. (This is the standard notation; there should be no possibility of confusion with the q for finite fields of Chapter 4.) Define

$$f_E(\tau) = \sum_{n=1}^{\infty} a_n q^n.$$

This is simply a generating function that encodes the number of points on E mod the various primes. It converges for $\tau \in \mathcal{H}$ and satisfies some amazing properties.

Let N be a positive integer and define

$$\Gamma_0(N) = \left\{ \begin{pmatrix} a & b \\ c & d \end{pmatrix} \in SL_2(\mathbf{Z}) \,\middle|\, c \equiv 0 \pmod{N} \right\}.$$

Then $\Gamma_0(N)$ is a subgroup of $SL_2(\mathbf{Z})$.

The following result was conjectured by Shimura and has been known by various names, for example, the **Weil conjecture**, the **Taniyama-Shimura-Weil conjecture**, and the **Taniyama-Shimura conjecture**. All three mathematicians played a role in its history.

THEOREM 12.4 (Breuil, Conrad, Diamond, Taylor, Wiles)

Let E be an elliptic curve defined over $\mathbf{Q}$. There exists an integer N such that, for all $\tau \in \mathcal{H}$,

1.

$$f_E\left(\frac{a\tau + b}{c\tau + d}\right) = (c\tau + d)^2 f_E(\tau) \quad \text{for all} \quad \begin{pmatrix} a & b \\ c & d \end{pmatrix} \in \Gamma_0(N)$$

2.

$$f_E(-1/(N\tau)) = \pm N\tau^2 f_E(\tau).$$

For a sketch of the proof of this result, see Chapter 13. The theorem (if we include statements about the behavior at cusps on the real axis) says that $f_E(\tau)$ is a cusp form of weight 2 and level N. The smallest possible N is called the **conductor** of E. A prime p divides this N if and only if E has bad reduction at p. When E has multiplicative reduction, p divides N only to the first power. If E has additive reduction and $p > 3$, then p^2 is the exact power of p dividing N. The formulas for $p = 2$ and 3 are slightly more complicated in this case. See [98].

The transformation law in (1) can be rewritten as

$$f_E\left(\frac{a\tau + b}{c\tau + d}\right) d\left(\frac{a\tau + b}{c\tau + d}\right) = f_E(\tau)\,d\tau$$

(this is bad notation: d represents both an integer and the differentiation operator; it should be clear which is which). Therefore,

$$f_E(\tau)\,d\tau$$

is a differential that is invariant under the action of $\Gamma_0(N)$.

Once we have the relation (1), the second relation of the theorem is perhaps not as surprising. Every function satisfying (1) is a sum of two functions satisfying (2), one with a plus sign and one with a minus sign (see Exercise 12.2). Therefore (2) says that f_E lies in either the plus space or the minus space.

Taniyama first suggested the existence of a result of this form in the 1950s. Eichler and Shimura then showed that if f is a cusp form (more precisely, a newform) of weight 2 (and level N for some N) such that all the coefficients a_n are integers, then there is an elliptic curve E with $f_E = f$. This is the converse of the theorem, but it gave the first real evidence that Taniyama's suggestion was reasonable. In 1967, Weil made precise what the integer N must be for any given elliptic curve. Since there are only finitely many modular forms f of a given level N that could arise from elliptic curves, this meant that the conjecture (Taniyama's suggestion evolved into a conjecture) could be investigated numerically. If the conjecture had been false for some explicit E, it could have been disproved by computing enough coefficients to see that f_E was not on the finite list of possibilities. Moreover, Weil showed that if functions like $L_E(s)$ (namely L_E and its twists) have analytic continuations and functional equations such as the one given in Corollary 12.5 below, then f_E must be a modular form. Since most people believe that naturally defined L-functions should have analytic continuations and functional equations, this gave the conjecture more credence. Around 1990, Wiles proved that there are infinitely many distinct E (that is, with distinct j-invariants) satisfying the theorem. In 1994, with the help of Taylor, he showed that the theorem is true for all E such that there is no additive reduction at any prime (but multiplicative reduction is allowed). Such curves are called **semistable**. Finally, in 2001, Breuil, Conrad, Diamond, and Taylor [15] proved the full theorem.

Let's assume Theorem 12.4 and show that $L_E(s)$ analytically continues and satisfies a functional equation. Recall that the **gamma function** is defined for $\Re(s) > 0$ by

$$\Gamma(s) = \int_0^\infty t^{s-1}e^t\,dt.$$

Integration by parts yields the relation $s\Gamma(s) = \Gamma(s+1)$, which yields the meromorphic continuation of $\Gamma(s)$ to the complex plane, with poles at the nonpositive integers. It also yields the relation $\Gamma(n) = (n-1)!$ for positive integers n.

COROLLARY 12.5
Let E and N be as in Theorem 12.4. Then

$$(\sqrt{N}/2\pi)^s\Gamma(s)L_E(s) = \mp(\sqrt{N}/2\pi)^{2-s}\Gamma(2-s)L_E(2-s)$$

for all $s \in \mathbf{C}$ (and both sides continue analytically to all of $\mathbf{C}$). The sign here is the opposite of the sign in (2) of Theorem 12.4.

PROOF   Using the definition of the gamma function, we have

$$(\sqrt{N}/2\pi)^s\Gamma(s)L_E(s) = \sum_{n=1}^{\infty} a_n(\sqrt{N}/2\pi n)^s \int_0^{\infty} t^{s-1}e^{-t}\,dt$$

$$= \sum_{n=1}^{\infty} a_n \int_0^{\infty} (u\sqrt{N})^s e^{-2\pi nu}\frac{du}{u} \quad \text{(let } t = 2\pi nu)$$

$$= \int_0^{\infty} (u\sqrt{N})^s f_E(iu)\frac{du}{u}$$

$$= \int_0^{1/\sqrt{N}} (u\sqrt{N})^s f_E(iu)\frac{du}{u} + \int_{1/\sqrt{N}}^{\infty} (u\sqrt{N})^s f_E(iu)\frac{du}{u}.$$

(The interchange of summation and integration to obtain the third equality is justified since the sum for $f(iu)$ converges very quickly near ∞.) Let ϵ be the sign in part (2) of Theorem 12.4. Then

$$f_E(i/(Nu)) = \epsilon(iu)^2 f_E(iu) = -\epsilon u^2 f_E(iu).$$

Therefore, let $u = 1/Nv$ to obtain

$$\int_0^{1/\sqrt{N}} (u\sqrt{N})^s f_E(iu)\frac{du}{u} = -\epsilon \int_{1/\sqrt{N}}^{\infty} (v\sqrt{N})^{2-s} f_E(iv)\frac{dv}{v}.$$

This implies that

$$(\sqrt{N}/2\pi)^s\Gamma(s)L_E(s) =$$
$$\int_{1/\sqrt{N}}^{\infty} (u\sqrt{N})^s f_E(iu)\frac{du}{u} - \epsilon \int_{1/\sqrt{N}}^{\infty} (v\sqrt{N})^{2-s} f_E(iv)\frac{dv}{v}.$$

Since $f(iu) \to 0$ exponentially as $u \to \infty$, it follows easily that both integrals converge and define analytic functions of s. Under $s \mapsto 2-s$, the right side, hence the left side, is multiplied by $-\epsilon$. This is precisely what the functional equation claims.   ∎

Example 12.2
Let E be the elliptic curve $y^2 + y = x^3 - x^2$ considered in the previous example. If we compute the number N_p of points on E mod p for various primes, we obtain, with $a_p = p + 1 - N_p$,

$$a_2 = -2, \quad a_3 = -1, \quad a_5 = 1, \quad a_7 = -2, \quad a_{13} = 4, \dots$$

(except for $p = 2, 3, 5$, the numbers a_p can be calculated using any of the equations in the previous example). The value

$$a_{11} = 1$$

is specified by the formulas for bad primes. We then calculate the coefficients for composite indices. For example,

$$a_6 = a_2 a_3 = 2, \quad a_4 = a_2^2 - 2 = 2$$

(see (12.2) and (12.1)). Therefore,

$$f_E(\tau) = q - 2q^2 - q^3 + 2q^4 + q^5 + 2q^6 - 2q^7 + \cdots.$$

It can be shown that

$$f(\tau) = q \prod_{j=1}^{\infty} (1 - q^j)^2 (1 - q^{11j})^2$$

is a cusp form of weight 2 and level $N = 11$. In fact, it is the only such form, up to scalar multiples. The product for f can be expanded into an infinite series

$$f(\tau) = q - 2q^2 - q^3 + 2q^4 + q^5 + 2q^6 - 2q^7 + \cdots.$$

It can be shown that $f = f_E$ (see [47]).
   The L-series for E satisfies the functional equation

$$(\sqrt{11}/2\pi)^s \Gamma(s) L_E(s) = +(\sqrt{11} 2\pi)^{2-s} \Gamma(2 - s) L_E(2 - s).$$

⧠

   In the early 1960s, Birch and Swinnerton-Dyer performed computer experiments to try to understand the relation between the number of points on an elliptic curve mod p as p ranges through the primes and the number of rational points on the curve. Ignoring the fact that the product for $L_E(s)$

doesn't converge at $s = 1$, let's substitute $s = 1$ into the product (we'll ignore the finitely many bad primes):

$$\prod_p \left(1 - a_p p^{-1} + p^{-1}\right)^{-1} = \prod_p \left(\frac{p - a_p + 1}{p}\right)^{-1} = \prod_p \frac{p}{N_p}.$$

If E has a lot of points mod p for many p, then many factors in the product are small, so we expect that $L_E(1)$ might be small. In fact, the data that Birch and Swinnerton-Dyer obtained led them to make the following conjecture.

CONJECTURE 12.6 (Conjecture of Birch and Swinnerton-Dyer, Weak Form)

Let E be an elliptic curve defined over $\mathbf{Q}$. The order of vanishing of $L_E(s)$ at $s = 1$ is the rank r of $E(\mathbf{Q})$. In other words, if $E(\mathbf{Q}) \simeq torsion \oplus \mathbf{Z}^r$, then $L_E(s) = (s - 1)^r g(s)$, with $g(1) \neq 0, \infty$.

One consequence of the conjecture is that $E(\mathbf{Q})$ is infinite if and only if $L_E(1) = 0$. This statement remains unproved, although there has been some progress. In 1977, Coates and Wiles showed that if E has complex multiplication and has a point of infinite order, then $L_E(1) = 0$. The results of Gross and Zagier on Heegner points (1983) imply that if E is an elliptic curve over $\mathbf{Q}$ such that $L_E(s)$ vanishes to order exactly 1 at $s = 1$, then there is a point of infinite order. However, if $L_E(s)$ vanishes to order higher than 1, nothing has been proved, even though there is conjecturally an abundance of points of infinite order. This is a common situation in mathematics. It seems that a solution is often easier to find when it is essentially unique than when there are many choices.

Soon, Conjecture 12.6 was refined to give not only the order of vanishing, but also the leading coefficient of the expansion at $s = 1$. To state the conjecture, we need to introduce some notation. If $P_1, \ldots, P_r$ form a basis for the free part of $E(\mathbf{Q})$, then

$$E(\mathbf{Q}) = E(\mathbf{Q})_{\text{torsion}} \oplus \mathbf{Z}\, P_1 \oplus \cdots \oplus \mathbf{Z}\, P_r.$$

Recall the height pairing $\langle P, Q \rangle$ defined in Section 8.5. We can form the $r \times r$ matrix $\langle P_i, P_j \rangle$ and compute its determinant to obtain what is known as the **elliptic regulator** for E. If $r = 0$, define this determinant to equal 1. Let ω_1, ω_2 be a basis of a lattice in $\mathbf{C}$ that corresponds to E by Theorem 9.19. We may assume that $\omega_2 \in \mathbf{R}$, by Exercise 9.4. If $E[2] \subset E(\mathbf{R})$, let $\Omega = 2\omega_2$. Otherwise, let $\Omega = \omega_2$. For each prime p, there are integers c_p that we won't define, except to say that if p is a prime of good reduction then $c_p = 1$. A formula for computing them is given [98]. Finally, recall that Ш is the (conjecturally finite) Shafarevich-Tate group of E.

CONJECTURE 12.7 (Conjecture of Birch and Swinnerton-Dyer)
Let E be an elliptic curve defined over **Q**. *Let r be the rank of* $E(\mathbf{Q})$. *Then*

$$L_E(s) = (s-1)^r \frac{\Omega\left(\prod_p c_p\right)(\#\text{III}_E)\det\langle P_i, P_j\rangle}{\#E(\mathbf{Q})^2_{\text{torsion}}} + (s-1)^{r+1}(b_{r+1} + \cdots).$$

This important conjecture combines most of the important information about E into one equation. When it was first made, there were no examples. As Tate pointed out in 1974 ([97, p. 198]),

> This remarkable conjecture relates the behavior of a function L at a point where it is not at present known to be defined to the order of a group III which is not known to be finite!

In 1986, Rubin gave the first examples of curves with finite III, and was able to compute the exact order of III in several examples. Since they were complex multiplication curves, $L_E(1)$ could be computed explicitly by known formulas (these had been used by Birch and Swinnerton-Dyer in their calculations), and this allowed the conjecture to be verified for these curves. Soon thereafter, Kolyvagin obtained similar results for elliptic curves satisfying Theorem 12.4 (which was not yet proved) such that $L_E(s)$ vanishes to order at most 1 at $s = 1$. Therefore, the conjecture is mostly proved (up to small rational factors) when $L_E(s)$ vanishes to order at most one at $s = 1$. In general, nothing is known when $L_E(s)$ vanishes to higher order. In fact, it is not ruled out (but most people believe it's very unlikely) that $L_E(s)$ could vanish at $s = 1$ to very high order even though $E(\mathbf{Q})$ has rank 0 or 1.

In 2000, the Clay Mathematics Institute listed the Conjecture of Birch and Swinnerton-Dyer as one of its million dollar problems. There are surely easier (but certainly less satisfying) ways to earn a million dollars.

For those who know some algebraic number theory, the conjecture is very similar to the analytic class number formula. For an imaginary quadratic field K, the zeta function of K satisfies

$$\zeta_K(s) = (s-1)^{-1}\frac{2\pi h}{w\sqrt{|d|}} + \cdots,$$

where h is the class number of K, d is the discriminant of K, and w is the number of roots of unity in K. Conjecture 12.7 for a curve of rank $r = 0$ predicts that

$$L_E(s) = \frac{\Omega\left(\prod_p c_p\right)\#\text{III}_E}{\#E(\mathbf{Q})^2_{\text{torsion}}} + \cdots.$$

The group III_E can be regarded as the analogue of the ideal class group, the number $\Omega\prod_p c_p$ plays the role of $2\pi/\sqrt{|d|}$, and $\#E(\mathbf{Q})_{\text{torsion}}$ is the analogue of w. Except for the square on the order of the torsion group, the two formulas for the leading coefficients have very similar forms.

Now let's look at real quadratic fields K. The class number formula says that

$$\zeta_K(s) = (s-1)^{-1}\frac{4h\log(\eta)}{2\sqrt{d}} + \cdots,$$

where h is the class number of K, d is the discriminant, and η is the fundamental unit. The Conjecture of Birch and Swinnerton-Dyer for a curve of rank $r = 1$, with generator P, predicts that

$$L_E(s) = (s-1)\frac{\Omega\left(\prod_p c_p\right)(\#\text{Ⅲ}_E)\,\hat{h}(P)}{\#E(\mathbf{Q})^2_{\text{torsion}}} + \cdots.$$

In this case, Ω is the analogue of $4/\sqrt{d}$ and $\#E(\mathbf{Q})_{\text{torsion}}$ plays the role of 2, which is the number of roots of unity in K. The height $\hat{h}(P)$ gives the size of P. Similarly, $\log(\eta)$ gives the size of η.

In general, we can write down a dictionary between elliptic curves and number fields:

$$\text{elliptic curves} \longleftrightarrow \text{number fields}$$
$$\text{points} \longleftrightarrow \text{units}$$
$$\text{torsion points} \longleftrightarrow \text{roots of unity}$$
$$\text{Shafarevich-Tate group} \longleftrightarrow \text{ideal class group}$$

This is not an exact dictionary, but it helps to interpret results in one area in terms of the other. For example, the Dirichlet unit theorem in algebraic number theory, which describes the group of units in a number field, is the analogue of the Mordell-Weil theorem, which describes the group of rational points on an elliptic curve. The finiteness of the ideal class group in algebraic number theory is the analogue of the conjectured finiteness of the Shafarevich-Tate group.

Exercises

12.1 Let $\mathbf{P}^1$ be one-dimensional projective space.

(a) Show that the number of points in $\mathbf{P}^1(\mathbf{F}_q)$ is $q+1$.

(b) Let $N_n = \#\mathbf{P}^1(\mathbf{F}_{q^n})$. Define the Z-function for $\mathbf{P}^1$ by

$$Z_{\mathbf{P}^1}(T) = \exp\left(\sum_{n=1}^{\infty}\frac{N_n}{n}T^n\right).$$

Show that

$$Z_{\mathbf{P}^1}(T) = \frac{1}{(1-T)(1-qT)}.$$

12.2 Let $M = \begin{pmatrix} a & b \\ c & d \end{pmatrix} \in GL_2(\mathbf{R})$ with $\det(M) > 0$. Define an action of M on functions on $\mathcal{H}$ by

$$(f|M)(z) = \det(M)(cz + d)^{-2} f(Mz),$$

where $Mz = \frac{az+b}{cz+d}$.

(a) Show that $(f|M_1)|M_2 = f|(M_1 M_2)$.

(b) Let $W = \begin{pmatrix} 0 & -1 \\ N & 0 \end{pmatrix}$. Show that $W \Gamma_0(N) W^{-1} = \Gamma_0(N)$.

(c) Suppose that f is a function with $f|M = f$ for all $M \in \Gamma_0(N)$. Let $g(z) = (f|W)(z)$. Show that $g|M = g$ for all $M \in \Gamma_0(N)$. (*Hint:* Combine parts (a) and (b).)

(d) Suppose that f is a function with $f|M = f$ for all $M \in \Gamma_0(N)$. Let $f^+ = \frac{1}{2}(f + f|W)$ and $f^- = \frac{1}{2}(f - f|W)$. Show that $f^+|W = f^+$ and $f^-|W = -f^-$. This gives a decomposition $f = f^+ + f^-$ in which f is written as a sum of two eigenfunctions for W.

12.3 It is well known that a product $\prod(1 + b_n)$ converges if $\sum |b_n|$ converges. Use this fact, plus Hasse's theorem, to show that the Euler product defining $L_E(s)$ converges for $\Re(s) > 3/2$.

Chapter 13

Fermat's Last Theorem

13.1 Overview

Around 1637, Fermat wrote in the margin of his copy of Diophantus's work that, when $n \geq 3$,

$$a^n + b^n = c^n, \quad abc \neq 0 \tag{13.1}$$

has no solution in integers a, b, c. This has become known as **Fermat's Last Theorem**. Note that it suffices to consider only the cases where $n = 4$ and where $n = \ell$ is an odd prime (since any $n \geq 3$ has either 4 or such an ℓ as a factor). The case $n = 4$ was proved by Fermat using his method of infinite descent (see Section 8.6). At least one unsuccessful attempt to prove the case $n = 3$ appears in Arab manuscripts in the 900s (see [26]). This case was settled by Euler (and possibly by Fermat). The first general result was due to Kummer in the 1840s: Define the Bernoulli numbers B_n by the power series

$$\frac{t}{e^t - 1} = \sum_{n=1}^{\infty} B_n \frac{t^n}{n!}.$$

For example,

$$B_2 = \frac{1}{6}, \quad B_4 = -\frac{1}{30}, \quad \ldots, \quad B_{12} = -\frac{691}{2730}.$$

Let ℓ be an odd prime. If ℓ does not divide the numerator of any of the Bernoulli numbers

$$B_2, B_4, \ldots, B_{\ell-3}$$

then (13.1) has no solutions for $n = \ell$. This criterion allowed Kummer to prove Fermat's Last Theorem for all prime exponents less than 100, except for $\ell = 37, 59, 67$. For example, 37 divides the numerator of the 32nd Bernoulli number, so this criterion does not apply. Using more refined criteria, based on the knowledge of which Bernoulli numbers are divisible by these exceptional ℓ, Kummer was able to prove Fermat's Last Theorem for the three remaining exponents. Refinements of Kummer's ideas by Vandiver and others, plus the advent of computers, yielded extensions of Kummer's results to many more exponents. For example, in 1992, Buhler, Crandall, Ernvall, and Metsänkylä proved Fermat's Last Theorem for all exponents less than 4×10^6. How could one check so many cases without seeing a pattern that would lead to a full proof? The reason is that these methods were a prime-by-prime check. For each prime ℓ, the Bernoulli numbers were computed mod ℓ. For around 61% of the primes, none of these Bernoulli numbers was divisible by ℓ, so Kummer's initial criterion yielded the result. For the remaining 39% of the primes, more refined criteria were used, based on the knowledge of which Bernoulli numbers were divisible by ℓ. For ℓ up to 4×10^6, these criteria sufficed to prove the theorem. But it was widely suspected that eventually there would be exceptions to these criteria, and hence more refinements would be needed. The underlying problem with this approach was that it did not include any conceptual reason for why Fermat's Last Theorem should be true. In particular, there was no reason why there couldn't be a few random exceptions.

In 1986, the situation changed. Suppose that

$$a^\ell + b^\ell = c^\ell, \quad abc \neq 0. \tag{13.2}$$

By removing common factors, we may assume that a, b, c are integers with $\gcd(a, b, c) = 1$, and by rearranging a, b, c and changing signs if necessary, we may assume that

$$b \equiv 0 \pmod 2, \quad a \equiv -1 \pmod 4. \tag{13.3}$$

Frey suggested that the elliptic curve

$$E_{\text{Frey}}: \quad y^2 = x(x - a^\ell)(x + b^\ell)$$

(this curve had also been considered by Hellegouarch) has such restrictive properties that it cannot exist, and therefore there cannot be any solutions to (13.2). As we'll outline below, subsequent work of Ribet and Wiles showed that this is the case.

When $\ell \geq 5$, the elliptic curve E_{Frey} has good or multiplicative reduction (see Exercise 2.19) at all primes (in other words, there is no additive reduction). Such an elliptic curve is called **semistable**. The discriminant of the cubic is the square of the product of the differences of the roots, namely

$$\left((a^\ell(-b^\ell)(a^\ell + b^\ell)) \right)^2 = (abc)^{2\ell}$$

(we have used (13.2)). Because of technicalities involving the prime 2 (related to the restrictions in (13.3)), the discriminant needs to be modified at 2 to yield what is known as the **minimal discriminant**

$$\Delta = 2^{-8}(abc)^{2\ell}$$

of E_{Frey}. A conjecture of Brumer and Kramer predicts that a semistable elliptic curve over **Q** whose minimal discriminant is an ℓth power will have a point of order ℓ. Mazur's Theorem (8.11) says that an elliptic curve over **Q** cannot have a point of order ℓ when $\ell \geq 11$. Moreover, if the 2-torsion is rational, as is the case with E_{Frey}, then there are no points of order ℓ when $\ell \geq 5$. Since Δ is almost an ℓth power, we expect E_{Frey} to act similarly to a curve that has a point of order ℓ. Such curves cannot exist when $\ell \geq 5$, so E_{Frey} should act like a curve that cannot exist. Therefore, we expect that E_{Frey} does not exist. The problem is to make these ideas precise.

Recall (see Chapter 12) that the L-series of an elliptic curve E over **Q** is defined as follows. For each prime p of good reduction, let

$$a_p = p + 1 - \#E(\mathbf{F}_p).$$

Then

$$L_E(s) = (*) \prod_p (1 - a_p p^{-s} + p^{1-2s})^{-1} = \sum_{n=1}^{\infty} \frac{a_n}{n^s},$$

where (*) represents the factors for the bad primes (see Section 12.2) and the product is over the good primes. Suppose $E(\mathbf{Q})$ contains a point of order ℓ. By Theorem 8.9, $E(\mathbf{F}_p)$ contains a point of order ℓ for all primes $p \neq \ell$ such that E has good reduction at p. Therefore, $\ell | \#E(\mathbf{F}_p)$, so

$$a_p \equiv p + 1 \pmod{\ell} \tag{13.4}$$

for all such p. This is an example of how the arithmetic of E is related to properties of the coefficients a_p. We hope to obtain information by studying these coefficients.

In particular, we expect a congruence similar to (13.4) to hold for E_{Frey}. In fact, a close analysis (requiring more detail than we give in Section 13.3) of Ribet's proof shows that E_{Frey} is trying to satisfy this congruence. However, the irreducibility of a certain Galois representation is preventing it, and this leads to the contradiction that proves the theorem.

The problem with this approach is that the numbers a_p at first seem to be fairly independent of each other as p varies. However, the Conjecture of Taniyama-Shimura-Weil (now Theorem 12.4) claims that, for an elliptic curve E over **Q**,

$$f_E(\tau) = \sum_{n=1}^{\infty} a_n q^n$$

(where $q = e^{2\pi i \tau}$) is a modular form for $\Gamma_0(N)$ for some N (see Section 12.2). In this case, we say that E is **modular**. This is a fairly rigid condition and can be interpreted as saying that the numbers a_p have some coherence as p varies. For example, it is likely that if we change one coefficient a_p, then the modularity will be lost. Therefore, modularity is a tool for keeping the numbers a_p under control. Frey predicted the following, which Ribet proved in 1986:

THEOREM 13.1

E_{Frey} *cannot be modular. Therefore, the conjecture of Taniyama-Shimura-Weil implies Fermat's Last Theorem.*

This result finally gave a theoretical reason for believing Fermat's Last Theorem. Then in 1994, Wiles proved

THEOREM 13.2

All semistable elliptic curves over $\mathbf{Q}$ *are modular.*

This result was subsequently extended to include all elliptic curves over $\mathbf{Q}$. See Theorem 12.4. Since the Frey curve is semistable, the theorems of Wiles and Ribet combine to show that E_{Frey} cannot exist, hence

THEOREM 13.3

Fermat's Last Theorem is true.

In the following three sections, we sketch some of the ideas that go into the proofs of Ribet's and Wiles's theorems.

13.2 Galois Representations

Let E be an elliptic curve over $\mathbf{Q}$ and let m be an integer. From Theorem 3.2, we know that

$$E[m] \simeq \mathbf{Z}_m \oplus \mathbf{Z}_m.$$

Let $\{\beta_1, \beta_2\}$ be a basis of $E[m]$ and let $\sigma \in G$, where

$$G = \text{Gal}(\overline{\mathbf{Q}}/\mathbf{Q}).$$

Since $\sigma\beta_i \in E[m]$, we can write

$$\sigma\beta_1 = a\beta_1 + c\beta_2, \quad \sigma\beta_2 = b\beta_1 + d\beta_2$$

with $a, b, c, d \in \mathbf{Z}_m$. We thus obtain a homomorphism

$$\rho_m : G \longrightarrow GL_2(\mathbf{Z}_m)$$
$$\sigma \longmapsto \begin{pmatrix} a & b \\ c & d \end{pmatrix}.$$

If $m = \ell$ is a prime, we call ρ_ℓ the mod ℓ Galois representation attached to E. We can also take $m = \ell^n$ for $n = 1, 2, 3, \ldots$. By choosing an appropriate sequence of bases, we obtain representations ρ_{ℓ^n} such that

$$\rho_{\ell^n} \equiv \rho_{\ell^{n+1}} \pmod{\ell^n}$$

for all n. These may be combined to obtain

$$\rho_{\ell^\infty} : G \longrightarrow GL_2(\mathcal{O}_\ell),$$

where $\mathcal{O}_\ell$ denotes any ring containing the ℓ-adic integers (see Appendix A). This is called the ℓ-adic Galois representation attached to E. An advantage of working with ρ_{ℓ^∞} is that the ℓ-adic integers have characteristic 0, so instead of congruences mod powers of ℓ, we can work with equalities.

Notation: Throughout this chapter, we will need rings that are finite extensions of the ℓ-adic integers. We'll denote such rings by $\mathcal{O}_\ell$. For many purposes, we can take $\mathcal{O}_\ell$ to equal the ℓ-adic integers, but sometimes we need slightly larger rings. Since we do not want to discuss the technical issues that arise in this regard, we simply use $\mathcal{O}_\ell$ to denote a varying ring that is large enough for whatever is required. The reader will not lose much by pretending that $\mathcal{O}_\ell$ is always the ring of ℓ-adic integers.

Suppose r is a prime of good reduction for E. There exists an element $\mathrm{Frob}_r \in G$ such that the action of Frob_r on $E(\overline{\mathbf{Q}})$ yields the action of the Frobenius ϕ_r on $E(\overline{\mathbf{F}}_r)$ when E is reduced mod r (the element Frob_r is not unique, but this will not affect us). In particular, when $\ell \neq r$, the matrices describing the actions of Frob_r and ϕ_r on the ℓ-power torsion are the same (use a basis and its reduction to compute the matrices). Let

$$a_r = r + 1 - \#E(\mathbf{F}_r).$$

From Proposition 4.11, we obtain that

$$\mathrm{Trace}(\rho_{\ell^n}(\mathrm{Frob}_r)) \equiv a_r \pmod{\ell^n}, \quad \det(\rho_{\ell^n}(\mathrm{Frob}_r)) \equiv r \pmod{\ell^n},$$

and therefore

$$\mathrm{Trace}(\rho_{\ell^\infty}(\mathrm{Frob}_r)) = a_r, \quad \det(\rho_{\ell^\infty}(\mathrm{Frob}_r)) = r.$$

Recall that the numbers a_r are used to produce the modular form f_E attached to E (see Section 12.2).

Suppose now that

$$\rho : G \longrightarrow GL_2(\mathcal{O}_\ell)$$

is a representation of G. Under certain technical conditions (namely, ρ is unramified at all but finitely many primes; see the end of this section), we may choose elements Frob_r (for the unramified primes) and define

$$a_r = \mathrm{Trace}(\rho(\mathrm{Frob}_r)).$$

This allows us to define a formal series

$$g = \sum_{n=1}^{\infty} a_n q^n.$$

We refer to g as the **potential modular form** attached to ρ. Of course, some conditions must be imposed on the a_r in order for this to represent a complex function (for example, the numbers $a_n \in \mathcal{O}_\ell$ must be identified with complex numbers), but we will not discuss this general problem here.

Let N be a positive integer. Recall that a modular form f of weight 2 and level N is a function analytic in the upper half plane satisfying

$$f\left(\frac{a\tau + b}{c\tau + d}\right) = (c\tau + d)^2 f(\tau) \tag{13.5}$$

for all

$$\begin{pmatrix} a & b \\ c & d \end{pmatrix} \in \Gamma_0(N)$$

(where $\Gamma_0(N)$ is the group of integral matrices of determinant 1 such that $c \equiv 0 \pmod{N}$). There are also technical conditions that we won't discuss for the behavior of f at the cusps. The **cusp forms** of weight 2 and level N, which we'll denote by $S(N)$, are those modular forms that take the value 0 at all the cusps. $S(N)$ is a finite dimensional vector space over $\mathbf{C}$. We represent cusp forms by their Fourier expansions:

$$f(\tau) = \sum_{n=1}^{\infty} b_n q^n,$$

where $q = e^{2\pi i \tau}$.

If $M|N$, then $\Gamma_0(N) \subseteq \Gamma_0(M)$, so a modular form of level M can be regarded as a modular form of level N. More generally, if $d|(N/M)$ and $f(\tau)$ is a cusp form of level M, then it can be shown that $f(d\tau)$ is a cusp form of level N. The subspace of $S(N)$ generated by such f, where M ranges through proper divisors of N and d ranges through divisors of N/M, is called the subspace of **oldforms**. There is a naturally defined inner product on $S(N)$, called the Petersson inner product. The space of **newforms** of level N is the perpendicular complement of the space of oldforms. Intuitively, the newforms are those that do not come from levels lower than N.

We now need to introduce the **Hecke operators**. Let r be a prime. Define

$$T_r\left(\sum_{n=1}^{\infty} b_n q^n\right) = \begin{cases} \sum_{n=1}^{\infty} b_{rn} q^n + \sum_{n=1}^{\infty} b_n q^{rn}, & \text{if } r \nmid N \\ \sum_{n=1}^{\infty} b_{rn} q^n, & \text{if } r \mid N. \end{cases} \tag{13.6}$$

It can be shown that T_r maps $S(N)$ into $S(N)$ and that the T_r's commute with each other. Define the **Hecke algebra**

$$\mathbf{T} = \mathbf{T}_N \subseteq \text{End}(S(N))$$

to be the image of $\mathbf{Z}[T_2, T_3, T_5, \dots]$ in the endomorphism ring of $S(N)$ (the endomorphism ring of $S(N)$ is the ring of linear transformations from the vector space $S(N)$ to itself).

A **normalized eigenform** of level N is a newform

$$f = \sum_{n=1}^{\infty} b_n q^n \in S(N)$$

of level N with $b_1 = 1$ and such that

$$T_r(f) = b_r f \quad \text{for all } r.$$

It can be shown that the space of newforms in $S(N)$ has a basis of normalized eigenforms. Henceforth, essentially all of the modular forms that we encounter will be normalized eigenforms of level N. Often, we shall refer to them simply as modular forms.

Let f be a normalized eigenform and suppose the coefficients b_n of f are rational integers. In this case, Eichler and Shimura showed that there is an elliptic curve E_f over $\mathbf{Q}$ such that

$$b_r = a_r$$

for all r (where $a_r = r + 1 - \#E_f(\mathbf{F}_r)$ for the primes of good reduction). In particular, the potential modular form f_{E_f} for E is the modular form f. Moreover, E_f has good reduction at the primes not dividing N. This result is, in a sense, a converse of the conjecture of Taniyama-Shimura-Weil. The conjecture can be restated as claiming that every elliptic curve E over $\mathbf{Q}$ arises from this construction. Actually, we have to modify this statement a little. Two elliptic curves E_1 and E_2 are called **isogenous** over $\mathbf{Q}$ if there is a nonconstant homomorphism $E_1(\overline{\mathbf{Q}}) \to E_2(\overline{\mathbf{Q}})$ that is described by rational functions over $\mathbf{Q}$ (see the end of Section 8.6). It can be shown that, in this case, $f_{E_1} = f_{E_2}$. Conversely, Faltings showed that if $f_{E_1} = f_{E_2}$ then E_1 and E_2 are isogenous. Since only one of E_1, E_2 can be the curve E_f, we must ask whether an elliptic curve E over $\mathbf{Q}$ is isogenous to one produced by the result of Eichler and Shimura. Theorem 12.4 says that the answer is yes.

If we have an elliptic curve E, how can we predict what N should be? The smallest possible N is called the **conductor** of E. For $E = E_f$, the primes dividing the conductor N are exactly the primes of bad reduction of E_f (these are also the primes of bad reduction of any curve isogenous to E_f over $\mathbf{Q}$). Moreover, $p|N$ and $p^2 \nmid N$ if and only if E_f has multiplicative reduction at p. Therefore, if E_f is semistable, then

$$N = \prod_{p|\Delta} p, \tag{13.7}$$

namely, the product of the primes dividing the minimal discriminant Δ. We see that N is squarefree if and only if E_f is semistable. Therefore, if E is an arbitrary modular semistable elliptic curve over $\mathbf{Q}$, then N is given by (13.7).

Combining the result of Eichler and Shimura with the Galois representations discussed above, we obtain the following. If $f = \sum b_n q^n$ is a normalized newform with rational integer coefficients, then there is a Galois representation

$$\rho_f : G \longrightarrow GL_2(\mathcal{O}_\ell)$$

such that

$$\text{Trace}(\rho_f(\text{Frob}_r)) = b_r, \quad \det(\rho_f(\text{Frob}_r)) = r \qquad (13.8)$$

for all $r \nmid \ell N$.

More generally, Eichler and Shimura showed that if $f = \sum b_n q^n$ is any normalized newform (with no assumptions on its coefficients), then there is a Galois representation

$$\rho_f : G \to GL_2(\mathcal{O}_\ell)$$

satisfying (13.8).

Returning to the situation where the coefficients b_n are in $\mathbf{Z}$, we let $\mathcal{M}$ be the kernel of the ring homomorphism

$$\mathbf{T} \longrightarrow \mathbf{F}_\ell$$
$$T_r \longmapsto b_r \pmod \ell.$$

Since the homomorphism is surjective (because 1 maps to 1) and $\mathbf{F}_\ell$ is a field, $\mathcal{M}$ is a maximal ideal of $\mathbf{T}$. Also, $\mathbf{T}/\mathcal{M} = \mathbf{F}_\ell$. Since $T_r - b_r \in \mathcal{M}$, the mod ℓ version of (13.8) says that

$$\text{Trace}(\rho_f(\text{Frob}_r)) \equiv T_r \mod \mathcal{M}, \qquad \det(\rho_f(\text{Frob}_r)) \equiv r \mod \mathcal{M}$$

for all $r \nmid \ell N$. This has been greatly generalized by Deligne and Serre:

THEOREM 13.4
Let $\mathcal{M}$ be a maximal ideal of $\mathbf{T}$ and let ℓ be the characteristic of $\mathbf{T}/\mathcal{M}$. There exists a semisimple representation

$$\rho_\mathcal{M} : G \longrightarrow GL_2(\mathbf{T}/\mathcal{M})$$

such that

$$\text{Trace}(\rho_\mathcal{M}(\text{Frob}_r)) \equiv T_r \mod \mathcal{M}, \quad \det(\rho_\mathcal{M}(\text{Frob}_r)) \equiv r \mod \mathcal{M}$$

for all primes $r \nmid \ell N$.

The semisimplicity of $\rho_\mathcal{M}$ means that either $\rho_\mathcal{M}$ is irreducible or it is the sum of two one-dimensional representations.

In general, let A be either $\mathcal{O}_\ell$ or a finite field. If

$$\rho : G \longrightarrow GL_2(A)$$

is a semisimple representation, then we say that ρ is **modular of level** N if there exists a homomorphism

$$\pi : \mathbf{T} \longrightarrow A$$

such that

$$\mathrm{Trace}(\rho(\mathrm{Frob}_r)) = \pi(T_r), \qquad \det(\rho_{\mathcal{M}}(\mathrm{Frob}_r)) = \pi(r)$$

for all $r \nmid \ell N$. This says that ρ is equivalent to a representation coming from one of the above constructions.

When $A = \mathbf{T}/\mathcal{M}$, the homomorphism π is the map $\mathbf{T} \to \mathbf{T}/\mathcal{M}$.

When $f = \sum b_n q^n$ is a normalized eigenform and $A = \mathcal{O}_\ell$, recall that $T_r(f) = b_r f$ for all r. This gives a homomorphism $\pi : \mathbf{T} \to \mathcal{O}_\ell$ (it is possible to regard the coefficients b_r as elements of a sufficiently large $\mathcal{O}_\ell$).

The way to obtain maximal ideals $\mathcal{M}$ of $\mathbf{T}$ is to use a normalized eigenform to get a map $\mathbf{T} \to \mathcal{O}_\ell$, then map $\mathcal{O}_\ell$ to a finite field. The kernel of the map from $\mathbf{T}$ to the finite field is a maximal ideal $\mathcal{M}$.

When A is a finite field, the level N of the representation ρ is not unique. In fact, a key result of Ribet (see Section 13.3) analyzes how the level can be changed. Also, in the definition of modularity in this case, we should allow modular forms of weight $k \geq 2$ (this means that the factor $(cz+d)^2$ in (13.5) is replaced by $(cz+d)^k$). However, this more general situation can be ignored for the present purposes.

If ρ is a modular representation of some level, and $c \in G$ is complex conjugation (regard $\overline{\mathbf{Q}}$ as a subfield of $\mathbf{C}$) then it can be shown that $\det(\rho(c)) = -1$. This says that ρ is an **odd** representation. A conjecture of Serre [87], which was a motivating force for much of the work described in this chapter, predicts that (under certain mild hypotheses) odd representations in the finite field case are modular (where we need to allow modular forms of weight $k \geq 2$ in the definition of modularity). Serre also predicts the level and the weight of a modular form that yields the representation.

Finally, there is a type of representation, called finite, that plays an important role in Ribet's proof. Let p be a prime. We can regard the Galois group for the p-adics as a subgroup of the Galois group for $\mathbf{Q}$:

$$G_p = \mathrm{Gal}(\overline{\mathbf{Q}}_p/\mathbf{Q}_p) \subset G = \mathrm{Gal}(\overline{\mathbf{Q}}/\mathbf{Q}).$$

There is a natural map from G_p to $\mathrm{Gal}(\overline{\mathbf{F}}_p/\mathbf{F}_p)$. The kernel is denoted I_p and is called the **inertia subgroup** of G_p:

$$G_p/I_p \simeq \mathrm{Gal}(\overline{\mathbf{F}}_p/\mathbf{F}_p). \tag{13.9}$$

A representation

$$\rho : G \to GL_2(\overline{\mathbf{F}}_\ell)$$

is said to be **unramified** at p if $\rho(I_p) = 1$, namely, I_p is contained in the kernel of ρ. If $p \neq \ell$ and ρ is unramified at p, then ρ is said to be **finite** at p.

If $p = \ell$, the definition of finite is much more technical (it involves finite flat group schemes) and we omit it. However, for the representation ρ_ℓ coming from an elliptic curve, there is the following:

PROPOSITION 13.5

Let E be an elliptic curve defined over $\mathbf{Q}$ and let Δ be the minimal discriminant of E. Let ℓ and p be primes (the case $p = \ell$ is allowed) and let ρ_ℓ be the representation of G on $E[\ell]$. Then ρ_ℓ is finite at p if and only if $v_p(\Delta) \equiv 0 \pmod{\ell}$, where v_p denotes the p-adic valuation (see Appendix A).

For a proof, see [87].

Consider the Frey curve. The minimal discriminant is

$$\Delta = 2^{-8}(abc)^{2\ell}.$$

Therefore, $v_p(\Delta) \equiv 0 \pmod{\ell}$ for all $p \neq 2$, so ρ_ℓ is finite at all odd primes. Moreover, ρ_ℓ is not finite at 2.

13.3 Sketch of Ribet's Proof

The key theorem that Ribet proved is the following.

THEOREM 13.6

Let $\ell \geq 3$ and let

$$\rho : G \to GL_2(\overline{\mathbf{F}}_\ell)$$

be an irreducible representation. Assume that ρ is modular of squarefree level N and that there exists a prime $q|N$, $q \neq \ell$, at which ρ is not finite. Suppose $p|N$ is a prime at which ρ is finite. Then ρ is modular of level N/p.

In other words, if ρ comes from a modular form of level N, then, under suitable hypotheses, it also comes from a modular form of level N/p.

COROLLARY 13.7

E_{Frey} cannot be modular.

PROOF   Since there are no solutions to the Fermat equation, and hence no Frey curves, when $\ell = 3$, we may assume $\ell \geq 5$. If E_{Frey} is modular, then the associated representation ρ_ℓ is modular of some level N. Since E_{Frey} is semistable, (13.7) says that

$$N = \prod_{p|abc} p.$$

It can be shown that ρ_ℓ is irreducible when $\ell \geq 5$ (see [87], where it is obtained as a corollary of Mazur's theorem (Theorem 8.11)). Let $q = 2$ in Ribet's theorem. As we showed at the end of Section 13.2, ρ_ℓ is not finite at 2 and is finite at all other primes. Therefore, Ribet's theorem allows us to remove the odd primes from N one at a time. We eventually find that ρ_ℓ is modular of level 2. This means that there is a normalized cusp form of weight 2 for $\Gamma_0(2)$ such that ρ_ℓ is the associated mod ℓ representation. But there are no nonzero cusp forms of weight 2 for $\Gamma_0(2)$, so we have a contradiction. Therefore, E_{Frey} cannot be modular.   ∎

COROLLARY 13.8

The Taniyama-Shimura-Weil conjecture (for semistable elliptic curves) implies Fermat's Last Theorem.

PROOF   We may restrict to prime exponents $\ell \geq 5$. If there is a nontrivial solution to the Fermat equation for ℓ, then the Frey curve exists. However, Corollary 13.7 and the Taniyama-Shimura-Weil conjecture imply that the Frey curve cannot exist. Therefore, there are no nontrivial solutions to the Fermat equation.   ∎

We now give a brief sketch of the proof of Ribet's theorem. The proof uses the full power of Grothendieck's algebraic geometry and is not elementary. Therefore, we give only a sampling of some of the ideas that go into the proof. For more details, see [74], [73], [69], [22].

We assume that ρ is as in Theorem 13.6 and that N is chosen so that

1. ρ is modular of squarefree level N,

2. both p and q divide N,

3. ρ is finite at p but is not finite at q.

The goal is to show that p can be removed from N. The main ingredient of the proof is a relation between Jacobians of modular curves and Shimura curves. In the following, we describe modular curves and Shimura curves and give a brief indication of how they occur in Ribet's proof.

Modular curves

Recall that $SL_2(\mathbf{Z})$ acts on the upper half plane $\mathcal{H}$ by linear fractional transformations:

$$\begin{pmatrix} a & b \\ c & d \end{pmatrix} \tau = \frac{a\tau + b}{c\tau + d}.$$

The fundamental domain $\mathcal{F}$ for this action is described in Section 9.3. The subgroup $\Gamma_0(N)$ (defined by the condition that $c \equiv 0 \pmod{N}$) also acts on $\mathcal{H}$. The **modular curve** $X_0(N)$ is defined over $\mathbf{C}$ by taking the upper half plane modulo the action of $\Gamma_0(N)$, and then adding finitely many points, called cusps, to make $X_0(N)$ compact. We obtain a fundamental domain $\mathcal{D}$ for $\Gamma_0(N)$ by writing

$$SL_2(\mathbf{Z}) = \cup_i \gamma_i \Gamma_0(N)$$

for some coset representatives γ_i and letting $\mathcal{D} = \cup_i \gamma_i^{-1} \mathcal{F}$. Certain edges of this fundamental domain are equivalent under the action of $\Gamma_0(N)$. When equivalent edges are identified, the fundamental domain gets bent around to form a surface. There is a hole in the surface corresponding to $i\infty$, and there are also finitely many holes corresponding to points where the fundamental domain touches the real axis. These holes are filled in by points, called **cusps**, to obtain $X_0(N)$. It can be shown that $X_0(N)$ can be represented as an algebraic curve defined over $\mathbf{Q}$.

Figure 13.1 gives a fundamental domain for $\Gamma_0(2)$. The three pieces are obtained as $\gamma_i^{-1}\mathcal{F}$, where

$$\gamma_1 = \begin{pmatrix} 1 & 0 \\ 0 & 1 \end{pmatrix}, \quad \gamma_2 = \begin{pmatrix} 0 & -1 \\ 1 & 0 \end{pmatrix}, \quad \gamma_3 = \begin{pmatrix} 1 & 1 \\ -1 & 0 \end{pmatrix}.$$

The modular curve $X_0(N)$ has another useful description, which works over arbitrary fields K with the characteristic of K not dividing N. Consider pairs (E, C), where E is an elliptic curve (defined over the algebraic closure $\overline{K}$) and C is a cyclic subgroup of $E(\overline{K})$ of order N. The set of such pairs is in one-to-one correspondence with the noncuspidal points of $X_0(N)(\overline{K})$. Of course, it is not obvious that this collection of pairs can be given the structure of an algebraic curve in a natural way. This takes some work.

Example 13.1
When $K = \mathbf{C}$, we can see this one-to-one correspondence as follows. An elliptic curve can be represented as

$$E_\tau = \mathbf{C}/(\mathbf{Z}\tau + \mathbf{Z}),$$

with $\tau \in \mathcal{H}$, the upper half plane. The set

$$C_\tau = \left\{ 0, \frac{1}{N}, \ldots, \frac{N-1}{N} \right\}$$

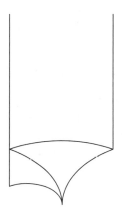

Figure 13.1

A Fundamental Domain for $\Gamma_0(2)$

is a cyclic subgroup of E_τ of order N. Let

$$\gamma = \begin{pmatrix} a & b \\ c & d \end{pmatrix} \in \Gamma_0(N)$$

and let

$$\gamma\tau = \frac{a\tau + b}{c\tau + d}.$$

Since

$$\mathbf{Z}\tau + \mathbf{Z} = \mathbf{Z}(a\tau + b) + \mathbf{Z}(c\tau + d) = (c\tau + d)(\mathbf{Z}\gamma\tau + \mathbf{Z}),$$

there is an isomorphism

$$f_\gamma : \ \mathbf{C}/(\mathbf{Z}\tau + \mathbf{Z}) \longrightarrow \mathbf{C}/(\mathbf{Z}\gamma\tau + \mathbf{Z})$$

given by

$$f_\gamma(z) = z/(c\tau + d).$$

This isomorphism between E_τ and $E_{\gamma\tau}$ maps the point k/N to

$$\frac{k}{N(c\tau + d)} = \frac{ka}{N} - k\frac{c}{N}\frac{a\tau + b}{c\tau + d}$$

$$\equiv \frac{ka}{N} \quad \mathrm{mod}\ \mathbf{Z}\gamma\tau + \mathbf{Z}$$

(we have used the fact that $c \equiv 0 \pmod{N}$). Therefore, the subgroup C_τ of E_τ is mapped to the corresponding subgroup $C_{\gamma\tau}$ of $E_{\gamma\tau}$, so f_γ maps the pair (E_τ, C_τ) to the pair $(E_{\gamma\tau}, C_{\gamma\tau})$. We conclude that if $\tau_1, \tau_2 \in \mathcal{H}$ are equivalent under the action of $\Gamma_0(N)$, then the corresponding pairs (E_{τ_j}, C_{τ_j}) are isomorphic. It is not hard to show that, conversely, if the pairs are isomorphic then the corresponding τ_j's are equivalent under $\Gamma_0(N)$. Moreover, every pair (E, C) of an elliptic curve over $\mathbf{C}$ and a cyclic subgroup C of order N is isomorphic to a pair (E_τ, C_τ) for some $\tau \in \mathcal{H}$. Therefore, the set of isomorphism classes of these pairs is in one-to-one correspondence with the points of $\mathcal{H}$ mod the action of $\Gamma_0(N)$. These are the noncuspidal points of $X_0(N)$.

Of course, over arbitrary fields, we cannot work with the upper half plane $\mathcal{H}$, and it is much more difficult to show that the pairs (E, C) can be collected together as the points on a curve $X_0(N)$. However, when this is done, it yields a convenient way to work with the modular curve $X_0(N)$ and its reductions mod primes.   ▯

For a nonsingular algebraic curve C over a field K, let $J(C)$ be the divisors (over $\overline{K}$) of degree 0 modulo divisors of functions. It is possible to represent $J(C)$ as an algebraic variety, called the **Jacobian** of C. When C is an elliptic curve E, we showed (Corollary 11.4; see also the sequence (9.3)) that $J(E)$ is a group isomorphic to $E(\overline{K})$. When $K = \mathbf{C}$, we thus obtained a torus. In general, if $K = \mathbf{C}$ and C is a curve of genus g, then $J(C)$ is isomorphic to a higher dimensional torus, namely, $\mathbf{C}^g$ mod a lattice of rank $2g$. The Jacobian of $X_0(N)$ is denoted $J_0(N)$.

The Jacobian $J_0(N)$ satisfies various functorial properties. In particular, a nonconstant map $\phi : X_0(N) \to E$ induces a map $\phi^* : E \to J_0(N)$ obtained by mapping a point P of E to the divisor on $X_0(N)$ formed by the sum of the inverse images of P minus the inverse images of $\infty \in E$:

$$\phi^* : P \longmapsto \sum_{\phi(Q)=P} [Q] - \sum_{\phi(R)=\infty} [R].$$

Therefore, we can map E to a subgroup of $J_0(N)$ (this map might have a nontrivial, but finite, kernel).

An equivalent formulation of the modularity of E is to say that there is a nonconstant map from $X_0(N)$ to E and therefore that E is isogenous to an elliptic curve contained in some $J_0(N)$.

If p is a prime dividing N, there are two natural maps $X_0(N) \to X_0(N/p)$. If (E, C) is a pair corresponding to a point in $X_0(N)$, then there is a unique subgroup $C' \subset C$ of order N/p. So we have a map

$$\alpha : (E, C) \longmapsto (E, C'). \tag{13.10}$$

However, there is also a unique subgroup $P \subset C$ of order p. It can be shown that E/P is an elliptic curve and therefore $(E/P, C/P)$ is a pair corresponding to a point on $X_0(N/p)$. This gives a map

$$\beta : (E, C) \longmapsto (E/P, C/P). \tag{13.11}$$

These two maps can be interpreted in terms of the complex model of $X_0(N)$. Since $\Gamma_0(N) \subset \Gamma_0(N/p)$, we can map $\mathcal{H}$ mod $\Gamma_0(N)$ to $\mathcal{H}$ mod $\Gamma_0(N/p)$ by mapping the equivalence class of τ mod $\Gamma_0(N)$ to the equivalence class of τ mod $\Gamma_0(N/p)$. This corresponds to the map α. The map β can be shown to correspond to the map $\tau \mapsto p\tau$. Note that these two maps represent the two methods of using modular forms for $\Gamma_0(N/p)$ to produce oldforms for $\Gamma_0(N)$.

The Hecke algebra $\mathbf{T}$ acts on $J_0(N)$. Let P be a point on $X_0(N)$. Recall that P corresponds to a pair (E, C), where E is an elliptic curve and C is a cyclic subgroup of order N. Let p be a prime. For each subgroup D of E of order p with $D \not\subseteq C$, we can form the pair $(E/D, (C + D)/D)$. It can be shown that E/D is an elliptic curve and $(C + D)/D$ is a cyclic subgroup of order N. Therefore, this pair represents a point on $X_0(N)$. Define

$$T_p([(E, C)]) = \sum_D [(E/D, (C + D)/D)] \in \mathrm{Div}(X_0(N)),$$

where the sum is over those D of order p with $D \not\subseteq C$ and where $\mathrm{Div}(X_0(N))$ denotes the divisors of $X_0(N)$ (see Chapter 11). It is not hard to show that this corresponds to the formulas for T_p given in (13.6). Clearly T_p maps divisors of degree 0 to divisors of degree 0, and it can be shown that it maps principal divisors to principal divisors. Therefore, T_p gives a map from $J_0(N)$ to itself. This yields an action of $\mathbf{T}$ on $J_0(N)$, and these endomorphisms are defined over $\mathbf{Q}$.

Let $\alpha \in \mathbf{T}$ and let $J_0(N)[\alpha]$ denote the kernel of α on $J_0(N)$. More generally, let I be an ideal of $\mathbf{T}$. Define

$$J_0(N)[I] = \bigcap_{\alpha \in I} J_0(N)[\alpha].$$

For example, when $I = n\mathbf{T}$ for an integer n, then $J_0(N)[I]$ is just $J_0(N)[n]$, the n-torsion on $J_0(N)$.

Now let's consider the representation ρ of Theorem 13.6. Since ρ is assumed to be modular, it corresponds to a maximal ideal $\mathcal{M}$ of $\mathbf{T}$. Let $\mathbf{F} = \mathbf{T}/\mathcal{M}$, which is a finite field. Then $W = J_0(N)[\mathcal{M}]$ has an action of $\mathbf{F}$, which means that it is a vector space over $\mathbf{F}$. Let ℓ be the characteristic of $\mathbf{F}$. Since $\ell = 0$ in $\mathbf{F}$, it follows that

$$W \subseteq J_0(N)[\ell],$$

the ℓ-torsion of $J_0(N)$. Since G acts on W, we see that W yields a representation ρ' of G over $\mathbf{F}$. It can be shown that ρ' is equivalent to ρ, so we can regard the representation space for ρ as living inside the ℓ-torsion of $J_0(N)$. This has great advantages. For example, if $M|N$ then there are natural maps $X_0(N) \to X_0(M)$. These yield (just as for the map $X_0(N) \to E$ above) maps $J_0(M) \to J_0(N)$. Showing that the level can be reduced from N to M is equivalent to showing that this representation space lives in these images of $J_0(M)$. Also, we are now working with a representation that lives inside a fairly concrete object, namely the ℓ-torsion of an abelian variety, rather than a more abstract situation, so we have more control over ρ.

Shimura curves

We now need to introduce what are known as Shimura curves. Recall that in Section 10.2 we defined quaternion algebras as (noncommutative) rings of the form

$$\mathcal{Q} = \mathbf{Q} + \mathbf{Q}\alpha + \mathbf{Q}\beta + \mathbf{Q}\alpha\beta,$$

where

$$\alpha^2, \beta^2 \in \mathbf{Q}, \quad \beta\alpha = -\alpha\beta.$$

We omit the requirement from Section 10.2 that $\alpha^2 < 0$ and $\beta^2 < 0$ since we want to consider indefinite quaternion algebras as well. Let r be a prime (possibly ∞) and let $\mathcal{Q}_r$ be the ring obtained by allowing r-adic coefficients in the definition of $\mathcal{Q}$. As we mentioned in Section 10.2, there is a finite set of primes r, called the ramified primes, for which $\mathcal{Q}_r$ has no zero divisors. On the other hand, when r is unramified, $\mathcal{Q}_r$ is isomorphic to $M_2(\mathbf{Q}_r)$, the ring of 2×2 matrices with r-adic entries.

Given two distinct primes p and q, there is a quaternion algebra $\mathcal{B}$ that is ramified exactly at p and q. In particular, $\mathcal{B}$ is unramified at ∞, so

$$\mathcal{B}_\infty = M_2(\mathbf{R}).$$

Corresponding to the integer $M = N/pq$, there is an order $\mathcal{O} \subset \mathcal{B}$, called an Eichler order of level M (an order in $\mathcal{B}$ is a subring of $\mathcal{B}$ that has rank 4 as an additive abelian group; see Section 10.2). Regarding $\mathcal{O}$ as a subset of $\mathcal{B}_\infty = M_2(\mathbf{R})$, define

$$\Gamma_\infty = \mathcal{O} \cap SL_2(\mathbf{R}).$$

Then Γ_∞ acts on $\mathcal{H}$ by linear fractional transformations. The **Shimura curve** C is defined to be $\mathcal{H}$ modulo Γ_∞.

There is another description of C, analogous to the one given above for $X_0(N)$. Let $\mathcal{O}_{\max}$ be a maximal order in $\mathcal{B}$. Consider pairs (A, B), where A is a two-dimensional abelian variety (these are algebraic varieties that, over $\mathbf{C}$, can be described as $\mathbf{C}^2$ mod a rank 4 lattice) and B is a subgroup of A isomorphic to $\mathbf{Z}_M \oplus \mathbf{Z}_M$. We restrict our attention to those pairs such that $\mathcal{O}_{\max}$ is contained in the endomorphism ring of A and such that $\mathcal{O}_{\max}$ maps B to B. When we are working over $\mathbf{C}$, such pairs are in one-to-one correspondence with the points on C. In general, over arbitrary fields, such pairs correspond in a natural way to points on an algebraic curve, which we again denote C.

Let J be the Jacobian of C. The description of C in terms of pairs (A, B) means that we can define an action of the Hecke operators on J, similarly to what we did for the modular curves.

Let $J[\ell]$ be the ℓ-torsion of the Jacobian J of C. It can be shown that the representation ρ occurs in $J[\mathcal{M}]$, so there is a space V isomorphic to the representation space W of ρ with

$$V \subseteq J[\mathcal{M}] \subseteq J[\ell].$$

We now have the representation ρ living in $J_0(N)[\ell]$ and in $J[\ell]$. The representation ρ can be detected using the reduction of $J_0(N)$ mod q and also using the reduction of J mod p, and Ribet uses a calculation with quaternion algebras to establish a relationship between these two reductions. This relationship allows him to show that p can be removed from the level N.

REMARK 13.9 A correspondence between modular forms for GL_2 and modular forms for the multiplicative group of a quaternion algebra plays a major role in work of Jacquet-Langlands. This indicates a relation between $J_0(N)$ and J. In fact, there is an surjection from $J_0(N)$ to J. However, this map is not being used in the present case since such a map would relate the reduction of $J_0(N)$ mod q to the reduction of J mod q. Instead, Ribet works with the reduction of $J_0(N)$ mod q and the reduction of J mod p. This switch between p and q is a major step in the proof of Ribet's theorem. ∎

13.4 Sketch of Wiles's Proof

In this section, we outline the proof that all semistable elliptic curves over $\mathbf{Q}$ are modular. For more details, see [22], [24], [99], [109]. Let E be a semistable elliptic curve and let

$$f_E = \sum_{n \geq 1} a_n q^n$$

be the associated potential modular form. We want to prove that f_E is a modular form (for some $\Gamma_0(N)$).

Suppose we have two potential modular forms

$$f = \sum_{n \geq 1} c_n q^n, \qquad g = \sum_{n \geq 1} c'_n q^n$$

arising from Galois representations $G \rightarrow GL_2(\mathcal{O}_p)$ (where $\mathcal{O}_p$ is some ring containing the p-adic integers. We assume that all of the coefficients c_n, c'_n are embedded in $\mathcal{O}_p$). Let $\tilde{p}$ be the prime above p in $\mathcal{O}_p$. (If $\mathcal{O}_p$ is the ring of p-adic integers, then $\tilde{p} = p$.) If $c_\ell \equiv c'_\ell \pmod{\tilde{p}}$ for almost all primes ℓ, then we write

$$f \equiv g \pmod{\tilde{p}}.$$

This means that the Galois representations mod $\tilde{p}$ associated to f and g are equivalent.

The following result of Langlands and Tunnell gives us a place to start.

THEOREM 13.10

Let E be an elliptic curve defined over $\mathbf{Q}$ and let $f_E = \sum_{n \geq 1} a_n q^n$ be the associated potential modular form. There exists a modular form

$$g_0 = \sum_{n \geq 1} b_n q^n$$

such that

$$a_\ell \equiv b_\ell \pmod{\tilde{3}}$$

for almost all primes ℓ (that is, with possibly finitely many exceptions), and where $\tilde{3}$ denotes a prime of $\mathcal{O}_3$.

Recall that $\mathcal{O}_3$ denotes an unspecified ring containing the 3-adic integers. If $\mathcal{O}_3$ is sufficiently large, the coefficients b_ℓ, which are algebraic integers, can be regarded as lying in $\mathcal{O}_3$.

The reason that 3 is used is that the group $GL_2(\mathbf{F}_3)$ has order 48, hence is solvable. The representation ρ_3 of G on $E[3]$ therefore has its image in a solvable group. The techniques of base change developed in the Langlands program apply to cyclic groups, hence to solvable groups, and these techniques are the key to proving the result. The groups $GL_2(\mathbf{F}_p)$ for $p \geq 5$ are not solvable, so the base change techniques do not apply. On the other hand, the representation ρ_2 for the Galois action on $E[2]$ is trivial for the Frey curves since the 2-torsion is rational for these curves. Therefore, it is not expected that ρ_2 should yield any information.

Note that the modular form g_0 does not necessarily have rational coefficients. Therefore, g_0 is not necessarily the modular form associated to an elliptic curve. Throughout Wiles's proof, Galois representations associated to arbitrary modular forms are used.

The result of Langlands and Tunnell leads us to consider the following.

GENERAL PROBLEM

Fix a prime p. Let $g = \sum_{n \geq 1} a_n q^n$ be a potential modular form (associated to a 2-dimensional Galois representation). Suppose there is a modular form $g_0 = \sum b_n q^n$ such that $g \equiv g_0 \pmod{p}$. Can we prove that g is a modular form?

The work of Wiles shows that the answer to the general problem is often yes. Let A be the set of all potential modular forms g with $g \equiv g_0 \pmod{p}$ (subject to certain restrictions). Let $M \subseteq A$ be the set of modular g's in A. We are assuming that $g_0 \in M$. The basic idea is the following. Let T_A be the tangent space to A at g_0 and let T_M be the tangent space to M at g_0. The goal is to show that $T_A = T_M$. Wiles shows that the spaces A and M are nice enough that the equality of tangent spaces suffices to imply that $A = M$.

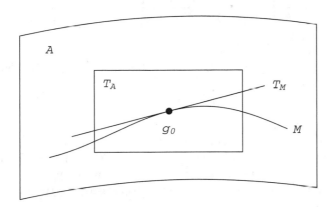

Figure 13.2
 Tangent Spaces

Example 13.2
Let E be given by

$$y^2 + xy + y = x^3 - x^2 - 171x + 1904.$$

This curve has multiplicative reduction at 17 and 37 and good reduction at all other primes. Therefore, E is semistable. The minimal discriminant of E is $\Delta = -17 \cdot 37^5$. Since E is semistable, the conductor of E is $N = 17 \cdot 37$. Therefore, we expect that g_E is a modular form for $\Gamma_0(17 \cdot 37)$. Counting points on $E \mod \ell$ for various ℓ yields the following values for a_ℓ (we ignore the bad prime 17):

ℓ	2	3	5	7	11	13	17	19	23
a_ℓ	-1	0	3	-1	-5	-2	$-$	1	-6

Therefore,

$$g_E = q - q^2 + 0 \cdot q^3 - q^4 + 3q^5 + \cdots.$$

There is a modular form

$$g_0 = \sum b_n q^n = q - q^2 + 0 \cdot q^3 - q^4 - 2q^5 + \cdots$$

for $\Gamma_0(17)$. The first few values of b_ℓ are as follows:

ℓ	2	3	5	7	11	13	17	19	23
b_ℓ	-1	0	-2	4	0	-2	$-$	-4	4

It can be shown that $a_\ell \equiv b_\ell \pmod 5$ for all $\ell \neq 17, 37$ (we ignore these bad primes), so

$$g_E \equiv g_0 \pmod 5.$$

Can we prove that g_E is a modular form?

Let A be the set of all potential modular forms g with $g \equiv g_0 \pmod 5$ and where the level N for g is allowed to contain only the primes $5, 17, 37$ in its factorization. There is also a technical condition, which we omit, on the ring generated by the coefficients of g. The subspace M of true modular forms contains g_0. Here are pictures of A and M:

$$A: \quad \bullet \quad \bullet$$
$$\quad g_0 \quad g_E$$

$$M: \quad \bullet \quad \text{or} \quad \bullet \quad \bullet$$
$$\qquad g_0 \qquad\qquad g_0 \quad g_E$$

Therefore, our intuitive picture given in Figure 13.2 is not quite accurate. In particular, the sets A and M are finite. However, by reinterpreting the geometric picture algebraically, we can still discuss tangent spaces.

Since the sets A and M are finite, why not count the elements in both sets and compare? First of all, this seems to be hard to do. Secondly, the tangent spaces yield enough information. Consider the following situation. Suppose you arrive at a train station in a small town. There are no signs telling you which town it is, but you know it must be either I or II. You have the maps given in Figure 13.3, where the large dot in the center indicates the station.

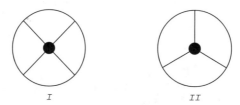

I   II

Figure 13.3
Two Small Towns

By counting the streets emanating from the station, you can immediately determine which town you are in. The reason is that you have a base point. If you didn't, then you might be on any of the vertices of I or II. You would not be able to count streets and identify the town. The configuration of streets at the station is the analogue of the tangent space at the base point. Of course, it is possible that two towns could have the same tangent spaces, but Wiles shows that this does not happen in his situation.

Tangent spaces

We now want to translate the notion of a tangent space into a useful algebraic formulation. Let $\mathbf{R}[x, y]$ be the ring of polynomials in two variables and let $f(x, y) \in \mathbf{R}[x, y]$. We can regard f as a function from the xy-plane to $\mathbf{R}$. Restricting f to the parabola $y = x^2 - 6x$, we obtain a function

$$f : \text{parabola} \longrightarrow \mathbf{R}.$$

If $g(x, y) \in \mathbf{R}[x, y]$, then f and g give the same function on the parabola if and only if $f - g$ is a multiple of $y + 6x - x^2$. For example, let $f = x^3 - y$ and $g = 6x + xy + 5x^2$. Then

$$f - g = -(x + 1)(y + 6x - x^2).$$

If we choose a point (a, b) on the parabola, then $b + 6a - a^2 = 0$, so

$$f(a, b) = g(a, b) - (a + 1)(b + 6a - a^2) = g(a, b).$$

Therefore, there is a one-to-one correspondence

$$\text{polynomial functions on the parabola} \quad \longleftrightarrow \quad \mathbf{R}[x, y]/(y + 6x - x^2).$$

The ring on the right consists of congruence classes of polynomials, where we say that two polynomials are congruent if their difference is a multiple of $y + 6x - x^2$. In this way, we have represented a geometric object, the parabola, by an algebraic object, the ring $\mathbf{R}[x, y]/(y + 6x - x^2)$.

Now let's consider the tangent line $y + 6x = 0$ at $(0, 0)$. It is obtained by taking the degree 1 terms in $y + 6x - x^2$. We can represent it by the set

$$\{ax + by \,|\, a, b \in \mathbf{R}\} \quad \text{mod } (y + 6x),$$

where we are taking all linear functions and regarding two of them as congruent if they differ by a multiple of $y + 6x$. Of course, we could have represented the tangent line by the ring $\mathbf{R}[x, y]/(y + 6x)$, but, since we already know that the tangent line is defined by a linear equation, we do not lose any information by replacing $\mathbf{R}[x, y]$ by the linear polynomials $ax + by$.

Now consider the surface

$$y - x^2 + xz + 6x + z = 0.$$

This surface contains the parabola $y = x^2 - 6x$, $z = 0$. The inclusion of the parabola in the surface corresponds to a surjective ring homomorphism

$$
\begin{array}{ccc}
\mathbf{R}[x, y, z]/(y - x^2 + xz + 6x + z) & \longrightarrow & \mathbf{R}[x, y]/(y + 6x - x^2) \\
f(x, y, z) & \longmapsto & f(x, y, 0).
\end{array}
$$

We also have a surjective map on the algebraic objects representing the tangent spaces

$$\{ax + by + cz\} \quad \mathrm{mod} \ (y + 6x + z) \quad \longrightarrow \quad \{ax + by\} \quad \mathrm{mod} \ (y + 6x)$$

corresponding to the inclusion of the tangent line to the parabola in the tangent plane for the surface at $(0, 0, 0)$. In this way, we can study relations between geometric objects by looking at the corresponding algebraic objects.

Wiles works with rings such as $\mathcal{O}_p[[x]]/(x^2 - px)$, where for simplicity we henceforth assume that $\mathcal{O}_p$ is the p-adic integers and where $\mathcal{O}_p[[x]]$ denotes power series with p-adic coefficients. The zeros of $x^2 - px$ are 0 and p, so this ring corresponds to the geometric object

$$
S_1 : \qquad \bullet \qquad \bullet
$$
$$
 0 \qquad p
$$

The tangent space is represented by the set obtained by looking only at the linear terms, namely $\{ax \mid a \in \mathcal{O}_p\} \quad \mathrm{mod} \ (px)$. Since

$$a_1 x \equiv a_2 x \quad \mathrm{mod} \ px \quad \Longleftrightarrow \quad a_1 \equiv a_2 \quad (\mathrm{mod} \ p),$$

the tangent space can be identified with $\mathbf{Z}_p$.

As another example, consider the ring $\mathcal{O}_p[[x]]/(x(x - p)(x - p^3))$, which corresponds to the geometric object

$$
S_2 : \qquad \bullet \qquad \bullet \qquad \bullet
$$
$$
 0 \qquad p \qquad p^3
$$

The tangent space is $\mathbf{Z}_{p^4}$.

There is an inclusion $S_1 \subset S_2$, which corresponds to the natural ring homomorphism

$$\mathcal{O}_p[[x]]/(x(x - p)(x - p^3)) \longrightarrow \mathcal{O}_p[[x]]/(x^2 - px).$$

The map on tangent spaces is the map from $\mathbf{Z}_{p^4}$ to $\mathbf{Z}_p$ that takes a number mod p^4 and reduces it mod p.

Now consider the ring $\mathcal{O}_p[[x, y]]/(x^2 - px, y^2 - py)$. In this case, we are looking at power series in two variables, and two power series are congruent if their difference is a linear combination of the form $A(x,y)(x^2-px)+B(x,y)(y^2-py)$ with $A, B \in \mathcal{O}_p[[x,y]]$. The corresponding geometric object is

$$S_3: \qquad \begin{matrix} \bullet & \bullet \\ (0,p) & (p,p) \\[1em] \bullet & \bullet \\ (0,0) & (p,0) \end{matrix}$$

It can be shown that two power series give the same function on this set of four points if they differ by a linear combination of $x^2 - px$ and $y^2 - py$. The tangent space is represented by

$$\{ax + by \mid a, b \in \mathcal{O}_p\} \quad \bmod (px, py),$$

which means we are considering two linear polynomials to be congruent if their difference is a linear combination of px and py. It is easy to see that

$$a_1 x + b_1 y \equiv a_2 x + b_2 y \quad \bmod (px, py) \quad \Longleftrightarrow \quad a_1 \equiv a_2, \quad b_1 \equiv b_2 \pmod{p}.$$

Therefore, the tangent space is isomorphic to $\mathbf{Z}_p \oplus \mathbf{Z}_p$.

The inclusion $S_1 \subset S_3$ corresponds to the ring homomorphism

$$\mathcal{O}_p[[x,y]]/(x^2 - px, \ y^2 - py) \longrightarrow \mathcal{O}_p[[x]]/(x^2 - px).$$

The map on tangent spaces is the map $\mathbf{Z}_p \oplus \mathbf{Z}_p \to \mathbf{Z}_p$ given by projection onto the first factor.

In all three examples above, the rings are given by power series over $\mathcal{O}_p$. The number of variables equals the number of relations and the resulting ring is a finitely generated $\mathcal{O}_p$-module (this is easily verified in the three examples). Such rings are called **local complete intersections**. For such rings, it is possible to recognize when a map is an isomorphism by looking at the tangent spaces.

Before proceeding, let's look at an example that is not a local complete intersection. Consider the ring

$$\mathcal{O}_p[[x,y]]/(x^2 - px, \ y^2 - py, \ xy).$$

The corresponding geometric object is

$$S_4: \qquad \begin{matrix} \bullet & \\ (0,p) & \\[1em] \bullet & \bullet \\ (0,0) & (p,0) \end{matrix}$$

There are two variables and three relations, so we do not have a complete intersection. The tangent space is $\mathbf{Z}_p \oplus \mathbf{Z}_p$. The inclusion $S_4 \subset S_3$ corresponds to the ring homomorphism

$$\mathcal{O}_p[[x,y]]/(x^2 - px, y^2 - py) \longrightarrow \mathcal{O}_p[[x,y]]/(x^2 - px, y^2 - py, xy)$$

and the map on tangent spaces is an isomorphism. However, $S_3 \neq S_4$. The problem is that the tangent space calculation does not notice the relation xy, which removed the point (p,p) from S_3 to get S_4. Therefore, the tangent space thinks this point is still there and incorrectly predicts an isomorphism between the three point space and the four point space.

The general fact we need is that if we have a surjective homomorphism of rings that are local complete intersections, and if the induced map on tangent spaces is an isomorphism, then the ring homomorphism is an isomorphism.

Deformations of Galois representations

Now let's return to our sets A and M. Corresponding to these two sets are rings R_A and R_M. We have $g_0 \in M \subseteq A$. Let T_A and T_M be the tangent spaces at g_0. In the examples above, the base point g_0 would correspond to $x = 0$ or to $(x, y) = (0, 0)$. Corresponding to the inclusion $M \subseteq A$, there are surjective maps

$$R_A \longrightarrow R_M, \quad T_A \longrightarrow T_M.$$

Therefore,

$$\#T_M \leq \#T_A.$$

The ring R_M can be constructed using the Hecke algebra and the ring R_A is constructed using results about representability of functors. In fact, it was shown that there is a representation

$$\rho_{\text{universal}} : G \longrightarrow GL_2(R_A)$$

with the following property. Let

$$\rho : G \longrightarrow GL_2(\mathcal{O}_p)$$

be a representation and let g be the potential modular form attached to ρ. Assume that ρ is unramified outside a fixed finite set of primes. If $g \equiv g_0$ $(\mathrm{mod}\ \tilde{p})$, then there exists a unique ring homomorphism

$$\phi : R_A \longrightarrow \mathcal{O}_p$$

such that the diagram

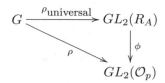

commutes.

The representations ρ such that $g \equiv g_0 \pmod{\tilde{p}}$ are examples of what are known as **deformations** of the Galois representation for g_0. The representation $\rho_{\text{universal}}$ is called a universal deformation.

Example 13.3

We continue with Example 13.2. Let $p = 5$ and take the fixed set of primes to be $\{5, 17, 37\}$. Then it can be shown that

$$R_A \simeq \mathcal{O}_5[[x]]/(x^2 - bx),$$

where $b/5$ is a 5-adic unit and $\mathcal{O}_5$ is the ring of 5-adic integers. This implies that $T_A = \mathbf{Z}_5$. The set A has two points, g_0 and g, corresponding to $x = 0$ and $x = b$.   □

There exists an integer n, defined below, such that

$$n \le \#T_M \le \#T_A.$$

Moreover, a result of Flach shows that $n \cdot T_A = 0$. If it can be shown that $n = \#T_A$, then $T_A = T_M$.

In our example, $n = 5$. Since we know that $T_A = \mathbf{Z}_5$, we have $n = \#T_A$. Therefore, $T_A = T_M$. It can be shown that R_A and R_M are local complete intersections. This yields $R_A = R_M$ and $A = M$. This implies that g is a modular form.

In general, recall that we started with a semistable elliptic curve E. Associated to E is the 3-adic Galois representation $\rho_{3\infty}$. The theorem of Langlands-Tunnell yields a modular form g_0, and therefore a Galois representation

$$\rho_0 : G \longrightarrow GL_2(\mathcal{O}_3).$$

We have

$$\rho_{3\infty} \equiv \rho_0 \pmod{\tilde{3}},$$

so the base point ρ_0 is modular and semistable mod $\tilde{3}$ (the notion of semistability can be defined for general Galois representations). Under the additional assumption that ρ_3 restricted to $\text{Gal}(\overline{\mathbf{Q}}/\mathbf{Q}(\sqrt{-3}))$ is absolutely irreducible, Wiles showed that if R_M is a local complete intersection then $n = \#T_A$ and the map $R_A \to R_M$ is an isomorphism of local complete intersections. Finally, in 1994, Wiles and Taylor used an ingenious argument to show that R_M is a local complete intersection, and therefore $A = M$.

What happens if ρ_3 does not satisfy the irreducibility assumption? Wiles showed that there is a semistable elliptic curve E' with the same mod 5 representation as E but whose mod 3 representation is irreducible. Therefore, E' is modular, so the mod 5 representation of E' is modular. This means that the mod 5 representation of E is modular. If the mod 5 representation, restricted to $\text{Gal}(\overline{\mathbf{Q}}/\mathbf{Q}(\sqrt{5}))$, is absolutely irreducible, then the above result of Wiles, with 5 in place of 3, shows that E is modular.

There are only finitely many elliptic curves over $\mathbf{Q}$ for which both the mod 3 representation (restricted to $\text{Gal}(\overline{\mathbf{Q}}/\mathbf{Q}(\sqrt{-3}))$) and the mod 5 representation (restricted to $\text{Gal}(\overline{\mathbf{Q}}/\mathbf{Q}(\sqrt{5}))$) are not absolutely irreducible. These finitely many exceptions can be proved to be modular individually.

Therefore, semistable elliptic curves over $\mathbf{Q}$ are modular. Eventually, the argument was extended by Breuil, Conrad, Diamond, and Taylor to include all elliptic curves over $\mathbf{Q}$ (Theorem 12.4).

The integer n is defined as follows. Let $g_0 = \sum b_m q^m$ and let

$$L(g_0, s) = \sum_{m=1}^{\infty} b_m m^{-s} = \prod_{\ell = \text{prime}} \left(1 - b_\ell \ell^{-s} + \ell^{1-2s}\right)^{-1}.$$

Write

$$1 - b_\ell X + \ell X^2 = (1 - \alpha_\ell X)(1 - \beta_\ell X).$$

The **symmetric square L-function** is defined to be

$$L(\text{Sym}^2 g_0, s) = \prod_{\ell \notin S} \left((1 - \alpha_\ell^2 \ell^{-s})(1 - \beta_\ell^2 \ell^{-s})(1 - \alpha_\ell \beta_\ell \ell^{-s})\right)^{-1},$$

where S is a finite set of bad primes (in our example, $S = \{5, 17, 37\}$). There exists a naturally defined transcendental number Ω (similar to the periods considered in Section 9.4), defined by a double integral, such that

$$\frac{L(\text{Sym}^2 g_0, 2)}{\Omega} = r = \text{a rational number.}$$

The number n is defined to be the p-part of r (that is, n is a power of p such that r equals n times a rational number with numerator and denominator prime to p).

The formula that Wiles proved is therefore that $L(\text{Sym}^2 g_0, 2)/\Omega$ equals $\#T_A$ times a rational number prime to p. This means that the order of an algebraic object, namely T_A, is expressed in terms of the value of an analytic function, in this case the symmetric square L-function. This formula is therefore of a nature similar to the analytic class number of algebraic number theory, which expresses the class number in terms of an L-series, and the conjecture of Birch and Swinnerton-Dyer (see Section 12.2), which expresses the order of the Shafarevich-Tate group of an elliptic curve in terms of the value of its L-series.

Appendix A

Number Theory

Basic results

Let n be a positive integer and let $\mathbf{Z}_n$ be the set of integers mod n. It is a group with respect to addition. We can represent the elements of $\mathbf{Z}_n$ by the numbers $0, 1, 2, \ldots, n-1$. Let

$$\mathbf{Z}_n^\times = \{a \mid 1 \le a \le n, \ \gcd(a, n) = 1\}.$$

Then $\mathbf{Z}_n^\times$ is a group with respect to multiplication mod n.

Let $a \in \mathbf{Z}_n^\times$. The **order of** a **mod** n is the smallest integer $k > 0$ such that $a^k \equiv 1 \pmod{n}$. The order of a mod n divides $\phi(n)$, where ϕ is the Euler ϕ-function.

Let p be a prime and let $a \in \mathbf{Z}_p^\times$. The order of a mod p divides $p-1$. A **primitive root** mod p is an integer g such that the order of g mod p equals $p-1$. If g is a primitive root mod p, then every integer is congruent mod p to 0 or to a power of g. For example, 3 is a primitive root mod 7 and

$$\{1, 3, 9, 27, 81, 243\} \equiv \{1, 3, 2, 6, 4, 5\} \pmod{7}.$$

There are $\phi(p-1)$ primitive roots mod p. In particular, a primitive root mod p always exists, so $\mathbf{Z}_p^\times$ is a cyclic group.

There is an easy criterion for deciding whether g is a primitive root mod p, assuming we know the factorization of $p-1$: If $g^{(p-1)/q} \not\equiv 1 \pmod{p}$ for all primes $q | p - 1$, then g is a primitive root mod p. This can be proved by noting that if g is not a primitive root, then its order is a proper divisor of $p-1$, hence divides $(p-1)/q$ for some prime q.

One way to find a primitive root for p, assuming the factorization of $p-1$ is known, is simply to test the numbers 2, 3, 5, 6, ... successively until a primitive root is found. Since there are many primitive roots, one should be found fairly quickly in most cases.

A very useful result in number theory is the following.

THEOREM A.1 (Chinese Remainder Theorem)
Let $n_1, n_2, \ldots, n_r$ be positive integers such that $\gcd(n_i, n_j) = 1$ when $i \neq j$.
Let $a_1, a_2, \ldots, a_r$ be integers. Then there exists an x such that

$$x \equiv a_i \pmod{n_i} \quad \text{for all } i.$$

The integer x is uniquely determined mod $n_1 n_2 \cdots n_r$.

For example, let $n_1 = 4$, $n_2 = 3$, $n_3 = 5$ and let $a_1 = 1$, $a_2 = 2$, $a_3 = 3$. Then $x = 53$ is a solution to the simultaneous congruences

$$x \equiv 1 \pmod 4, \quad x \equiv 2 \pmod 3, \quad x \equiv 3 \pmod 5,$$

and any solution x satisfies $x \equiv 53 \pmod{60}$.

Another way to state the Chinese Remainder Theorem is to say that if $\gcd(n_i, n_j) = 1$ for $i \neq j$, then

$$\mathbf{Z}_{n_1 n_2 \cdots n_r} \simeq \mathbf{Z}_{n_1} \oplus \cdots \oplus \mathbf{Z}_{n_r}$$

(see Appendix B for the definition of $\oplus$). This is an isomorphism of additive groups. It is also an isomorphism of rings.

p-adic numbers

Let p be a prime number and let x be a nonzero rational number. Write

$$x = p^r \frac{a}{b},$$

where a, b are integers such that $p \nmid ab$. Then r is called the *p*-**adic valuation** of x and is denoted by

$$r = v_p(x).$$

Define $v_p(0) = \infty$. The *p*-**adic absolute value** of x is defined to be

$$|x|_p = p^{-r}.$$

Define $|0|_p = 0$.

For example,

$$\left| \frac{12}{35} \right|_2 = \frac{1}{4}, \quad \left| \frac{11}{250} \right|_5 = 125, \quad \left| \frac{1}{2} - 41 \right|_3 = \frac{1}{81}.$$

The last example says that $1/2$ and 41 are close 3-adically. Note that two integers are close *p*-adically if and only if they are congruent mod a large power of p.

The *p*-**adic integers** are most easily regarded as sums of the form

$$\sum_{n=0}^{\infty} a_n p^n, \quad a_n \in \{0, 1, 2, \ldots, p-1\}.$$

Such infinite sums do not converge in the real numbers, but they do make sense with the *p*-adic absolute value since $|a_n p^n|_p \to 0$ as $n \to \infty$.

Arithmetic operations are carried out just as with finite sums. For example, in the 3-adic integers,

$$(1 + 2 \cdot 3 + 0 \cdot 3^2 + \cdots) + (1 + 2 \cdot 3 + 1 \cdot 3^2 + \cdots) = 2 + 4 \cdot 3 + 1 \cdot 3^2 + \cdots$$
$$= 2 + 1 \cdot 3 + 2 \cdot 3^2 + \cdots$$

(where we wrote $4 = 1 + 3$ and regrouped, or "carried," to obtain the last expression). If

$$x = a_k p^k + a_{k+1} p^{k+1} + \cdots$$

with $a_k \neq 0$, then

$$-x = (p - a_k)p^k + (p - 1 - a_{k+1})p^{k+1} + (p - 1 - a_{k+2})p^{k+2} + \cdots \quad \text{(A.1)}$$

(use the fact that $p^{k+1} + (p-1)p^{k+1} + (p-1)p^{k+2} + \cdots = 0$ because the sum telescopes, so all the terms cancel). Therefore, *p*-adic integers have additive inverses. It is not hard to show that the *p*-adic integers form a ring.

Any rational number with denominator not divisible by p is a *p*-adic integer. For example, in the 3-adics,

$$\frac{1}{2} = \frac{-1}{1-3} = -(1 + 3 + 3^2 + \cdots) = 2 + 3 + 3^2 + \cdots,$$

where we used (A.1) for the last equality. In fact, it can be shown that if $x = \sum_{n=0}^{\infty} a_n p^n$ is a *p*-adic integer with $a_0 \neq 0$, then $1/x$ is a *p*-adic integer.

The *p*-**adic rationals**, which we denote by $\mathbf{Q}_p$, are sums of the form

$$y = \sum_{n=m}^{\infty} a_n p^n, \quad \text{(A.2)}$$

with m positive or negative or zero and with $a_n \in \{0, 1, \ldots, p-1\}$. If $y \in \mathbf{Q}_p$, then $p^k y$ is a *p*-adic integer for some integer k. The *p*-adic rationals form a field, and every rational number lies in $\mathbf{Q}_p$. If $a_m \neq 0$ in (A.2), then we define

$$v_p(y) = m, \quad |y|_p = p^{-m}.$$

This agrees with the definitions of the *p*-adic valuation and absolute value defined above when y is a rational number.

Another way to look at *p*-adic integers is the following. Consider sequences of integers $x_1, x_2, \ldots$ such that

$$x_m \equiv x_{m+1} \pmod{p^m} \quad \text{(A.3)}$$

for all $m \geq 1$. Since $x_m \equiv x_k \pmod{p^m}$ for all $k \geq m$, the base p expansions for all x_k with $k \geq m$ must agree through the p^{m-1} term. Therefore, the sequence of integers x_m determines an expression of the form

$$\sum_{n=0}^{\infty} a_n p^n,$$

where

$$x_m \equiv \sum_{n=0}^{m-1} a_n p^n \pmod{p^m}$$

for all m. In other words, the sequence of integers determines a p-adic integer. Conversely, the partial sums of a p-adic integer determine a sequence of integers satisfying (A.3).

Let's use these ideas to show that -1 is a square in the 5-adic integers. Let $x_1 = 2$, so

$$x_1^2 \equiv -1 \pmod{5}.$$

Suppose we have defined x_m such that

$$x_m^2 \equiv -1 \pmod{5^m}.$$

Let $x_{m+1} = x_m + b5^m$, where

$$b \equiv \frac{-1 - x_m^2}{2 \cdot 5^m x_m} \pmod{5}.$$

Note that $x_m^2 \equiv -1 \pmod{5^m}$ implies that the right side of this last congruence is defined mod 5. A quick calculation shows that

$$x_{m+1}^2 \equiv -1 \pmod{5^{m+1}}.$$

Since (A.3) is satisfied, there is a 5-adic integer x with $x \equiv x_m \pmod{5^m}$ for all m. Moreover,

$$x^2 \equiv -1 \pmod{5^m}$$

for all m. This implies that $x^2 = -1$.

In general, this procedure leads to the following very useful result.

THEOREM A.2 (Hensel's Lemma)
Let $f(X)$ be a polynomial with coefficients that are p-adic integers and suppose x_1 is an integer such that

$$f(x_1) \equiv 0 \pmod{p}.$$

If

$$f'(x_1) \not\equiv 0 \pmod{p},$$

then there exists a p-adic integer x with $x \equiv x_1$ (mod p) and

$$f(x) = 0.$$

COROLLARY A.3
Let p be an odd prime and suppose b is a p-adic integer that is a nonzero square mod p. Then b is the square of a p-adic integer.

The corollary can be proved by exactly the same method that was used to prove that -1 is a square in the 5-adic integers. The corollary can also be deduced from the theorem as follows. Define $f(X) = X^2 - b$ and let $x_1^2 \equiv b$ (mod p). Then $f(x_1) \equiv 0$ (mod p) and

$$f'(x_1) = 2x_1 \not\equiv 0 \quad (\text{mod } p)$$

since p is odd and $x_1 \not\equiv 0$ by assumption. Hensel's Lemma shows that there is a p-adic integer x with $f(x) = 0$. This means that $x^2 = b$, as desired.

When $p = 2$, the corollary is not true. For example, 5 is a square mod 2 but is not a square mod 8, hence is not a 2-adic square. However, the inductive procedure used above yields the following:

PROPOSITION A.4
If b is a 2-adic integer such that $b \equiv 1$ (mod 8) then b is the square of a 2-adic integer.

Appendix B

Groups

Basic definitions

Since most of the groups in this book are additive abelian groups, we'll use additive notation for the group operations in this appendix. Therefore, a group G has a binary operation $+$ that is associative. There is an additive identity that we'll call 0 satisfying

$$0 + g = g + 0 = g$$

for all $g \in G$. Each $g \in G$ is assumed to have an additive inverse $-g$ satisfying

$$(-g) + g = g + (-g) = 0.$$

If n is a positive integer, we let

$$ng = g + g + \cdots + g \quad (n \text{ summands}).$$

If $n < 0$, we let $ng = -(|n|g) = -(g + \cdots + g)$.

Almost all of the groups in this book are abelian, which means that $g + h = h + g$ for all $g, h \in G$.

If G is a finite group, the **order** of G is the number of elements in G. The **order of an element** $g \in G$ is the smallest integer $k > 0$ such that $kg = 0$. If k is the order of g, then

$$g^i = g^j \iff i \equiv j \pmod{k}.$$

The basic result about orders is the following.

THEOREM B.1 (Lagrange's Theorem)
Let G be a finite group.

1. *Let H be a subgroup of G. Then the order of H divides the order of G.*

2. *Let $g \in G$. Then the order of g divides the order of G.*

A **cyclic** group is a group isomorphic to either $\mathbf{Z}$ or $\mathbf{Z}_n$ for some n. These groups have the property that they can be generated by one element. For example, $\mathbf{Z}_4$ is generated by 1, and it is also generated by 3 since $\{0, 3, 3 + 3, 3 + 3 + 3\}$ is all of $\mathbf{Z}_4$. The following result says that the converse of Lagrange's theorem holds for finite cyclic groups.

THEOREM B.2

Let G be a finite cyclic group of order n. Let $d > 0$ divide n.

1. *G has a unique subgroup of order d.*

2. *G has d elements of order dividing d, and G has $\phi(d)$ elements of order exactly d (where $\phi(d)$ is Euler's ϕ-function).*

For example, $\mathbf{Z}_6$ contains the subgroup $\{0, 2, 4\}$ of order 3. The elements $2, 4 \in \mathbf{Z}_6$ have order 3.

The **direct sum** of two groups G_1 and G_2 is defined to be the set of ordered pairs formed from elements of G_1 and G_2:

$$G_1 \oplus G_2 = \{(g_1, g_2) \mid g_1 \in G_1,\, g_2 \in G_2\}.$$

Ordered pairs can be added componentwise:

$$(g_1, g_2) + (h_1, h_2) = (g_1 + h_1, g_2 + h_2).$$

This makes $G_1 \oplus G_2$ into a group with $(0, 0)$ as the identity element. A similar definition holds for the direct sum of more than two groups. We write G^r for the direct sum of r copies of G. In particular, $\mathbf{Z}^r$ denotes the set of r-tuples of integers, which is a group under addition.

Structure theorems

THEOREM B.3

A finite abelian group is isomorphic to a group of the form

$$\mathbf{Z}_{n_1} \oplus \mathbf{Z}_{n_2} \oplus \cdots \oplus \mathbf{Z}_{n_s}$$

with $n_i \mid n_{i+1}$ for $i = 1, 2, \ldots, s - 1$. The integers n_i are uniquely determined by G.

An abelian group G is called **finitely generated** if there is a finite set $\{g_1, g_2, \ldots, g_k\}$ contained in G such that every element of G can be written (not necessarily uniquely) in the form

$$m_1 g_1 + \cdots + m_k g_k$$

with $m_i \in \mathbf{Z}$.

THEOREM B.4

A finitely generated abelian group is isomorphic to a group of the form

$$\mathbf{Z}^r \oplus \mathbf{Z}_{n_1} \oplus \mathbf{Z}_{n_2} \oplus \cdots \oplus \mathbf{Z}_{n_s}$$

with $r \geq 0$ and with $n_i | n_{i+1}$ for $i = 1, 2, \ldots, s-1$. The integers r and n_i are uniquely determined by G.

The subgroup of G isomorphic to

$$\mathbf{Z}_{n_1} \oplus \mathbf{Z}_{n_2} \oplus \cdots \oplus \mathbf{Z}_{n_s}$$

is called the **torsion subgroup** of G. The integer r is called the **rank** of G.

Appendix C

Fields

Let K be a field. There is a ring homomorphism $\psi : \mathbf{Z} \to K$ that sends $1 \in \mathbf{Z}$ to $1 \in K$. If ψ is injective, then we say that K has **characteristic** 0. Otherwise, there is a smallest positive integer p such that $\psi(p) = 0$. In this case, we say that K has **characteristic** p. If p factors as ab with $1 < a \le b < p$, then $\psi(a)\psi(b) = \psi(p) = 0$, so $\psi(a) = 0$ or $\psi(b) = 0$, contradicting the minimality of p. Therefore, p is prime.

When K has characteristic 0, the field $\mathbf{Q}$ of rational numbers is contained in K. When K has characteristic p, the field $\mathbf{F}_p$ of integers mod p is contained in K.

Let K and L be fields with $K \subseteq L$. If $\alpha \in L$, we say that α is **algebraic** over K if there exists a nonconstant polynomial

$$f(X) = X^n + a_{n-1}X^{n-1} + \cdots + a_0$$

with $a_0, \ldots, a_{n-1} \in K$ such that $f(\alpha) = 0$. We say that L is an **algebraic** over K, or that L is an **algebraic extension** of K, if every element of L is algebraic over K. An **algebraic closure** of a field K is a field $\overline{K}$ containing K such that

1. $\overline{K}$ is algebraic over K.

2. Every nonconstant polynomial $g(X)$ with coefficients in $\overline{K}$ has a root in $\overline{K}$ (this means that $\overline{K}$ is algebraically closed).

If $g(X)$ has degree n and has a root $\alpha \in \overline{K}$, then we can write $g(X) = (X - \alpha)g_1(X)$ with $g_1(X)$ of degree $n - 1$. By induction, we see that $g(X)$ has exactly n roots (counting multiplicity) in $\overline{K}$.

It can be shown that every field K has an algebraic closure, and that any two algebraic closures of K are isomorphic. Throughout the book, we implicitly assume that a particular algebraic closure of a field K has been chosen, and we refer to it as the algebraic closure of K.

When $K = \mathbf{Q}$, the algebraic closure $\overline{\mathbf{Q}}$ is the set of complex numbers that are algebraic over $\mathbf{Q}$. When $K = \mathbf{C}$, the algebraic closure is $\mathbf{C}$ itself, since the fundamental theorem of algebra states that $\mathbf{C}$ is algebraically closed.

Finite fields

Let p be a prime. The integers mod p form a field $\mathbf{F}_p$ with p elements. It can be shown that the number of elements in a finite field is a power of a prime, and for each power p^n of a prime p, there is a unique (up to isomorphism) field with p^n elements. (*Note:* The ring $\mathbf{Z}_{p^n}$ is not a field when $n \geq 2$ since p does not have a multiplicative inverse; in fact, p is a zero divisor since $p \cdot p^{n-1} \equiv 0$ (mod p^n).) In this book, the field with p^n elements is denoted $\mathbf{F}_{p^n}$. Another notation that appears often in the literature is $GF(p^n)$. It can be shown that

$$\mathbf{F}_{p^m} \subseteq \mathbf{F}_{p^n} \quad \Longleftrightarrow \quad m|n.$$

The algebraic closure of $\mathbf{F}_p$ can be shown to be

$$\overline{\mathbf{F}}_p = \bigcup_{n \geq 1} \mathbf{F}_{p^n}.$$

THEOREM C.1

Let $\overline{\mathbf{F}}_p$ be the algebraic closure of $\mathbf{F}_p$ and let $q = p^n$. Then

$$\mathbf{F}_q = \{\alpha \in \overline{\mathbf{F}}_p \,|\, \alpha^q = \alpha\}.$$

PROOF   The group $\mathbf{F}_q^\times$ of nonzero elements of $\mathbf{F}_q$ forms a group of order $q - 1$, so $\alpha^{q-1} = 1$ when $0 \neq \alpha \in \mathbf{F}_q$. Therefore, $\alpha^q = \alpha$ for all $\alpha \in \mathbf{F}_q$.

Recall that a polynomial $g(X)$ has multiple roots if and only if $g(X)$ and $g'(X)$ have a common root. Since

$$\frac{d}{dX}(X^q - X) = qX^{q-1} - 1 = -1$$

(since $q = p^n = 0$ in $\mathbf{F}_p$), the polynomial $X^q - X$ has no multiple roots. Therefore, there are q distinct $\alpha \in \overline{\mathbf{F}}_p$ such that $\alpha^q = \alpha$.

Since both sets in the statement of the theorem have q elements and one is contained in the other, they are equal.   ∎

Define the q-th power **Frobenius automorphism** ϕ_q of $\overline{\mathbf{F}}_q$ by the formula

$$\phi_q(x) = x^q \quad \text{for all } x \in \overline{\mathbf{F}}_q.$$

PROPOSITION C.2

Let q be a power of the prime p.

1. $\overline{\mathbf{F}}_q = \overline{\mathbf{F}}_p$.

2. ϕ_q *is an automorphism of* $\overline{\mathbf{F}}_q$. *In particular,*

$$\phi_q(x+y) = \phi_q(x) + \phi_q(y), \qquad \phi_q(xy) = \phi_q(x)\phi_q(y)$$

for all $x, y \in \overline{\mathbf{F}}_q$.

3. *Let* $\alpha \in \overline{\mathbf{F}}_q$. *Then*

$$\alpha \in \mathbf{F}_{q^n} \iff \phi_q^n(\alpha) = \alpha.$$

PROOF   Part (1) is a special case of a more general fact: If $K \subseteq L$ and every element of L is algebraic over K, then $\overline{L} = \overline{K}$. This can be proved as follows. If α is algebraic over L and L is algebraic over K, then a basic property of algebraicity is that α is then algebraic over K. Therefore, $\overline{L}$ is algebraic over K and is algebraically closed. Therefore, it is an algebraic closure of K.

Part (3) is just a restatement of Theorem C.1, with q^n in place of q.

We now prove part (2). If $1 \le j \le p-1$, the binomial coefficient $\binom{p}{j}$ has a factor of p in its numerator that is not canceled by the denominator, so

$$\binom{p}{j} \equiv 0 \pmod{p}.$$

Therefore,

$$(x+y)^p = x^p + \binom{p}{1} x^{p-1} y + \binom{p}{2} x^{p-2} y^2 + \cdots + y^p$$
$$= x^p + y^p$$

since we are working in characteristic p. An easy induction yields that

$$(x+y)^{p^n} = x^{p^n} + y^{p^n}$$

for all $x, y \in \overline{\mathbf{F}}_p$. This implies that $\phi_q(x+y) = \phi_q(x) + \phi_q(y)$. The fact that $\phi_q(xy) = \phi_q(x)\phi_q(y)$ is clear. This proves that ϕ_q is a homomorphism of fields. Since a homomorphism of fields is automatically injective (see the discussion preceding Proposition C.5), it remains to prove that ϕ_q is surjective. If $\alpha \in \overline{\mathbf{F}}_p$, then $\alpha \in \mathbf{F}_{q^n}$ for some n, so $\phi_q^n(\alpha) = \alpha$. Therefore, α is in the image of ϕ_q, so ϕ_q is surjective. Therefore, ϕ_q is an automorphism.   ∎

In Appendix A, it was pointed out that $\mathbf{F}_p^\times = \mathbf{Z}_p^\times$ is a cyclic group, generated by a primitive root. More generally, it can be shown that $\mathbf{F}_q^\times$ is a cyclic group. A useful consequence is the following.

PROPOSITION C.3
Let m *be a positive integer with* $p \nmid m$ *and let* μ_m *be the group of* m*th roots of unity. Then*

$$\mu_m \subseteq \mathbf{F}_q^\times \iff m \,|\, q-1.$$

PROOF   By Lagrange's theorem (see Appendix B), if $\mu_m \subseteq \mathbf{F}_q^\times$, then $m | q - 1$. Conversely, suppose $m | q - 1$. Since $\mathbf{F}_q^\times$ is cyclic of order $q - 1$, it has a subgroup of order m (see Appendix B). By Lagrange's theorem, the elements of this subgroup must satisfy $x^m = 1$, hence they must be the m elements of μ_m. ∎

Let $\mathbf{F}_q \subseteq \mathbf{F}_{q^n}$ be finite fields. We can regard $\mathbf{F}_{q^n}$ as a vector space of dimension n over $\mathbf{F}_q$. This means that there is a basis $\{\beta_1, \ldots, \beta_n\}$ of elements of $\mathbf{F}_{q^n}$ such that every element of $\mathbf{F}_{q^n}$ has a unique expression of the form

$$a_1 \beta_1 + \cdots + a_n \beta_n$$

with $a_1, \ldots, a_n \in \mathbf{F}_q$. The next result says that it is possible to choose a basis of a particularly nice form, sometimes called a **normal basis**.

PROPOSITION C.4
There exists $\beta \in \mathbf{F}_{q^n}$ such that

$$\{\beta, \phi_q(\beta), \ldots, \phi_q^{n-1}(\beta)\}$$

is a basis of $\mathbf{F}_{q^n}$ as a vector space over $\mathbf{F}_q$.

An advantage of a normal basis is that the qth power map becomes a shift operator on the coordinates: Let

$$x = a_1 \beta + a_2 \phi_q(\beta) + \cdots + a_n \phi_q^{n-1}(\beta),$$

with $a_i \in \mathbf{F}_q$. Then $a_i^q = a_i$ and $\phi_q^n(\beta) = \beta$, so

$$\begin{aligned} x^q &= a_1 \beta^q + a_2 \phi_q(\beta^q) + \cdots + a_n \phi_q^{n-1}(\beta^q) \\ &= a_n \phi_q^n(\beta) + a_1 \phi_q(\beta) + \cdots + a_{n-1} \phi_q^{n-1}(\beta) \\ &= a_n \beta + a_1 \phi_q(\beta) + \cdots + a_{n-1} \phi_q^{n-1}(\beta). \end{aligned}$$

Therefore, if x has coordinates $(a_1, \ldots, a_n)$ with respect to the normal basis, then x^q has coordinates $(a_n, a_1, \ldots, a_{n-1})$. Therefore, the computation of qth powers is very fast and requires no calculation in $\mathbf{F}_{q^n}$. This has great computational advantages.

Embeddings and automorphisms

Let K be a field of characteristic 0, so $\mathbf{Q} \subseteq K$. An element $\alpha \in K$ is called **transcendental** if it is not the root of any nonzero polynomial with rational coefficients, that is, if it is not algebraic over $\mathbf{Q}$. A set of elements $S = \{\alpha_i\} \subseteq K$ (with i running through some (possibly infinite) index set I) is called **algebraically dependent** if there are n distinct elements $\alpha_1, \ldots, \alpha_n$ of S, for some $n \geq 1$, and a nonzero polynomial $f(X_1, \ldots, X_n)$ with rational coefficients such that $f(\alpha_1, \ldots, \alpha_n) = 0$. The set S is called **algebraically independent** if it is not algebraically dependent. This means that there is no polynomial relation among the elements of S. A maximal algebraically independent subset of K is called a **transcendence basis** of K. The **transcendence degree** of K over $\mathbf{Q}$ is the cardinality of a transcendence basis (the cardinality is independent of the choice of transcendence basis). If every element of K is algebraic over $\mathbf{Q}$, then the transcendence degree is 0. The transcendence degree of $\mathbf{C}$ over $\mathbf{Q}$ is infinite, in fact, uncountably infinite.

Let K be a field of characteristic 0, and choose a transcendence basis S. Let F be the field generated by $\mathbf{Q}$ and the elements of S. The maximality of S implies that every element of K is algebraic over F. Therefore, K can be obtained by starting with $\mathbf{Q}$, adjoining algebraically independent transcendental elements, then making an algebraic extension.

Let K and L be fields and let $f : K \to L$ be a homomorphism of fields. We always assume f maps $1 \in K$ to $1 \in L$. Then f is injective. One way to see this is to note that if $0 \neq x \in K$, then $1 = f(x)f(x^{-1}) = f(x)f(x)^{-1}$; since $f(x)$ has a multiplicative inverse, it cannot be 0.

The following result is very useful. It is proved using Zorn's Lemma (see [56]).

PROPOSITION C.5

Let K and L be fields. Assume that L is algebraically closed and that there is a field homomorphism

$$f : K \longrightarrow L.$$

Then there is a homomorphism $\tilde{f} : \overline{K} \to L$ such that $\tilde{f}$ restricted to K is f.

Proposition C.5 has the following nice consequence.

COROLLARY C.6

Let K be a field of characteristic 0. Assume that K has finite transcendence degree over $\mathbf{Q}$. Then there is a homomorphism $K \to \mathbf{C}$. Therefore, K can be regarded as a subfield of $\mathbf{C}$.

PROOF   Choose a transcendence basis $S = \{\alpha_1, \ldots, \alpha_n\}$ of K and let F be the field generated by $\mathbf{Q}$ and S. Since $\mathbf{C}$ has uncountable transcendence degree over $\mathbf{Q}$, we can choose n algebraically independent elements $\tau_1, \ldots, \tau_n \in \mathbf{C}$. Define $f : F \to \mathbf{C}$ by making f the identity map on $\mathbf{Q}$ and setting $f(\alpha_j) = \tau_j$ for all j. The proposition says that f can be extended to $\tilde{f} : \overline{F} \to \mathbf{C}$. Since K is an algebraic extension of F, we have $K \subseteq \overline{F}$. Restricting $\tilde{f}$ to K yields the desired homomorphism from $K \to \mathbf{C}$. Since a homomorphism of fields is injective, K is isomorphic to its image under this homomorphism. Therefore, K is isomorphic to a subfield of $\mathbf{C}$. ∎

The proposition also holds, with a similar proof, if the transcendence degree of K is at most the cardinality of the real numbers, which is the cardinality of a transcendence basis of $\mathbf{C}$.

If $\alpha \in \mathbf{C}$ is algebraic over $\mathbf{Q}$, then $f(\alpha) = 0$ for some nonzero polynomial with rational coefficients. Let $\mathrm{Aut}(\mathbf{C})$ be the set of field automorphisms of $\mathbf{C}$ and let $\sigma \in \mathrm{Aut}(\mathbf{C})$. Then $\sigma(1) = 1$, from which it follows that σ is the identity on $\mathbf{Q}$. Therefore,

$$0 = \sigma(f(\alpha)) = f(\sigma(\alpha)),$$

so $\sigma(\alpha)$ is one of the finitely many roots of $f(X)$. The next result gives a converse to this fact.

PROPOSITION C.7

Let $\alpha \in \mathbf{C}$. If the set
$$\{\sigma(\alpha) \,|\, \sigma \in \mathit{Aut}(\mathbf{C})\},$$
where σ runs through all automorphisms of $\mathbf{C}$, is finite, then α is algebraic over $\mathbf{Q}$.

PROOF   Suppose α is transcendental. There is a transcendence basis S of $\mathbf{C}$ with $\alpha \in S$. Then $\mathbf{C}$ is algebraic over the field F generated by $\mathbf{Q}$ and S.

The map

$$\sigma : F \longrightarrow F$$
$$\alpha \longmapsto \alpha + 1$$
$$\beta \longmapsto \beta \quad \text{when } \beta \in S,\ \beta \neq \alpha$$

defines an automorphism of F. By Proposition C.7, σ can be extended to a map $\tilde{\sigma} : \mathbf{C} \to \mathbf{C}$. We want to show that $\tilde{\sigma}$ is an automorphism, which means that we must show that $\tilde{\sigma}$ is surjective. Let $y \in \mathbf{C}$. Since y is algebraic over F, there is a nonzero polynomial $g(X)$ with coefficients in F such that $g(y) = 0$. Let $g^{\sigma^{-1}}$ denote the result of applying σ^{-1} to all of the coefficients of g (note that we know σ^{-1} exists on F because we already know that σ is an automorphism of F). For any root r of $g^{\sigma^{-1}}$, we have

$$0 = \tilde{\sigma}\left(g^{\sigma^{-1}}(r)\right) = g(\tilde{\sigma}(r)).$$

Therefore, $\tilde{\sigma}$ maps the roots of $g^{\sigma^{-1}}$ to roots of g. Since the two polynomials have the same number of roots, $\tilde{\sigma}$ gives a bijection between the two sets of roots. In particular, $\tilde{\sigma}(r) = y$ for some r. Therefore, y is in the image of $\tilde{\sigma}$. This proves that $\tilde{\sigma}$ is surjective, so $\tilde{\sigma}$ is an automorphism of $\mathbf{C}$.

Since

$$\tilde{\sigma}^j(\alpha) = \alpha + j,$$

the set of images of α under automorphisms of $\mathbf{C}$ is infinite, in contradiction to our assumption. Therefore, α cannot be transcendental, hence must be algebraic. ∎

References

[1] MFIPS 186-2. *Digital signature standard*. Federal Information Processing Standards Publication 186. U. S. Dept. of Commerce/National Institute of Standards and Technology, 2000.

[2] L. V. Ahlfors. *Complex analysis*. McGraw-Hill Book Co., New York, third edition, 1978. An introduction to the theory of analytic functions of one complex variable, International Series in Pure and Applied Mathematics.

[3] M. F. Atiyah and C. T. C. Wall. Cohomology of groups. In *Algebraic Number Theory (Proc. Instructional Conf., Brighton, 1965)*, pages 94–115. Thompson, Washington, D.C., 1967.

[4] A. O. L. Atkin and F. Morain. Elliptic curves and primality proving. *Math. Comp.*, 61(203):29–68, 1993.

[5] A. O. L. Atkin and F. Morain. Finding suitable curves for the elliptic curve method of factorization. *Math. Comp.*, 60(201):399–405, 1993.

[6] R. Balasubramanian and N. Koblitz. The improbability that an elliptic curve has subexponential discrete log problem under the Menezes-Okamoto-Vanstone algorithm. *J. Cryptology*, 11(2):141–145, 1998.

[7] I. F. Blake, G. Seroussi, and N. P. Smart. *Elliptic curves in cryptography*, volume 265 of *London Mathematical Society Lecture Note Series*. Cambridge University Press, Cambridge, 2000. Reprint of the 1999 original.

[8] D. Boneh. The decision Diffie-Hellman problem. In *Algorithmic number theory (Portland, OR, 1998)*, volume 1423 of *Lecture Notes in Comput. Sci.*, pages 48–63. Springer-Verlag, Berlin, 1998.

[9] D. Boneh and N. Daswani. Experimenting with electronic commerce on the PalmPilot. In *Financial Cryptography '99*, volume 1648 of *Lecture Notes in Comput. Sci.*, pages 1–16. Springer-Verlag, Berlin, 1999.

[10] D. Boneh and M. Franklin. Identity based encryption from the Weil pairing. In *Advances in Cryptology, Crypto 2001 (Santa Barbara, CA)*, volume 2139 of *Lecture Notes in Comput. Sci.*, pages 213–229. Springer-Verlag, Berlin, 2001.

[11] A. I. Borevich and I. R. Shafarevich. *Number theory.* Translated from the Russian by N. Greenleaf. Pure and Applied Mathematics, Vol. 20. Academic Press, New York, 1966.

[12] J. M. Borwein and P. B. Borwein. *Pi and the AGM.* Canadian Mathematical Society Series of Monographs and Advanced Texts. John Wiley & Sons Inc., New York, 1987. A study in analytic number theory and computational complexity, A Wiley-Interscience Publication.

[13] W. Bosma and H. W. Lenstra, Jr. Complete systems of two addition laws for elliptic curves. *J. Number Theory*, 53(2):229–240, 1995.

[14] R. P. Brent, R. E. Crandall, K. Dilcher, and C. van Halewyn. Three new factors of Fermat numbers. *Math. Comp.*, 69(231):1297–1304, 2000.

[15] C. Breuil, B. Conrad, F. Diamond, and R. Taylor. On the modularity of elliptic curves over **Q**: wild 3-adic exercises. *J. Amer. Math. Soc.*, 14(4):843–939 (electronic), 2001.

[16] K. S. Brown. *Cohomology of groups*, volume 87 of *Graduate Texts in Mathematics*. Springer-Verlag, New York, 1994. Corrected reprint of the 1982 original.

[17] J. W. S. Cassels. *Lectures on elliptic curves*, volume 24 of *London Mathematical Society Student Texts*. Cambridge University Press, Cambridge, 1991.

[18] C. H. Clemens. *A scrapbook of complex curve theory.* Plenum Press, New York, 1980. The University Series in Mathematics.

[19] H. Cohen. *A course in computational algebraic number theory*, volume 138 of *Graduate Texts in Mathematics*. Springer-Verlag, Berlin, 1993.

[20] H. Cohen, A. Miyaji, and T. Ono. Efficient elliptic curve exponentiation using mixed coordinates. In *Advances in cryptology—ASIACRYPT'98 (Beijing)*, volume 1514 of *Lecture Notes in Comput. Sci.*, pages 51–65. Springer-Verlag, Berlin, 1998.

[21] I. Connell. *Elliptic curve handbook.* Course Notes. McGill University, August, 1996.

[22] G. Cornell, J. H. Silverman, and G. Stevens, editors. *Modular forms and Fermat's last theorem.* Springer-Verlag, New York, 1997. Papers from the Instructional Conference on Number Theory and Arithmetic Geometry held at Boston University, Boston, MA, August 9–18, 1995.

[23] D. A. Cox. The arithmetic-geometric mean of Gauss. *Enseign. Math. (2)*, 30(3-4):275–330, 1984.

[24] H. Darmon, F. Diamond, and R. Taylor. Fermat's last theorem. In *Current developments in mathematics, 1995 (Cambridge, MA)*, pages 1–154. Internat. Press, Cambridge, MA, 1994.

[25] M. Deuring. Die Typen der Multiplikatorenringe elliptischer Funktionenkörper. *Abh. Math. Sem. Hamburg*, 14:197–272, 1941.

[26] L. E. Dickson. *History of the theory of numbers. Vol. II: Diophantine analysis*. Chelsea Publishing Co., New York, 1966.

[27] N. D. Elkies. The existence of infinitely many supersingular primes for every elliptic curve over **Q**. *Invent. Math.*, 89(3):561–567, 1987.

[28] A. Enge. *Elliptic curves and their applications to cryptography: An introduction*. Kluwer Academic Publishers, Dordrecht, 1999.

[29] G. Frey, M. Müller, and H.-G. Rück. The Tate pairing and the discrete logarithm applied to elliptic curve cryptosystems. *IEEE Trans. Inform. Theory*, 45(5):1717–1719, 1999.

[30] G. Frey and H.-G. Rück. A remark concerning m-divisibility and the discrete logarithm in the divisor class group of curves. *Math. Comp.*, 62(206):865–874, 1994.

[31] W. Fulton. *Algebraic curves*. Advanced Book Classics. Addison-Wesley Publishing Company Advanced Book Program, Redwood City, CA, 1989. An introduction to algebraic geometry, Notes written with the collaboration of R. Weiss, Reprint of 1969 original.

[32] S. Galbraith, K. Harrison, and D. Soldera. Implementing the Tate pairing. In *Algorithmic number theory (Sydney, Australia, 2002)*, volume 2369 of *Lecture Notes in Comput. Sci.*, pages 324–337. Springer-Verlag, Berlin, 2002.

[33] S. D. Galbraith and N. P. Smart. A cryptographic application of Weil descent. In *Cryptography and coding (Cirencester, 1999)*, volume 1746 of *Lecture Notes in Comput. Sci.*, pages 191–200. Springer-Verlag, Berlin, 1999.

[34] P. Gaudry, F. Hess, and N. P. Smart. Constructive and destructive facets of Weil descent on elliptic curves. *J. Cryptology*, 15(1):19–46, 2002.

[35] S. Goldwasser and J. Kilian. Primality testing using elliptic curves. *J. ACM*, 46(4):450–472, 1999.

[36] R. Hartshorne. *Algebraic geometry*. Springer-Verlag, New York, 1977. Graduate Texts in Mathematics, No. 52.

[37] O. Herrmann. Über die Berechnung der Fourierkoeffizienten der Funktion $j(\tau)$. *J. reine angew. Math.*, 274/275:187–195, 1975. Collection of articles dedicated to Helmut Hasse on his seventy-fifth birthday, III.

[38] E. W. Howe. The Weil pairing and the Hilbert symbol. *Math. Ann.*, 305(2):387–392, 1996.

[39] D. Husemoller. *Elliptic curves*, volume 111 of *Graduate Texts in Mathematics*. Springer-Verlag, New York, 1987. With an appendix by R. Lawrence.

[40] H. Ito. Computation of the modular equation. *Proc. Japan Acad. Ser. A Math. Sci.*, 71(3):48–50, 1995.

[41] H. Ito. Computation of modular equation. II. *Mem. College Ed. Akita Univ. Natur. Sci.*, (52):1–10, 1997.

[42] M. J. Jacobson, N. Koblitz, J. H. Silverman, A. Stein, and E. Teske. Analysis of the xedni calculus attack. *Des. Codes Cryptogr.*, 20(1):41–64, 2000.

[43] A. Joux. A one round protocol for tripartite Diffie-Hellman. In *Algorithmic Number Theory (Leiden, The Netherlands, 2000)*, volume 1838 of *Lecture Notes in Comput. Sci.*, pages 385–394. Springer-Verlag, Berlin, 2000.

[44] A. Joux. The Weil and Tate pairings as building blocks for public key cryptosystems. In *Algorithmic number theory (Sydney, Australia, 2002)*, volume 2369 of *Lecture Notes in Comput. Sci.*, pages 20–32. Springer-Verlag, Berlin, 2002.

[45] A. Joux and R. Lercier. Improvements to the general number field sieve for discrete logarithms in prime fields. *To appear in Math. Comp.*, 2002.

[46] E. Kani. Weil heights, Néron pairings and V-metrics on curves. *Rocky Mountain J. Math.*, 15(2):417–449, 1985. Number theory (Winnipeg, Man., 1983).

[47] A. W. Knapp. *Elliptic curves*, volume 40 of *Mathematical Notes*. Princeton University Press, Princeton, NJ, 1992.

[48] N. Koblitz. CM-curves with good cryptographic properties. In *Advances in cryptology—CRYPTO '91 (Santa Barbara, CA, 1991)*, volume 576 of *Lecture Notes in Comput. Sci.*, pages 279–287. Springer-Verlag, Berlin, 1992.

[49] N. Koblitz. *Introduction to elliptic curves and modular forms*, volume 97 of *Graduate Texts in Mathematics*. Springer-Verlag, New York, second edition, 1993.

[50] N. Koblitz. *A course in number theory and cryptography*, volume 114 of *Graduate Texts in Mathematics*. Springer-Verlag, New York, second edition, 1994.

[51] N. Koblitz. *Algebraic aspects of cryptography*, volume 3 of *Algorithms and Computation in Mathematics*. Springer-Verlag, Berlin, 1998. With an appendix by A. J. Menezes, Y.-H. Wu, and R. J. Zuccherato.

[52] D. S. Kubert. Universal bounds on the torsion of elliptic curves. *Proc. London Math. Soc. (3)*, 33(2):193–237, 1976.

[53] S. Lang. *Elliptic curves: Diophantine analysis*, volume 231 of *Grundlehren der Mathematischen Wissenschaften*. Springer-Verlag, Berlin, 1978.

[54] S. Lang. *Abelian varieties*. Springer-Verlag, New York, 1983. Reprint of the 1959 original.

[55] S. Lang. *Elliptic functions*, volume 112 of *Graduate Texts in Mathematics*. Springer-Verlag, New York, second edition, 1987. With an appendix by J. Tate.

[56] S. Lang. *Algebra*, volume 211 of *Graduate Texts in Mathematics*. Springer-Verlag, New York, third edition, 2002.

[57] H. Lange and W. Ruppert. Addition laws on elliptic curves in arbitrary characteristics. *J. Algebra*, 107(1):106–116, 1987.

[58] G.-J. Lay and H. G. Zimmer. Constructing elliptic curves with given group order over large finite fields. In *Algorithmic number theory (Ithaca, NY, 1994)*, volume 877 of *Lecture Notes in Comput. Sci.*, pages 250–263. Springer-Verlag, Berlin, 1994.

[59] H. W. Lenstra, Jr. Elliptic curves and number-theoretic algorithms. In *Proceedings of the International Congress of Mathematicians, Vol. 1, 2 (Berkeley, Calif., 1986)*, pages 99–120, Providence, RI, 1987. Amer. Math. Soc.

[60] H. W. Lenstra, Jr. Factoring integers with elliptic curves. *Ann. of Math. (2)*, 126(3):649–673, 1987.

[61] E. Liverance. *Heights of Heegner points in a family of elliptic curves*. PhD thesis, Univ. of Maryland, 1993.

[62] B. Mazur. Rational isogenies of prime degree (with an appendix by D. Goldfeld). *Invent. Math.*, 44(2):129–162, 1978.

[63] H. McKean and V. Moll. *Elliptic curves. Function theory, geometry, arithmetic*. Cambridge University Press, Cambridge, 1997.

[64] A. Menezes. *Elliptic curve public key cryptosystems*, volume 234 of *The Kluwer International Series in Engineering and Computer Science*. Kluwer Academic Publishers, Boston, MA, 1993. With a foreword by N. Koblitz.

[65] A. Menezes and M. Qu. Analysis of the Weil descent attack of Gaudry, Hess and Smart. In *Topics in cryptology—CT-RSA 2001 (San Francisco, CA)*, volume 2020 of *Lecture Notes in Comput. Sci.*, pages 308–318. Springer-Verlag, Berlin, 2001.

[66] A. J. Menezes, T. Okamoto, and S. A. Vanstone. Reducing elliptic curve logarithms to logarithms in a finite field. *IEEE Trans. Inform. Theory*, 39(5):1639–1646, 1993.

[67] A. J. Menezes, P. C. van Oorschot, and S. A. Vanstone. *Handbook of applied cryptography*. CRC Press Series on Discrete Mathematics and its Applications. CRC Press, Boca Raton, FL, 1997. With a foreword by R. L. Rivest.

[68] T. Nagell. Sur les propriétés arithmétiques des cubiques planes du premier genre. *Acta Math.*, 52:93–126, 1929.

[69] J. Oesterlé. Nouvelles approches du "théorème" de Fermat. *Astérisque*, (161-162):Exp. No. 694, 4, 165–186 (1989). Séminaire Bourbaki, Vol. 1987/88.

[70] IEEE P1363-2000. *Standard specifications for public key cryptography*.

[71] J. M. Pollard. Monte Carlo methods for index computation (mod p). *Math. Comp.*, 32(143):918–924, 1978.

[72] V. Prasolov and Y. Solovyev. *Elliptic functions and elliptic integrals*, volume 170 of *Translations of Mathematical Monographs*. American Mathematical Society, Providence, RI, 1997. Translated from the Russian manuscript by D. Leites.

[73] K. A. Ribet. From the Taniyama-Shimura conjecture to Fermat's last theorem. *Ann. Fac. Sci. Toulouse Math. (5)*, 11(1):116–139, 1990.

[74] K. A. Ribet. On modular representations of $\text{Gal}(\overline{\mathbf{Q}}/\mathbf{Q})$ arising from modular forms. *Invent. Math.*, 100(2):431–476, 1990.

[75] A. Robert. *Elliptic curves*. Springer-Verlag, Berlin, 1973. Lecture Notes in Mathematics, Vol. 326.

[76] A. Rosing. *Implementing Elliptic Curve Cryptography*. Manning Publications Company, 1999.

[77] H.-G. Rück. A note on elliptic curves over finite fields. *Math. Comp.*, 49(179):301–304, 1987.

[78] T. Satoh. On p-adic point counting algorithms for elliptic curves over finite fields. In *Algorithmic number theory (Sydney, Australia, 2002)*, volume 2369 of *Lecture Notes in Comput. Sci.*, pages 43–66. Springer-Verlag, Berlin, 2002.

[79] T. Satoh and K. Araki. Fermat quotients and the polynomial time discrete log algorithm for anomalous elliptic curves. *Comment. Math. Univ. St. Paul.*, 47(1):81–92, 1998. Errata: 48 (1999), 211-213.

[80] R. Schoof. Elliptic curves over finite fields and the computation of square roots mod p. *Math. Comp.*, 44(170):483–494, 1985.

[81] R. Schoof. Nonsingular plane cubic curves over finite fields. *J. Combin. Theory Ser. A*, 46(2):183–211, 1987.

[82] R. Schoof and L. Washington. Untitled. In *Dopo le parole (aangeboden aan Dr. A. K. Lenstra.* verzameld door H. W. Lenstra, Jr., J. K. Lenstra en P. van Emde Boas, Amsterdam, 16 mei, 1984. 1 page.

[83] A. Selberg and S. Chowla. On Epstein's zeta-function. *J. reine angew. Math.*, 227:86–110, 1967.

[84] I. A. Semaev. Evaluation of discrete logarithms in a group of p-torsion points of an elliptic curve in characteristic p. *Math. Comp.*, 67(221):353–356, 1998.

[85] J.-P. Serre. Complex multiplication. In *Algebraic Number Theory (Proc. Instructional Conf., Brighton, 1965)*, pages 292–296. Thompson, Washington, D.C., 1967.

[86] J.-P. Serre. *A course in arithmetic.* Springer-Verlag, New York, 1973. Translated from the French, Graduate Texts in Mathematics, No. 7.

[87] J.-P. Serre. Sur les représentations modulaires de degré 2 de $\mathrm{Gal}(\overline{\mathbf{Q}}/\mathbf{Q})$. *Duke Math. J.*, 54(1):179–230, 1987.

[88] D. Shanks. Class number, a theory of factorization, and genera. In *1969 Number Theory Institute (Proc. Sympos. Pure Math., Vol. XX, State Univ. New York, Stony Brook, NY, 1969)*, pages 415–440. Amer. Math. Soc., Providence, RI, 1971.

[89] G. Shimura. *Introduction to the arithmetic theory of automorphic functions*, volume 11 of *Publications of the Mathematical Society of Japan.* Princeton University Press, Princeton, NJ, 1994. Reprint of the 1971 original, Kanô Memorial Lectures, 1.

[90] J. H. Silverman. *The arithmetic of elliptic curves*, volume 106 of *Graduate Texts in Mathematics.* Springer-Verlag, New York, 1986.

[91] J. H. Silverman. The difference between the Weil height and the canonical height on elliptic curves. *Math. Comp.*, 55(192):723–743, 1990.

[92] J. H. Silverman. *Advanced topics in the arithmetic of elliptic curves*, volume 151 of *Graduate Texts in Mathematics.* Springer-Verlag, New York, 1994.

[93] J. H. Silverman. The xedni calculus and the elliptic curve discrete logarithm problem. *Des. Codes Cryptogr.*, 20(1):5–40, 2000.

[94] J. H. Silverman and J. Suzuki. Elliptic curve discrete logarithms and the index calculus. In *Advances in cryptology—ASIACRYPT '98 (Beijing, China)*, volume 1514 of *Lecture Notes in Comput. Sci.*, pages 110–125. Springer-Verlag, Berlin, 1998.

[95] J. H. Silverman and J. Tate. *Rational points on elliptic curves.* Undergraduate Texts in Mathematics. Springer-Verlag, New York, 1992.

[96] N. P. Smart. The discrete logarithm problem on elliptic curves of trace one. *J. Cryptology*, 12(3):193–196, 1999.

[97] J. Tate. The arithmetic of elliptic curves. *Invent. Math.*, 23:179–206, 1974.

[98] J. Tate. Algorithm for determining the type of a singular fiber in an elliptic pencil. In *Modular functions of one variable, IV (Proc. Internat. Summer School, Univ. Antwerp, Antwerp, 1972)*, pages 33–52. Lecture Notes in Math., Vol. 476. Springer-Verlag, Berlin, 1975.

[99] R. Taylor and A. Wiles. Ring-theoretic properties of certain Hecke algebras. *Ann. of Math. (2)*, 141(3):553–572, 1995.

[100] E. Teske. Speeding up Pollard's rho method for computing discrete logarithms. In *Algorithmic number theory (Portland, OR, 1998)*, volume 1423 of *Lecture Notes in Comput. Sci.*, pages 541–554. Springer-Verlag, Berlin, 1998.

[101] W. Trappe and L. Washington. *Introduction to cryptography with coding theory.* Prentice Hall, Upper Saddle River, NJ, 2002.

[102] J. B. Tunnell. A classical Diophantine problem and modular forms of weight 3/2. *Invent. Math.*, 72(2):323–334, 1983.

[103] S. Wagstaff. *Cryptanalysis of number theoretic ciphers.* Computational Mathematics Series. Chapman and Hall/CRC, Boca Raton, 2003.

[104] D. Q. Wan. On the Lang-Trotter conjecture. *J. Number Theory*, 35(3):247–268, 1990.

[105] L. Washington. Wiles' strategy. In *Cuatrocientos años de matemáticas en torno al Último Teorema de Fermat (ed. by C. Corrales Rodrigáñez and C. Andradas)*, pages 117–136. Editorial Complutense, Madrid, 1999. Section 13.4 of the present book is a reworking of much of this article.

[106] L. C. Washington. *Introduction to cyclotomic fields*, volume 83 of *Graduate Texts in Mathematics*. Springer-Verlag, New York, second edition, 1997.

[107] W. C. Waterhouse. Abelian varieties over finite fields. *Ann. Sci. École Norm. Sup. (4)*, 2:521–560, 1969.

[108] E. Weiss. *Cohomology of groups.* Pure and Applied Mathematics, Vol. 34. Academic Press, New York, 1969.

[109] A. Wiles. Modular elliptic curves and Fermat's last theorem. *Ann. of Math. (2)*, 141(3):443–551, 1995.

[110] H. C. Williams. A $p+1$ method of factoring. *Math. Comp.*, 39(159):225–234, 1982.

Index